The Chern Symposium 1979

Proceedings of the International Symposium
on Differential Geometry in honor of
S.-S. Chern, held in Berkeley, California,
June 1979

Edited by

W.-Y. Hsiang, S. Kobayashi, I. M. Singer,
A. Weinstein, J. Wolf, H.-H. Wu

With Contributions by

M. F. Atiyah   Raoul Bott   Eugenio Calabi
Mark Green   Phillip Griffiths   F. Hirzebruch
Nicolaas H. Kuiper   J. Moser   Louis Nirenberg
Robert Osserman   Wu Wen-tsün   Chen Ning Yang
Shing-Tung Yau

Springer-Verlag
New York   Heidelberg   Berlin

W.-Y. Hsiang   S. Kobayashi
I. M. Singer   J. Wolf   H.-H. Wu
Department of Mathematics
University of California
Berkeley, CA 94720
USA

A. Weinstein
Department of Mathematics
California Institute of Technology
Pasadena, CA 91125
USA

AMS Subject Classifications (1980): 00A10, 51Mxx, 53-XX, 55Rxx, 81E10, 83C95, 83C99

Library of Congress Cataloging in Publication Data

Proceedings of the International Symposium on
   Differential Geometry in Honor of S.-S.
   Chern, held in Berkeley, California, June
   1979.
   The Chern Symposium 1979.

Bibliography: p.
   1.   Global analysis (mathematics)—Congresses.
   2.   Geometry, Differential—Congresses.
   I.   Hsiang, Wu Yi, 1937   II.   Title.
   QA614.I57   1979   514'.74   80-26330

9 8 7 6 5 4 3 2 1

ISBN   0-387-90537-5   Springer-Verlag   New York   Heidelberg   Berlin
ISBN   3-540-90537-5   Springer-Verlag   Berlin   Heidelberg   New York

Preface

This volume attests to the vitality of differential geometry as it probes deeper into its internal structure and explores ever widening connections with other subjects in mathematics and physics.

To most of us Professor S. S. Chern *is* modern differential geometry, and we, his students, are grateful to him for leading us to this fertile landscape.

The aims of the symposium were to review recent developments in geometry and to expose and explore new areas of research. It was our way of honoring Professor Chern upon the occasion of his official retirement as Professor of Mathematics at the University of California.

This book is a record of the scientific events of the symposium and reflects Professor Chern's wide interest and influence. The conference also reflected Professor Chern's personality. It was a serious occasion, active yet relaxed, mixed with gentleness and good humor. We wish him good health, a long life, happiness, and a continuation of his extraordinarily deep and original contributions to mathematics.

I. M. Singer

Contents

Real and Complex Geometry in Four Dimensions

M. F. Atiyah*

1. Introduction

Some fifty years ago Einstein's theory of general relativity provided a great stimulus for different geometry, but after a period of fruitful interaction the interests of geometers and physicists diverged. Within the past few years there has been a resurgence of geometrical ideas in physics, arising partly from the popularity of gauge theories in elementary-particle physics and partly from the work of Hawking and Penrose on black holes. A characteristic feature in both cases has been the significance of global or topological properties, and it is this feature which has particularly attracted mathematicians.

The ideas and problems arising in this way have had a considerable impact on geometry, especially the differential geometry of four dimensions. Through the remarkable twistor theory of Penrose [13] there is also an intimate connection with complex analytic geometry. In this lecture I shall attempt to survey these mathematical developments with especial emphasis on Penrose's ideas and their application.

As is well known, Einstein's theory involves a Minkowski-type metric of signature $(1, 3)$ on space-time, and this is quite different in its global properties from conventional Riemannian geometry with the usual positive definite metric. One of the interesting developments is that, for quantum-mechanical reasons, physicists have now become interested in the analytic continuation from the Minkowski to the Euclidean region. This leads to global problems of the type conventionally studied by geometers, and it is this aspect on which I shall be mainly reporting, although many of the mathematical ideas and techniques apply also in the Minskowski framework.

2. Special Features of Four Dimensions

Mathematicians have a natural tendency to generality, developing methods which will apply over a wide front. Geometrically this means that they do not normally assign special importance to any particular dimension, though, of

* Mathematical Institute, Oxford University, 24-29 St. Giles, Oxford OX1 3LB, England.

course, dimensions one and two are clearly special for a variety of reasons. In physics, on the other hand, four dimensions are of fundamental importance, and techniques which exploit special features of four dimensions are much sought after. Now it so happens that four-dimensional space does have some remarkable properties which distinguish it from spaces of higher dimension, and I shall first describe some of these properties.

From a group-theoretical point of view we have the remarkable fact that the rotation group $SO(n)$ is a simple Lie group for all $n \neq 4$ (for $n = 2$ the group is abelian but still "simple"). For $n = 4$ we have the local isomorphism

$$SO(4) \approx SO(3) \times SO(3) \tag{2.1}$$

or, for the simply-connected groups, the isomorphism

$$Spin(4) \cong SU(2) \times SU(2). \tag{2.2}$$

Closely associated wlth (2.1) is the existence of the algebra H of quaternions— one of the great discoveries of Hamilton. We can identify $SU(2)$ with the group of quaternions of unit norm, and (2.2) corresponds to the fact that left and right multiplication by unit quaternions generates all rotations in $H = R^4$.

In terms of representations (2.2) corresponds to the isomorphism

$$R^4 \otimes C \cong S^+ \otimes S^-, \tag{2.3}$$

where S^+ and S^- are the basic 2-dimensional representations of the two copies of $SU(2)$: they are also the two half-spin representations of $Spin(4)$.

The imaginary quaternions are spanned by i, j, k and an imaginary quaternion x of unit norm

$$x = x_1 i + x_2 j + x_3 k, \qquad \sum x_i^2 = 1,$$

has the property that $x^2 = -1$. Thus the points of the 2-sphere $\sum_{i=1}^{3} x_i^2 = 1$ parametrize the complex structures on R^4 (i.e., isomorphisms $R^4 \cong C^2$) which preserve length and orientation. This can also be seen from (2.3): by fixing a nonzero vector in S^+ we can identify R^4 with S^-.

In terms of Lie algebras the decomposition (2.1) can also be described as follows. We consider the exterior square $\Lambda^2(R^4)$, which may be viewed as the Lie algebra of $SO(4)$, and we introduce the duality operator $*$:

$$* : \Lambda^2(R^4) \to \Lambda^2(R^4) \tag{2.4}$$

which sends $e_i \wedge e_j$ into $e_k \wedge e_l$ where e_i, e_j, e_k, e_l is an orthonormal oriented base of R^4. Since $(*)^2 = 1$, we can decompose $\Lambda^2(R^4)$ into the ± 1-eigenspaces of $*$:

$$\Lambda^2(R^4) \cong \Lambda_+^2 \oplus \Lambda_-^2. \tag{2.5}$$

Each of the pieces $\Lambda_\pm^2$ is three-dimensional and can be identified with the Lie algebra of $SO(3)$, so that (2.5) corresponds to (2.1). Note that $*$ can be defined on the exterior algebra of R^n, for any n, but only if $n = 4$ does $*$ induce a map of 2-forms into 2-forms.

The decomposition (2.5) is particularly significant in differential geometry because curvature is a 2-form (with values in some other Lie algebra). Thus the curvature of any principal bundle with connection over a Riemannian 4-manifold has an intrinsic decomposition into *self-dual* and *anti-self-dual* parts according to (2.5).

If we multiply the metric on R^n by a scalar factor λ, then $*$, on $\Lambda^p(R^n)$, gets multiplied by $\lambda^{n-p/2}$. Hence if $n = 2p$, $*$ is conformally invariant on $\Lambda^p(R^n)$. In particular, $*$ is conformally invariant on 2-forms in R^4. This is of considerable physical significance because many of the basic equations in physics are conformally invariant.

3. Geometry of Twistors

Motivated by a variety of physical arguments, R. Penrose [13] has proposed that space-time points should not be regarded as fundamental and that instead, one should formulate physics in a different space—the space of (projective) "twistors". Mathematically one can view this procedure as a kind of transform (cf. the Fourier transform) and it turns out to have some remarkable properties. Although Penrose is primarily interested in the Minkowski space, I shall describe the twistor construction for Euclidean space. This is somewhat easier to see and is also more relevant to the geometrical problems I shall discuss.

One possible mathematical motivation for twistors can be put as follows. It is well known that many interesting problems in two dimensions are best tackled by using complex-variable methods. Unfortunately in four dimensions the naive introduction of two complex variables does not seem very sensible, since we have no preferred choice. As we saw in Section 2, the admissible isomorphisms $R^4 \cong C^2$ are parametrized by points of a 2-sphere. We might, therefore, try to use *all* such complex coordinates simultaneously, in which case we need one other complex variable (parametrizing $S^2 = C \cup \infty$) to label our two complex variables. This suggests we should somehow be in a space of three complex variables. In fact, this idea works and is best described in geometric terms as follows.

We start with the complex projective 3-space $P_3(C)$ with complex homogeneous coordinates (x, y, u, v). Let us remove the complex projective line given by $u = v = 0$. Then any line in $P_3(C) - P_1(C)$ is given by a pair of equations:

$$x = au + bv,$$
$$y = cu + dv. \tag{3.1}$$

In particular, we shall be interested in those lines for which

$$c = -\bar{b}, \qquad d = \bar{a}. \tag{3.2}$$

Note that the determinant of (3.1) then becomes

$$\Delta = |a|^2 + |b|^2, \tag{3.3}$$

and so the line (3.1) never intersects the line $x = y = 0$ (unless of course it coincides with it). Similarly, no two lines of our family intersect, and they fill out the whole of $P_3(C) - P_1(C)$. Thus we have a fibration

$$P_3(C) - P_1(C)$$
$$\downarrow$$
$$R^4$$

obtained by assigning to each point of $P_3(C) - P_1(C)$ the four coordinates $\{\text{Re}(a), \text{Im}(a), \text{Re}(b), \text{Im}(b)\}$. If we restrict this fibration to a plane of the form

$$\alpha u + \beta v = 0,$$

we get an isomorphism $C^2 \to R^4$, and this depends on the ratio $(\alpha, \beta) \in P_1(C)$. Thus our picture embodies the idea of introducing complex coordinates outlined earlier.

Our fibration turns out to depend in an essential manner only on the conformal structure of R^4. It can, therefore, be extended to the one-point compactification S^4 of R^4, and we get a fibration

$$P_3(C)$$
$$\downarrow$$
$$S^4$$

in which the line $u = v = 0$, which we earlier excluded, sits over the point at ∞ of $S^4 = R^4 \cup \infty$. This fibration arises naturally if we use quaternions to identify C^4 with H^2 and S^4 with $P_1(H)$, the quaternion projective line. We simply assign to every complex one-dimensional subspace of C^4 the quaternion one-dimensional subspace of H which it generates.

The map $\sigma : P_3(C) \to P_3(C)$ defined by

$$\sigma(x, y, u, v) = (-\bar{y}, \bar{x}, -\bar{v}, \bar{u})$$

plays an important role. It preserves the fibration (because of (3.2)) and induces the antipodal map on each fibre.

In principle, now we can lift problems from S^4 (or R^4) to $P_3(C)$ and try to use complex methods. That this is remarkably successful for many physically interesting problems will be demonstrated in the following sections. These applications provide a mathematical justification of the Penrose theory.

4. Linear Field Equations

The simplest and most basic differential equation in R^4 is the Laplace equation $\Delta\phi = 0$. Naively, one might hope to obtain solutions ϕ from holomorphic functions on $P_3(C) - P_1(C)$. However, $P_3(C) - P_1(C)$ contains many projective lines, namely all those with equations (3.1) (without the reality constraint (3.2)). Any holomorphic function has to be constant on a projective line (Liouville's theorem), and since there are so many lines, it follows easily that a holomorphic function on $P_3(C) - P_1(C)$ has to be a constant. This is because we have not allowed poles or other singularities. Suppose now that we cover $P_3(C) - P_1(C)$ by two open sets U_0, U_∞ where U_0 meets each fibre of our fibration in a disc (near "0") and U_∞ in another disc (near "∞"). Suppose next that $\Phi(x, y, u, v)$ is holomorphic, defined in $U_0 \cap U_\infty$ and homogeneous of degree -2. Then restricting to any fibre P_ξ, we get Φ_ξ defined in an annular strip near the equator. Being homogenous of degree -2, Φ can be viewed naturally as a differential on P_ξ: if we use inhomogeneous coordinates putting $v = 1$, so that u is taken as parameter on P_ξ, our differential is just $\Phi_\xi(x, y, u, 1)\,du$ (x, y being replaced by the functions of u given by (3.1), (3.2), and $\xi = (a, b)$). We can now define a function ϕ on R^4 by

$$\phi(\xi) = \frac{1}{2\pi i} \int_{|u|=1} \Phi_\xi \, du. \qquad (4.1)$$

The first basic result of the twistor theory is that ϕ, given by (4.1), is a solution of the Laplace equation and that essentially every solution is obtained in this way. Of course, Φ is not unique, since it can be clearly modified to $\Phi + \Phi_0 - \Phi_\infty$, whose Φ_0 is holomorphic throughout U_0 and Φ_∞ is holomorphic in U_∞, without affecting the Cauchy integral (4.1). However, this is the only ambiguity in the choice of Φ. More formally, in the language of sheaf cohomology we have (Atiyah [1], Hitchin [9], Wells [16])

$$H^1(P_3(C) - P_1(C), \mathbb{O}(-2)) \cong \text{space of solutions of } \Delta\phi = 0 \text{ on } R^4. \quad (4.2)$$

It is important to know that the argument leading to (4.2) is purely local in R^4. Thus (4.2) continues to hold for an open set U of R^4 and its inverse image $\tilde{U}$ in $P_3(C) - P_1(C)$. It also holds on the compactified spaces $P_3(C) \to S^4$ provided we replace the flat Laplacian by the conformally invariant Laplacian $\Delta + R/6$, where R is the scalar curvature (and Δ is the positive Laplacian, i.e., $-\sum \partial^2/\partial x_i^2$).

Remark. As mentioned before, Penrose is primarily interested in the Minkowski space case where Δ is the hyperbolic wave operator. The analogue of (3.2) then involves boundary-value considerations (Wells [16]). On the other hand, for a metric of signature $(2, 2)$ Δ becomes ultrahyperbolic, and the counterpart of (4.2) is then the much older result of F. John [10]. The connection between John's result and (4.2) is explained in [1, Chapter VI, Section 5].

Not surprisingly, perhaps, if we replace the homogeneity -2 in (3.2) by $-m$, we get a corresponding statement for solutions of the linear equations for differing values of the spin. Thus for $m = 0$ and 4 we get the two halves (self-dual and anti-self-dual) of Maxwell's equations in vacuo, while $m = 1, 3$ give similarly the two halves of the massless Dirac equation. Again, everything is local and can be made conformally invariant.

It is a significant feature of the Penrose transform that it treats the two types of field (self-dual or anti-self-dual) differently. The whole theory depends in an important fashion on the initial choice of orientation of R^4.

5. The Yang–Mills Equations

We recall that Yang–Mills theory (for a given compact Lie group G) is a nonabelian generalization of Maxwell theory. One might, therefore, hope for a nonabelian analogue of the results described in the preceding section. This turns out to be the case, providing we consider only the self-dual Yang–Mills equations (as suggested by the remarks at the end of Section 4).

We recall that a Yang-Mills potential A_μ, with values in the Lie algebra of G, defines the Yang–Mills field or curvature

$$F_{\mu\nu} = \partial_\mu A_\nu - \partial_\nu A_\mu + \left[A_\mu, A_\nu \right].$$

The Yang–Mills Lagrangian is

$$- \int \text{trace}\, F_{\mu\nu} F^{\mu\nu} = - \int \text{trace}(F_\wedge {}^* F)$$

where on the left we sum over all μ, ν and on the right we view F as a matrix valued 2-form (we consider G as a group of unitary matrices for simplicity). The Yang–Mills equation (in Euclidean space) is then a second-order nonlinear differential equation for A_μ, namely

$$\sum_\mu \nabla_\mu F_{\mu\nu} = 0, \tag{5.1}$$

where ∇_μ denotes the covariant derivative, so that $\nabla_\mu F_{\mu\nu} = \partial_\mu F_{\mu\nu} + [A_\mu, F_{\mu\nu}]$. Now, in view of the Bianchi identity, (5.1) is automatically satisfied if

$$F = \pm {}^* F, \tag{5.2}$$

i.e., if F is self-dual (or anti-self-dual).

If we now lift a Yang–Mills potential (i.e., connection in differential-geometric terms) up to the Penrose twistor space, it turns out that (5.2) with the appropriate sign is equivalent to the lifted bundle having a *holomorphic* structure. This was first noted by R. S. Ward [15] and is explained in detail in [1, 3].

Applying this basic observation globally over S^4, the problem of constructing *instantons* (i.e., self-dual solutions of the Yang–Mills equation over the whole of S^4) converts into a problem concerning algebraic bundles over $P_3(C)$. This problem has been effectively solved (Atiyah et al. [4]), and a full account can be found in [1].

Besides working on the whole of S^4 there are other global problems of interest. In particular, there is interest in studying time-independent solutions on R^4, since these have an interpretation in terms of magnetic monopoles (in Minkowski space). The Penrose theory leads to an interesting problem concerning holomorphic bundles which deserves further investigation. Because of time independence it turns out that the base space of the bundle can be taken as the complex analytic surface of tangent vectors to P_1.

6. The Einstein Equations

The Penrose theory also applies, in the self-dual setup, to Einstein's vacuum equations. We recall that the Riemann curvature tensor breaks up in the form

$$\text{Riemann} = \text{Ricci} \oplus \text{Weyl}, \tag{6.1}$$

where the Weyl tensor is the conformally invariant part of the curvature. The Einstein equation is

$$\text{Ricci} = 0. \tag{6.2}$$

In the Riemannian 4-dimensional case we can further decompose the Weyl tensor W:

$$W = W^+ \oplus W^- \tag{6.3}$$

according to the *-operator. The Penrose theory applies in a very beautiful fashion when we put $W^- = 0$. One can then construct a 3-dimensional complex manifold Z, fibered over our 4-dimensional Riemannian manifold M, generalizing the fibration $P_3(C) \to S^4$. This depends only on the conformal structure of M. To get an Einstein metric in this conformal class then corresponds to giving a *holomorphic* fibration $Z \to P_1$ together with an isomorphism $K \cong H^{-4}$ (here K is the canonical line bundle of Z, and H is the Hopf bundle of P_1 pulled up to Z). The flat example is the projection $P_3(C) - P_1(C) \to P_1(C)$ given by $(x, y, u, v) \to (u, v)$.

Thus a self-dual Einstein metric on M^4 gives rise to the following picture:

$$
\begin{array}{ccc}
Z & \overset{\beta}{\to} & M^4 \\
{\scriptstyle \alpha}\downarrow & & \\
P_1 & &
\end{array}
$$

where α is holomorphic, but β is only real differentiable. Differentiably (α, β) give an isomorphism

$$Z \to P_1 \times M.$$

The fibres of β are holomorphic sections of α and define a real 4-parameter family of such sections. Conversely these data, i.e., the holomorphic fibration α together with the 4-parameter family of sections, enable one to reconstruct the Einstein metric on M (the parameter space of the sections or equivalently any one of the fibres of α) (Penrose [14], Hitchin [8, Section 2]).

An interesting concrete application of these ideas arises in connection with the ALE-manifolds of Gibbons and Hawking [6]. These are complete self-dual Einstein 4-manifolds which asymptotically are locally Euclidean, i.e., they look like Euclidean space modulo a finite group Γ near infinity. Gibbons and Hawking construct such manifolds explicitly when Γ is cyclic of order k and Hitchin [8] has given the Penrose description of these manifolds. He considers the equation

$$xy = \prod_{i=1}^{k} (z - p_i(u, v)) \tag{6.4}$$

where $p_i(u, v) = a_i u^2 + 2b_i uv - \bar{a}_i v^2$ (b real). If we assign weights $(k, k, 2, 1, 1)$ to the five variables (x, y, z, u, v), then (6.4) is homogeneous and so defines a 3-dimensional complex space Z' in the "weighted projective space" $C^5 - \{0\}/C^*$, where $\lambda \in C^*$ acts by

$$\lambda(x, y, z, u, v) = (\lambda^k x, \lambda^k y, \lambda^2 z, \lambda u, \lambda v).$$

The space Z is then obtained from Z' by removing the curve $u = v = 0$ and resolving (judiciously) the finite number of quadratic singularities arising from $p_i = p_j$. The projection $Z \to p_1$ is given by $(x, y, z, u, v) \to (u, v)$, and the 4-

parameter family of sections are given by

$$z = au^2 + 2buv - \bar{a}v^2 \qquad (b \text{ real}),$$

$$x = A \prod (u - \alpha_i v),$$

(6.5)

$$y = B \prod (u - \beta_i v),$$

where α_i, β_i are the two roots of

$$(a - a_i)u^2 + 2(b - b_i)u - (\bar{a} - \bar{a}_i) = 0.$$

(6.6)

Note that the discriminant Δ_i of (6.6) is positive and so we can systematically distinguish between α_i and β_i. The parameters A, B have to satisfy the constraints

$$AB = \prod (a - a_i), \qquad A\bar{A} = \prod ((b - b_i) + \Delta_i),$$

so that (6.5) depends finally on the four real parameters $(\text{Re}(a), \text{Im}(a), t, \arg A)$.

This description of the ALE spaces has been used [2] to construct explicitly the Green's function on M^4, i.e., the kernel of the inverse of the Laplace–Beltrami operator. A direct computation of this Green's function by more conventional methods was given by Page [12].

The Penrose approach suggests strongly that it should be possible to construct ALE-manifolds for all the finite subgroups of SU(2), corresponding to the symmetries of the regular polygon and the regular solids in 3-space. Hitchin (unpublished) has carried this out for the dihedral case. In all cases, one can start from the classical equation of the algebraic surface C^2/Γ given by the relation between the Γ-invariant polynomials. The theory, and in particular the deformation theory, of these singularities has been extensively studied, and from this one can write down an equation generalizing (6.4). The most difficult and as yet unsolved problem is to find the counterpart of the equations (6.5) describing the rational sections. The manifold M is always the standard resolution of C^2/Γ, and from Yau's work on the Calabi conjecture [18] one can deduce the existence of a complete self-dual Einstein metric on these manifolds. What is not entirely clear from this point of view is whether this metric will be asymptotically Euclidean.

Once one has the Penrose description of these manifolds, the method developed in [2] shows that the Green's function is necessarily rational and its explicit determination is a purely algebraic problem.

If we are interested in compact self-dual Einstein manifolds, then Yau's work shows that a quartic surface in $P_3(C)$ always has such a metric. The explicit description of this metric from the algebra has not yet been found. It is a more difficult problem than the one studied by Hitchin for rational surfaces, and its answer is likely to involve some interesting transcendental functions.

7. Conclusions

As the preceding sections will have made clear, the ideas and problems arising from physics have led to some extremely interesting mathematical work having intimate relations with classical algebraic geometry. As I have indicated, there are still a number of further problems in this direction which remain to be solved. Specifically, we can list

(a) Magnetic Monopoles

An outstanding problem here is to find if there are SU(2) magnetic monopoles of higher magnetic charge. This can be converted into a problem of complex analytic geometry, but this problem is more difficult than the one corresponding to instantons and has not yet been solved.[1]

(b) ALE-Manifolds

The explicit construction of the ALE-metrics for the groups corresponding to the cube, tetrahedron, and icosahedron has yet to be carried out. The Green's functions should then be computed by the method of [2]. Also the "moduli space," i.e., the parameter space of the metrics, should be found.

(c) Quartic Surfaces

Find the Yau metric on a quartic surface. One might begin by considering very special quartics such as $\sum x_i^4 = 0$ or the classical Kummer surface.

(d) Instantons

Although the instanton problem has in principle been solved in [3], our understanding of the moduli space is still very limited. For example, we do not even know if it is connected.

(e) The Full Yang–Mills Equation

The full (second-order) Yang–Mills equation (4.1) has been given a twistor interpretation by Witten [17] and Green et al. [7]. This has not yet been exploited. For example, one might hope to prove (for SU(2)) that there are no solutions on S^4 except the self-dual or anti-self-dual solutions.

(f) The Full Einstein Equations

Inspired by (e), one might be ambitious and look for a twistor interpretation of the full Einstein equations. In this direction one might note the intriguing remark made by Hawking that the Schwarzschild metric can, in suitable coordinates, be viewed as the sum of a self-dual metric and its complex conjugate.

I would now like to make some general remarks concerning the Penrose transform. As we have seen, it is remarkably successful in solving certain interesting differential equations, including some nonlinear ones. It may usefully be compared with the Fourier transform, which is, of course, designed to solve linear differential equations (with constant coefficients) by reducing them to algebra. The Penrose theory is less universal, in the sense that it is restricted to four dimensions and only deals with certain types of differential equation. On the

[1] Note added in proof: Recent progress on this problem has been made by R. S. Ward (to appear in *Comm. Math. Phys.*).

other hand, it has the advantage of dealing with some nonlinear problems and with variable coefficients. It is also invariant under a large group, namely the conformal group.

The Fourier transform is nonlocal in the sense that local behavior of a function is reflected in asymptotic properties of its Fourier transform. The Penrose transform is "semilocal" in the following sense. We recall the isomorphism (2.3) which factorizes 4-space into a product of two spin spaces. Now, as I pointed out earlier, the Penrose theory distinguishes between one type of spin and its opposite: one is treated locally and the other nonlocally. Thus, in the linear case explained in Section 4, a local solution in a small region of R^4 corresponds to a 1-dimensional sheaf cohomology class in the neighborhood of a line in $P_3(C)$: it is already "global" in one direction, namely along the line. The global instanton problem of Section 5 corresponds to a problem in the whole of $P_3(C)$, and this is now global in all directions.

To conclude, I would like to say how pleased I am to be able to report on these topics at this symposium in honor of Professor Chern. No one has done more than he, over the past few decades, to encourage the vigorous development of both real and complex geometry. He has also played an important role in the current dialogue between geometers and physicists.

REFERENCES

[1] M. F. Atiyah, Classical geometry of Yang–Mills fields. Fermi Lectures, Scuola Normale Pisa, 1980.

[2] M. F. Atiyah, Green's functions in self-dual 4-manifolds. *Adv. Math.*, to appear.

[3] M. F. Atiyah, N. J. Hitchin, and I. M. Singer, Self-duality in four-dimensional Riemannian geometry. *Proc. R. Soc. Lond. A* **362**, 425–461 (1978).

[4] M. F. Atiyah, N. J. Hitchin, V. G. Drinfeld, and Yu. I. Manin, Construction of instantons, *Phys. Lett.* **65A**, 285–287 (1978).

[5] M. F. Atiyah and R. S. Ward, Instantons and algebraic geometry. *Commun. Math. Phys.* **55**, 117–126 (1977).

[6] G. W. Gibbons and S. W. Hawking, Gravitational multi-instantons. *Phys. Lett.* **78B**, 430 (1978).

[7] P. S. Green, J. Isenberg, and P. B. Yasskin, Non-dual Yang–Mills fields, *Phys. Lett.* **78B**, 462–464 (1978).

[8] N. J. Hitchin, Polygons and gravitons. *Math. Proc. Camb. Phil. Soc.* **85**, 465–476 (1979).

[9] N. J. Hitchin, Field equations on self-dual spaces. *Proc. R. Soc. Lond.*, to appear.

[10] F. John, The ultrahyperbolic differential equation in four independent variables. *Duke Math. J.* **4**, 300–322 (1938).

[11] D. E. Lerner and P. D. Sommers (eds.), *Complex Manifold Techniques in Theoretical Physics.* Pitman, 1979.

[12] D. Page, Green's Functions for Gravitational Multi-Instantons.

[13] R. Penrose, The twistor programme. *Reports on Math. Phys.* **12**, 65–76 (1977).

[14] R. Penrose, Non-linear gravitons and curved twistor theory. *Gen. Relativ. Grav.* **7**, 31–52 (1976).

[15] R. S. Ward, On the self-dual Yang–Mills equations. *Phys. Lett.* **61A**, 81–82 (1977).

[16] R. O. Wells, Jr., Complex manifolds and mathematical physics. *Bull. Amer. Math. Soc.* **1**, 296–336 (1979).

[17] E. Witten, An interpretation of classical Yang–Mills theory. *Phys. Lett.* **77B**, 394–398 (1978).

[18] S. T. Yau, On the Ricci curvature of a compact Kähler manifold and the complex Monge-Ampère equations I. *Comm. Pure Appl. Math.* **31**, 339–411 (1978).

Equivariant Morse Theory and the Yang–Mills Equation on Riemann Surfaces[1]

Raoul Bott*

It is a great pleasure to address this symposium in honor of my dear friend, teacher, and collaborator. I first met Chern in 1950, when he dropped in to visit Princeton for just one day and I sat near him at lunch. I don't suppose that you remember this occasion, my dear friend, though I am sure I contrived to attract your attention by some impertinence or other. For I was immediately captivated by what you said and how you said it.

Later of course it was my privilege to see much of you and even to do joint work with you. And just as for Michael Atiyah, the geometric theory of characteristic classes both primary and secondary, which you have shaped for us, has been one of the great loves and inspirations of my mathematical work.

In a way what I will have to say today is also related to refined characteristic classes, but only indirectly, and rather than emphasizing this aspect I would like to present it here as yet another testament to the unity of mathematics, relating as it does Yang–Mills theory (in what the physicists like to call the trivial two-dimensional case) via Morse theory to some famous results in number theory.

This trip which Michael Atiyah and I stumbled onto two years ago is part fact and part surmise. But that is really how Michael and I like our mathematics—nowadays; I suppose because of the company we have been keeping.

To start off let me remind you of the Jacobian of a curve, which was also the starting point of Phil Griffiths' lecture; however, I will define it as follows:

Definition. The Jacobian $J(M)$ of a compact Riemann Surface M is the set of isomorphism classes of holomorphic line bundles L over M with 1st Chern class 0. The set $J(M)$ carries a natural complex and in fact algebraic structure.

Now this $J(M)$ can also be described in a purely "Riemannian" terms and is then seen to be related to the Yang–Mills functional.

Let me explain this in some detail. Recall first that a complex structure on a curve defines a conformal class of Reimann metrics on M, and let $\tilde{M}$ denote M

[1] This work supported in part through funds provided by the National Science Foundation under grant 33-966-7566-2.

* Department of Mathematics, Harvard University, Cambridge, MA 02138, USA.

endowed with one of these. Next construct the space

$$P = \tilde{M} \times S^1, \qquad S^1 = \{z \,|\, |z| = 1\} \tag{1}$$

considered as the trivial circle bundle over $\tilde{M}$ under the natural group action:

$$(m, s) \cdot s' = (m, ss'), \qquad m \in M, \quad s \in S^1.$$

We also consider S^1 as the Riemann manifold with unit 1-form $d\theta/2\pi$ ($z = e^{i\theta}$).

We may then consider the *space of Riemann structures on P*, which

(a) are invariant under the action of S^1,
(b) restrict to the given structure on each fibre S^1
(c) agree with the structure of the base on any normal tangent space to S^1 in $\tilde{M}$.

We call the space of these Riemann structures, $\mathcal{R}(P)$, and it should be clear after a little reflection that $\mathcal{R}(P)$ can also be thought of as the "space of connections" for P in the usual sense. That is $\mathcal{R}(P)$ is also the space of functions which assign to each point $p \in P$ a subspace $\mathcal{H}_p$ of the tangent space to P at p, subject to two conditions:

(a') $\mathcal{H}_p$ is transversal to the fibre S^1 through p.
(b') The $\mathcal{H}_p$ are preserved by the action of S^1.

Again a little reflection shows that such a connection is now completely determined by a 1-form ω in $\Omega^1(P)$, subject to:

(a'') ω restricts to $d\theta/2\pi$ on any fibre.
(b'') ω is invariant under the action of S^1.

From these two properties it follows immediately that for two such ω's, say ω and ω':

$$\omega - \omega' = \pi^*\alpha, \tag{2}$$

where $\alpha \in \Omega^1(\tilde{M})$, and *this* π denotes the projection

$$\begin{matrix} P \\ \downarrow{\scriptstyle\pi} \\ M \end{matrix} \, . \tag{3}$$

Note that the other projection of the product

$$S^1 \xleftarrow{\sigma} P \tag{4}$$

allows one to pull the form $d\theta/2\pi$ back to P and that it then clearly staisfies requirements (a) and (b). Furthermore

$$d\sigma^* \frac{d\theta}{2\pi} \equiv 0. \tag{5}$$

From (1) and (2) it then follows, by setting $\omega' = \sigma^* \, d\theta/2\pi$, that

$$d\omega = \pi^* \, d\alpha. \tag{6}$$

Now by definition $d\omega = \pi^* F$, where $F \in \Omega^2(M)$ is the *curvature of the connection* ω; and the "Yang–Mills functional"

$$S(\omega) : \mathcal{R}(P) \to R \tag{7}$$

is therefore simply the assignment

$$\omega \mapsto \int_{\tilde{M}} \|d\alpha\|^2,$$

the norm being taken in the Hodge sense, i.e. in terms of the metric of $\tilde{M}$.

Thus *in the abelian case, the Yang–Mills functional reduces to the norm in the Hodge sense on the 2-forms on the base.*

Now the extrema of this norm are of course well known. In each homology class this norm has a unique minimum, the harmonic form of that class. Thus in our case, the class of $d\alpha$ being zero, this harmonic form is identically 0.

Recapitulating, we have found the following state of affairs.

After a choice $(\sigma^* d\theta/2\pi)$ the space $\mathcal{C}(P)$ can be identified with the 1-form $\Omega^1(\tilde{M})$ on the base, and the Yang–Mills equations have as their solutions precisely the closed 1-forms $Z^1(\tilde{M}) \subset \Omega^1(\tilde{M})$.

Now this is of course not very interesting. However, what *is* of interest, both in physics and in mathematics, is the equivalence class of the solutions of the Yang–Mills equation under the group $\mathcal{G}(P)$ of gauge transformations of P, because $\mathcal{G}(P)$ is a natural symmetry of the functional S.

Precisely, a diffeomorphism f of P is a gauge transformation if it

(a) preserves the fibres,
(b) commutes with the action of S^1.

For such an f, it is clear that $f^*\omega$ will have the properties (a), (b) of a connection if ω has, and hence that

$$f^*\omega - \omega = d\pi^*\alpha, \qquad \alpha \in \Omega^1(M), \tag{8}$$

whence $F(f^*\omega) = F(\omega)$.

Thus $\mathcal{G}(P)$ acts naturally on $\mathcal{C}(P)$ preserving S, and hence S descends to a function

$$\underline{S} : \mathcal{C}(P)/\mathcal{G}(P) \to R, \tag{9}$$

and properly speaking we are interested in the stationary set of $\underline{S}$ rather than S.

Now in our case I claim that this critical set of $\underline{S}$ is a real torus $J(\tilde{M})$ of dimension $2 \times \text{genus}(M)$ endowed with a natural Riemann and compatible complex (in fact algebraic) structure, and so considered may be identified with the Jacobian $J(M)$.

I have time for only two hints on this quite easy equivalence. One first observes that in our case

$$\mathcal{G}(P) = \text{Maps}(M; S^1) \tag{10}$$

and as such, $\mathcal{G}(P)$ falls into components which are in one-to-one correspondence with

$$\text{Hom}\big(Z, H^1(M; Z)\big) \simeq Z^{2g}. \tag{11}$$

Next one observes that the action of the identity component $\mathcal{G}_0(P)$ on $Z^1(M)$ yields $H^1(M; R)$ as quotient, and that finally

$$Z^1(M)/\mathcal{G}(P) \simeq H^1(M; R)/H^1(M; Z) \tag{12}$$

is a torus.

The second hint concerns the complex structure on the result. This structure arises in the following manner.

We complexify S^1 in the natural manner to obtain $C^* = \{C - 0\}$ and correspondingly complexify P to

$$P_C = M \times C^*. \tag{13}$$

Now corresponding to a connection form ω on P there is a unique connection form ω_C which extends ω to P_C. Namely, if $\omega = \sigma^* \, d\theta/2\pi + \pi^*\alpha$, set

$$\omega_C = \sigma^* \frac{1}{2\pi i} \frac{dz}{z} + \pi^*\alpha^{1,0} \tag{14}$$

and define the almost complex structure on P_C, by declaring that its forms of type (1,0), i.e. $\Omega^{1,0}(P_C)$, are to be generated by $\pi^*\Omega^{1,0}(M)$ *and* ω_C. For dimensional reasons (on the base) the Nirenberg–Newlander conditions of integrability:

$$d\Omega^{1,0} \subset \Omega^{1,0} + \Omega^{1,1}, \tag{15}$$

are then easily checked. □

Let us return to algebraic geometry for a moment. The definition of the Jacobian in that category clearly motivates higher-order analogues. Namely, one is tempted to define $J_n(M)$ as the isomorphism classes of n-dimensional holomorphic vector bundles. Actually this generalization fails in the sense that this $J_n(M)$ would be a very bad set, i.e. with no good algebraic properties. However, if one restricts ones self to a subclass of holomorphic vector bundles, the so-called "semistable bundles" of Mumford,[2] then their isomorphism classes have a natural projective structure, and are the proper generalizations of the Jacobian. However the topological structure of these higher J_k's is much more complicated than that of $J_1(M)$, and in fact still largely unknown.

On the other hand the relation of $J_k(M)$ to the Yang–Mills theory is perfect, in the sense that:

$J_k(M)$ is the minimum set of $\underline{S}$ on $\mathcal{C}(P)/\mathcal{G}(P)$, where $\underline{S}$ is the Yang–Mills functional for the trivial principal bundle $P = U(n) \times M$ over M.

This result is an immediate consequence of fine theorems of Narasimhan and Sechadri according to which the space $J_k(M)$ is identified with conjugacy classes of homomorphisms of $\pi_1(M)$ into $U(k)$. But the minimum of $\underline{S}$ clearly corresponds to a connection with curvature $F \equiv 0$, that is to *flat bundles*, and these are classified by their holonomy maps which again are the conjugacy classes of homomorphisms of $\pi_1(M)$ into U_k. □

This and other evidence led M. Atiyah and myself to conjecture that in some appropriate sense the Morse theory should be applicable to $\underline{S}$ and that such an

---

[2] For $n = 2$, the stability condition of Mumford is the following one: E is *semistable* if and only if for each holomorphic line subbundle $L \subset E$, $c_1(L) \leqslant \frac{1}{2}c_1(E)$. Bundles which satisfy $c_1(L) < \frac{1}{2}c_1(E)$ are called *stable*, and their equivalence classes form a nonsingular variety, which however need not be compact.

application could yield new information about the $J_k(M)$'s. To explain this development let me therefore give a very quick review of the Morse theory, before applying it in our circumstance.

If f is a smooth function on a compact, finite-dimensional manifold W, a critical point of f is of course a point p where all the partials $\partial f/\partial x^i$ relative to local coordinates x^i vanish; and at such a point the matrix $\partial^2 f/\partial x^i\,\partial x^j|_p$ acquires certain invariant aspects. For instance the number λ_p of negative eigenvalues and the number ν_p of zero eigenvalues are intrinsic (i.e. independent of the local coordinates chosen) and are called the *index and nullity of f at p* respectively. When all the nullities of f are zero, f is called *nondegenerate*, and in that case Morse attaches to f the series

$$\mathfrak{M}_t = \sum t^{\lambda_p}, \qquad p \text{ critical for } f, \tag{16}$$

and shows that this series *dominates* the Poincaré series of (M),

$$P_t(M) = \dim H^i(M; R), \tag{17}$$

in a certain sense:

$$\mathfrak{M}_t(f) \geqslant P_t(M). \tag{18}$$

In particular each coefficient of $\mathfrak{M}_t$ is greater than or equal to the corresponding coefficient of P_t. We will therefore say that f is a *perfect Morse function* if

$$\mathfrak{M}_t(f) = P_t(M). \tag{19}$$

For example, in the Riemann surface of Figure 1, the z coordinate has critical points as indicated with indices as shown. Thus

$$\mathfrak{M}_t(z) = 1 + 4t + t^2. \tag{20}$$

This polynomial agrees with $P_t(M) = 1 + 2gt + t^2$, and hence z is a perfect Morse function on M.

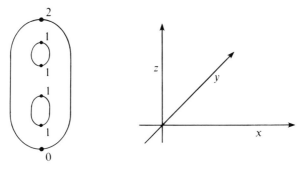

Figure 1.

We will also need a slight generalization of this concept on nondegeneracy. Namely, suppose f has a critical set which is itself a manifold N, but with $\nu_p = \dim N$ for each p in N. Under these conditions λ_p will be a constant λ_N, along N, and for functions having only such critical sets N, we define[3]

$$\mathfrak{M}_t(f) = \sum_N P_t(N) t^{\lambda_N}. \tag{21}$$

This granted, the Morse inequality will still hold:

$$\mathfrak{M}_t(f) \geqslant P_t(M), \tag{22}$$

and so here we may again speak of a perfect Morse function.

Thus *the perfect Morse functions exhibit the minimal critical behavior compatible with the topology of their domain.* On the other hand their critical set can be large. For example, the constant function is perfect in this sense. For a more typical example consider a torus T lying on a table. Then the height above that table defines a perfect Morse function on T. Its Morse polynomial is

$$\mathfrak{M}_t(f) = (1 + t) + t(1 + t), \tag{23}$$

the two terms corresponding to the minimum circle with index 0 and the maximum circle, which has index 1.

With these notions understood, I can finally explain to you what Michael Atiyah and I believe to be the case. Essentially we assert that the Yang–Mills functional is a perfect Morse function on $\mathcal{C}(P)/\mathcal{G}(P)$ when the base is a Riemann surface M, and P is any unitary principal bundle over M.

Now to prove such an assertion one of course has to extend all the concepts introduced to mapping spaces etc., and there are a legion of technical questions to establish. But above and beyond these there is the problem that the space $\mathcal{C}(P)/\mathcal{G}(P)$ will certainly *not* be a manifold in general because the action of $\mathcal{G}(P)$ on $\mathcal{C}(P)$ will not be free. Let me therefore explain next how we get around this problem, but again only in a finite-dimensional setting.

Suppose then that f is defined on W as before, and that in addition f is invariant under the action of a compact group G. When is f then to be considered a perfect equivariant Morse function? If G acts freely, then of course this should mean that the induced function $\underline{f}$ on W/G is perfect in the usual sense. But what if the action of G is not free? At this point a page from the book of homotopy theory is advisable. Namely, in that discipline there is a construction which converts all actions into free actions, *without changing the homotopy of the space on which the group acts.* This construction depends on the fact that a topological group G admits a *free* action on a contractible space EG, and that such actions are essentially unique. These actions are hard to find, and usually EG is a huge space, but still tractable if G is. For example let $G = S^1$. Then S^1 acts freely on $S^{2n+1} = \{\sum |z_i|^2 = 1\}$ by sending $z_k \to e^{i\theta} z_k$. However S^{2n+1} is not contractible. It is only in the limit $n \to \infty$ that we obtain ES^1 as the unit sphere in Hilbert space, which is known to be contractible.

The quotient of EG by the action of G is called the *classifying space of G* and is denoted by BG:

$$BG = EG/G. \tag{24}$$

---

[3] Properly speaking, $P_t(N) = \sum \dim H^i(N)$ when 0 is the orientation sheaf of the "negative" normal bundle to N.

For instance in the above example this clearly yields

$$BS^1 = CP^\infty, \tag{25}$$

the infinite complex projective space. Recall also that the Poincaré series of this space is given by

$$P_t(CP^\infty) = 1 + t^2 + t^4 + \cdots = \frac{1}{1 - t^2}. \tag{26}$$

More generally BU_n is the Grassmannian of n-planes in Hilbert space and

$$P_t(BU_n) = \frac{1}{1 - t^2} \cdot \frac{1}{1 - t^4} \cdots \frac{1}{1 - t^{2n}}.$$

The construction G to EG and BG is functorial, and the *topology* of BG in some sense reflects the *algebraic structure* of G. This understood, it is now clear how the homotopy theorist makes actions free. Namely given an action of G on W, he simply passes to the diagonal action of G on

$$W \times EG \tag{27}$$

and he then defines the *homotopy quotient* $W//G$ of the original action by

$$W//G = W \times EG/G. \tag{28}$$

This homotopy quotient has the following elementary properties:

$$pt//G \cong BG, \tag{29a}$$

$$W//G \cong W/G \qquad \text{if the action is free,} \tag{29b}$$

$$W//G \text{ is a fibre bundle over } BG \text{ with fibre } W. \tag{29c}$$

Thus in particular, if W is *contractible*, then

$$W//G \cong BG.$$

Here of course $\cong$ means homotopy type, and in (29a) pt means the trivial action on a point.

This homotopy wisdom now turns out to blend very nicely with the Morse theory. Indeed, if f is G-invariant on W, then the pullback of f to $W \times EG$ is invariant under the diagonal action and hence descends to a function

$$f_G : W//G \to R. \tag{30}$$

We therefore propose to call f a *perfect G-invariant Morse function* on W iff f_G is a perfect Morse function on $W//G$ in the usual sense.

In putting this definition to use, we must still remark on the very nice dictionary which relates the critical points of f on W to those of f_G on $W//G$. It goes as follows:

If $N \subset W$ is a nondegenerate critical manifold of f which consists of a single G-orbit, with stability group H:

$$N = G/H, \tag{31}$$

then corresponding to N, f_G will have a nondegenerate critical manifold BH of the same index as N.

As an example, consider the height function $f \equiv z$, on the 2-sphere $x^2 + y^2 + z^2 = 1$. It is clearly invariant under the rotations about the z-axis, which form an S^1. The Morse function of f is $1 + t^2$. As the critical points of f are both fixed points of S^1 (i.e., $H = G$) they both contribute critical manifolds BS^1 to f_G, and so the Morse "series" of f_G is given by

$$\mathfrak{M}_t(f_G) = \frac{1}{1 - t^2} + \frac{t^2}{1 - t^2} . \tag{32}$$

It is easy to compute $H^*(S^2 // S^1)$ in this case, and one then finds that

$$P_t(S^2 // S^1) = \frac{1 + t^2}{1 - t^2} ,$$

so that f_G is indeed a perfect S-equivariant Morse function. Note by the way that in this example $P_t(S^2/S^1) = 1$ and so is clearly extraneous to the problem.

We are now ready to return to the Yang–Mills situation. Precisely then our conjecture-theorm is that on principal $U(n)$ bundles P over Riemann surfaces,

The Yang–Mills Functional

$$S : \mathcal{C}(P) \to R$$

induces a perfect $\mathcal{G}(P)$-equivariant Morse function on $\mathcal{C}(P) // \mathcal{G}(P) \simeq B\mathcal{G}(P)$, so that in particular

$$P_t(B\mathcal{G}) = \mathfrak{M}_t(S_{\mathcal{G}(P)}). \tag{33}$$

Note that the homotopy quotient of $\mathcal{C}(P)$ by $\mathcal{G}(P)$ is of course $B\mathcal{G}(P)$ because $\mathcal{C}(P)$ is contractible (corresponding to the property (3) of the homotopy quotient construction) so that in the left-hand side of (33) the space of connections has completely disappeared. On the other hand, by our dictionary the right-hand side is related to the Yang–Mills extrema on $\mathcal{C}(P)$.

Next we observe that the left-hand side is computable. This follows first of all from the following essentially well-known description of $B\mathcal{G}$:

$$B\mathcal{G}(P) = \mathrm{Map}_P(M; BG). \tag{34}$$

Here P has G for its structure group, and is therefore induced by a map

$$\mu_P : M \to BG, \tag{35}$$

and on the right-hand side we mean the component of the mapping space of all maps of M to BG containing μ_P. Using this relation (34) with $G = U(n)$ and M a compact Riemann Surface of genus g, one then obtains the following formula:

$$P_t(B\mathcal{G}) = \frac{\left\{ (1 + t)(1 + t^3) \cdots (1 + t^{2n-1}) \right\}^{2g}}{\left\{ (1 - t^2)(1 - t^4) \cdots (1 - t^{2n}) \right\}^2 (1 - t^{2n})} . \tag{36}$$

This formula follows from essentially standard results of homotopy theory and is valid for $\mathcal{G}$ the group of gauge transformations $\mathcal{G}(P)$ of *any* U_n bundle P over M. It turns out furthermore that the spaces $B\mathcal{G}$ are *torsion free*.

The right-hand side of (33) needs considerably more work for its computation. First of all we need to understand all the critical points of S on $\mathcal{C}(P)$. We also need to compute their indices, and finally their stability groups under the action of $\mathcal{G}(P)$. Note by the way that the stability group H_ω of a connection ω in $\mathcal{C}(P)$ is always a *compact* Lie group, because it is a closed subgroup of a group of diffeomorphisms of P, preserving the Riemann structure induced by ω on P.

To give the experts at least the flavor of some of these computations, let me now quickly describe the index formula for a critical connection $\omega \in \mathcal{C}(P)$ in the general setting.

Consider then a G-bundle (G-compact) over M, and let $\mathcal{C}(P)$ denote the space of connections on P. Thus $\omega \in \mathcal{C}(P)$ is a 1-form on P with values in $\mathfrak{g}$, the Lie algebra of G, subject to the two conditions

(a) ω restricts to the identity in the vertical directions,
(b) $g^*\omega = \operatorname{Ad} g \cdot \omega$,

where $g \in G$, g^* is induced by right translation of P by g, and Ad is the adjoint representation.

From these formulae it follows that the 2-form

$$d\omega + \tfrac{1}{2}[\omega, \omega]$$

naturally descends to a 2-form F on M with values in the bundle Ad P associated to P via the adjoint representation:

$$F \in \Omega^2(M; \operatorname{Ad} P).$$

This is the curvature of ω, and the Yang–Mills functional is then

$$S(\omega) = \int_M \|F\|^2. \tag{37}$$

At an extremum ω of S the Yang–Mills equations take the form

$$d_\omega \cdot F = 0, \qquad d_\omega * F = 0 \tag{38}$$

where d_ω denotes the differential operator induced on $\Omega^*(M; \operatorname{Ad} P)$, so that the first of these equations is the Banchi identity. When dim $M = 2$, $* F$ is simply a section of Ad P, and hence the second equation expresses the fact that $* F$ is *covariant constant*.

It follows that the bracket with $(i/2\pi) * F$ decomposes the complexification of Ad P into three pieces:

$$\operatorname{Ad} P_C = \operatorname{Ad}^+ P \oplus \operatorname{Ad}^0 P \oplus \operatorname{Ad}^- P \tag{39}$$

according to the positive, zero, and negative eigenvalues of $(i/2\pi) * F$.

These bundles all turn out to have natural complex structures over M, and in terms of these we have the following formulae:

$$\operatorname{index} \omega = 2 \cdot \dim H^1(M; \operatorname{Ad}^- P),$$

$$\operatorname{nullity} \omega = 2 \cdot \dim H^1(M; \operatorname{Ad}^0 P). \tag{40}$$

Furthermore the index is *stable* in the sense that the above formula can also be put in purely C^∞ terms:

$$\operatorname{index} \omega = 2\{c_1(\operatorname{Ad}^+ P) + \dim(\operatorname{Ad}^+ P)(g - 1)\}. \tag{41}$$

On the other hand, the nullity is not stable in this sense.

So much for the experts. For one and all let me just say that these formulae arise from a spectral estimate for the eigenvalues of the elliptic problem arising in the second variation of the Yang–Mills functional, and that finally, equipped with this information, one can compute $\mathfrak{M}_t(S_g)$ purely in terms of the Poincaré series of the *absolute minimum* of $\underline{S}$, i.e. the space of semistable bundles over $\mathfrak{M}$. For instance, when $n = 2$ and $c_1(P) = 1$, one obtains the following series for $\mathfrak{M}_t$:

$$\mathfrak{M}_t = \frac{P_t(J_2(M;1))}{1-t^2} + \frac{t^{2g+4}(1+t^2)^{4g}}{(1-t^2)^2(1-t^4)}. \tag{42}$$

Here the first term describes the Poincaré series of the absolute minimum of S_G as a product of BS^1 with the space of semistable bundles. This is legitimate because $c_1(P) = 1$ implies that the semistable bundles are stable (Chern classes are integer-valued) and hence form a *nonsingular variety*. Note that the stability group of J_2 is S^1, precisely the center of $U(2)$, and this accounts for the $1/(1-t^2)$ factor in this term.

The next term should be thought of as the sum

$$\frac{t^{2g}(1+t)^{4g}}{(1-t^2)^2} + \frac{t^{2g+4}(1+t)^{4g}}{(1-t^2)^2} + \cdots \tag{43}$$

corresponding to the higher critical points of $\underline{S}$. These correspond precisely to the case when P splits into the direct product of two circle bundles P_1 and P_2 each endowed with a critical connection of its own. Thus each of these gives rise to a whole $2g$-dimensional torus of critical points, accounting for the $(1+t)^{4g}$ factor.

These critical points are in one-to-one correspondence with the possible splittings of P, and these can be labeled by the first Chern classes of the P_i, that is, by pairs of integers $n_1 \leqslant n_2$ with $n_1 + n_2 = 1$. If L_i denotes the line bundle associated to P_i in the usual manner, then one finds furthermore that "at" (n_1, n_2)

$$\operatorname{Ad} P_C \simeq (L_1^* + L_2^*) \otimes (L_1 + L_2) \tag{44}$$

with

$$\operatorname{Ad}_+ P \simeq L_1^* \otimes L_2. \tag{45}$$

This bundle has $c_1 = n_2 - n_1 = 2n_2 - 1$, so that the corresponding index is given by

$$2\{(2n_2 - 1) + (g - 1)\} = 4(n_2 - 1) + 2g. \tag{46}$$

Finally note that the stability group of all of these critical sets is $S^1 \times S^1$, accounting for the factor $1/(1-t^2)^2$. Putting these facts together and summing over n_2 therefore yields the second term.

In this case our assertion (33) therefore reads

$$\frac{\{(1+t)(1+t^3)\}^{2g}}{(1-t^2)^2(1-t^4)} = \frac{P_t(J_2)}{(1-t^2)} + \frac{t^{2g}(1+t)^{4g}}{(1-t^2)^2(1-t^4)}. \tag{47}$$

When Michael Atiyah and I first wrote down this formula we hardly believed that when solving for $P_t(J_2)$ we would find a polynomial. But we did, and to our delight found that it gave the correct answer for the Betti numbers of the space J_2 of stable bundles of Chern class 1, as computed by Newstead, who had treated this case by direct topological means (see [4]). Later however we were amazed to find precisely the formula (47) in a paper by Harder, arrived at this time from a completely different point of view (see [2]).

Harder, writing at a time where the Weil conjectures were still unproved, planned a nontrivial check of them via a setting of the stable-bundle problem in characteristic p. Using a former result which goes back to C. L. Siegel, he first found a formula for the number of rational points on J_2 mod p^q and finally applied the Weil conjectures to obtain (47). Later he and others extended the formula to bundles of every dimension, and obtained precisely the formula (33) in general. Thus, as the Weil conjectures are now resolved, we actually *know* that (33) is correct, at least over the rationals. (This restriction must be made because the Weil conjectures only deal with the rational homology.)

Atiyah and I of course have different reasons for believing that S is a perfect $\mathcal{G}(P)$-invariant Morse function, and thus to establish (33), as well as prove that in the nonsingular case the higher J's have no torsion. Our reasons for believing this are Morse-theoretic and depend in particular on the completion principle which Hans Samelson and I used a long time ago to prove that certain other functions associated with Lie groups were perfect (see [1]). For functions f with only nondegenerate critical points, this principle asserts that f is a perfect Morse function on W if for each critical point p we can find an orientable manifold M_p and a smooth map

$$\mu : M_p \to W$$

such that

(a) $\mu_p^* f$ has a single nondegenerate maximum of $\mu_p^{-1}(p)$,
(b) $\dim M_p = \text{index } f$ at p.

Indeed if such M_p's (called completing manifolds) can be found for all p, then their fundamental classes map into a set of generators for $H_*(W)$ under μ_p and W is free of torsion.

Now each critical point of the Yang–Mills functional, in the present context, comes equipped with its own completing manifold. Let me illustrate this in our 2-dimensional example. The minimum of S is of course its own completing manifold. At a higher-index critical point the complex bundle E_ω associated to ω over M splits into two line bundles:

$$E = L_1 \oplus L_2. \tag{48}$$

Now the extensions of L_2 by L_1 naturally form a family of bundles parameterized by a complex projective space of $\dim H^1(M; L_2^* \otimes L_1)$, i.e. precisely of the dimension we found for the Yang–Mills index at ω'. At the ω' corresponding to a nontrivial extension one then checks that the Yang–Mills index has *decreased*, and so these projective space have the desired properties.

This then brings us to the end of our excursion, which has already turned out to be more strenuous then I had intended. Clearly there are many details to be filled in this approach to stable bundles—details with which Michael and I would appreciate some help. The details of the index formula and the completion argument are about half written up and should be available next fall.

REFERENCES

[1] R. Bott and H. Samelson, Applications of the theory of Morse to symmetric spaces. *Amer. J. Math.* **80**, 964–1029 (1968).

[2] G. Harder, Eine Bemerkung zu einer Arbeit von P. E. Newstead. *J. für Math.* **242**, 16–25 (1970).

[3] M. S. Narasimhan and C. S. Seshadri, Stable and unitary vector bundles on a compact Riemann surface. *Ann. of Math.* **82**, 540–567 (1965).

[4] P. E. Newstead, Stable bundles of rank 2 and odd degree over a curve of genus 2. *Topology* **7**, 205–215 (1968).

Isometric Families of Kähler Structures[1]

Eugenio Calabi*

1. Introduction

The present study is motivated by the classical problem, loosely stated, of determining and describing, for any given family of complex analytic structures on a fixed, underlying real manifold, which subfamily consists of algebraic manifolds. This problem, which is probably inaccessible in its full generality, even when correctly stated with its missing qualifications, is treated here in a very restricted case, where each of the complex structures of the families under consideration admits a Kähler metric, such that the complex manifolds of the same family are isometric (though not complex analytically, in general). Clearly, this situation is highly restrictive compared to the general problem; however the information that one can extract from the two typical classes of such isometric families that can occur provides some interesting results on the global structure of some moduli spaces of nonalgebraic varieties, such as the $K3$-surfaces, the complex tori, and a special type of $2n$-dimensional, rational affine variety.

2. The Holonomy Group

Let (M, g) be a connected, $2n$-dimensional Riemannian manifold, and for some fixed $x \in M$ let H_x denote the linear holonomy group, based at x, relative to the Levi–Civita connection of (M, g). Then $H_x \subset O(M_x)$, where M_x denotes the tangent space of M at x and $O(M_x)$ is the group of linear transformations of M_x, isomorphic to $O(2n)$, that leaves the quadratic form $g(x)$ invariant. It is well known (Lichnérowicz [12], Newlander and Nirenberg [13]) that (M, g) is the Riemannian structure underlying a Kähler structure $(M; g, J)$ if and only if there exists an imbedding of the linear group $U(n)$ in $O(M_x)$, its image denoted by $U(M_x)$, such that $H_x \subset U(M_x) \subset O(M_x)$. The imbedding is uniquely determined by the image $J(x) \in O(M_x)$ of the element $\sqrt{-1}\, 1 \in U(n)$ ($1 =$ identity); the element $J(x)$ is, *a priori*, an arbitrary element of $O(M_x)$ satisfying $(J(x))^2$

[1] Supported in part by NSF grant No. MCS 78-02285.
*Department of Mathematics, University of Pennsylvania, Philadephia PA 19104, USA, and Institute for Advanced Study, Princeton, NJ 08540, USA.

$= -1$, and the group $U(M_x)$ is then the centralizer of $J(x)$ in $O(M_x)$. The necessary and sufficient condition for (M, g) to be "Kählerizable" is that the centralizer $Z(H_x)$ of H_x in $O(M_x)$ contain an element $J(x)$ with $J(x)^2 = -1$. In this case the parallel transport of $J(x)$ from x to any other point of M is independent of the path and thus generates the structure tensor J whose covariant derivative is zero, so that the almost complex structure defined by J is integrable.

If there is one tensor J with the properties just described, then also $-J$ satisfies the same conditions, corresponding to the conjugate complex structure. The interesting cases occur when the centralizer of H_x contains a continuum of elements $J(x) \in O(M_x)$ satisfying $(J(x))^2 = -1$. For such manifolds the centralizer group of H_x in $O(M_x)$ is a compact group with more than one dimension. Keeping in mind Marcel Berger's classification of the admissible holonomy groups of locally irreducible, non-symmetric Riemannian manifolds [4] and subsequent later contributions (Brown and Gray [5], Alekseevskiĭ [1]) to that topic, the only irreducible holonomy groups that can occur, with some isolated exceptions, are the compact groups $Sp(n)$ acting in $4n$-space, endowing it with a structure of a quaternionic module. We are thus limited to examining only two typical cases of manifolds admitting Kählerian structures in infinitely many complex structures with mutually isometric Kähler metrics: the locally flat ones, and the $Sp(n)$-manifolds. Among the former, we consider only the tori, because of the importance of the function theory in abelian varieties and other, nonalgebraic, tori; among the latter, we deal especially with the $K3$-surfaces, since they are the only compact manifolds of $4n$ real dimensions admitting $Sp(n)$ as holonomy group, according to a recent result announced by Bogomolov [6]. In higher dimensions we consider only one isolated case in each dimensionality $4n$, the manifold that can be described briefly as the cotangent vector bundle over the complex projective n-space; we include it because it helps in describing an interesting "vanishing cycle" that occurs in the resolution of a certain type of singularity.

3. Complex Structures in $2n$-Dimensional Tori

We recall the classical theorem on the period relations for n-dimensional abelian varieties; we restate it here in the form used by A. Weil [15]. A $2n$-dimensional torus T (i.e. the affine space R^{2n} modulo the discrete translation group Z^{2n}) with a translation-invariant complex structure J, abbreviated the pair (T, J), is an algebraic manifold (indeed, after Kodaira, a projective algebraic manifold), if and only if there exists a 2-dimensional, integral cohomology class on T, uniquely representable by a translation invariant, exterior 2-form ω, with the following properties. If $(z_1, \ldots, z_n)$ is any complex system of complex-valued, translation-flat, local coordinates on T that is holomorphic relative to J, then the form ω is purely of bidegree $(1, 1)$ with respect to $(z_1, \ldots, z_n; \bar{z}_1, \ldots, \bar{z}_n)$, i.e.

$$\omega = (2\pi i)^{-1} \sum_{\alpha,\beta=1}^{n} a_{\alpha\bar{\beta}} \, dz^{\alpha} \wedge dz^{\bar{\beta}}, \tag{3.1}$$

and the matrix $(a_{\alpha\bar{\beta}})$ of coefficients of $2\pi i\omega$, necessarily Hermitian, is positive definite.

If we disregard, for the moment, the complex structure J, the integral cohomology class ω is simply one whose nth cup power is not zero: such a class is called a *polarization* of T. While the set of all translation-invariant complex structures J on T is equivalent to the homogeneous space $GL(2n, R)/GL(n, C)$ the set of those J's for which the translation-invariant form ω representing a given polarization is of bidegree $(1, 1)$, called complex structures compatible with ω, belong to $n + 1$ disjunct, $n(n + 1)$-dimensional (over R) components, according to the signature of the matrix $(a_{\alpha\bar{\beta}})$ occurring in (3.1); the component where $(a_{\alpha\bar{\beta}})$ is of index j $(0 \leqslant j \leqslant n)$ is equivalent to the complex symmetric space $Sp(n, R)/U(n; j)$, where $U(n; j)$ denotes the subgroup of $GL(n, C)$ that leaves invariant a Hermitian form of the type $-\sum_{\alpha=1}^{j} |z_\alpha|^2 + \sum_{\alpha=j+1}^{n} |z_\alpha|^2$.

While the generic structure in the components corresponding to the index j $(1 \leqslant j < n)$ is not algebraic,[2] there is some interest in the particular complex structures arising there, because such families of complex tori, or some holomorphic subfamilies of them, occur naturally as intermediate Jacobi varieties of certain appropriate families of higher-dimensional algebraic manifolds.

Every n-dimensional complex torus (T, J) admits a family of locally flat Kähler metrics g equivalent to the space $GL(n, C)/U(n)$ of all positive definite Hermitian forms. Disregarding again the complex structure, any such metric is merely one of the set of translation-invariant Riemannian metrics on T, the total set being equivalent to the symmetric space $GL(2n, R)/O(2n)$; however, for each such metric g, the set of complex structures J that are *compatible with g*, i.e. with respect to which the symmetric bilinear form derived from g is the real part of a Hermitian form, is equivalent to the compact, complex symmetric space $O(2n)/U(n)$.

We now look at the combined problem; we assume that we are given a translation-invariant Riemannian metric g on T and a polarization, represented by a unique real-valued, translation-invariant form

$$\omega = \sum_{1 \leqslant i < j \leqslant 2n} a_{ij}\, dx_i \wedge dx_j,$$

with a nonsingular coefficient matrix (a_{ij}), unrelated to g. We pose the question of describing the set of complex structures J that are compatible simultaneously with g and with ω, and how this set is partitioned according to the index of ω with respect to J.

Theorem 3.1. *Given a $2n$-dimensional torus T with an arbitrary translation-invariant Riemannian metric g and a polarization represented by a translation-invariant, nondegenerate, exterior 2-form ω, for each integer j $(0 \leqslant j \leqslant n)$ there are, in general, exactly $\binom{n}{j}$ complex structures J on T with respect to which g is Hermitian and ω is a form of bidegree purely $(1, 1)$ and of index j. More precisely, there is always a unique J corresponding to a positive polarization $(j = 0)$; then the conjugate complex structure $-J$ is the unique one corresponding to the index*

---

[2] I owe this remark to P. Griffiths.

$j = n$; for the remaining, intermediate values of j, the set $\mathcal{J}(g, \omega, j)$ of complex structures J compatible with both g and ω, for which ω is of index j, depends in a certain way on a certain partition of n, $n = \sum_{\rho=1}^{r} \{ m_\rho \mid 1 \leqslant m_1 \leqslant \cdots \leqslant m_r \leqslant n \}$, depending on the pair (g, ω), namely, $\mathcal{J}(g, \omega; j)$ is the disjoint union of the product of real Grassmannians as follows:

$$\mathcal{J}(g, \omega; j) = \bigcup_{j \,:\, 0 < j_\rho \leqslant m_\rho} \left(\prod_{\rho=1}^{r} \mathrm{Gr}(m_\rho; j_\rho) \right) \quad \text{with} \quad \sum_{\rho=1}^{r} j_\rho = j, \qquad (3.2)$$

where $\mathrm{Gr}(m; k)$ denotes the Grassmann manifold describing the set of all k-dimensional vector subspaces of R^m.

Proof. We identify each translation-invariant complex structure J on T with the one induced by J, and also denoted by J, on the tangent space V at any point of T; V in turn is naturally identified with the universal covering group of the translation group of T. Following the viewpoint described by Atiyah [2], each complex structure J on V is uniquely defined by any n-dimensional (over C) vector subspace W of $V_C = V \otimes_R C$ that is transversal to V (equivalently, transversal to the complex conjugate subspace $\overline{W}$). Then V_C naturally splits as a direct sum $W \oplus \overline{W}$, and the complex structure induced by W on V is the one defined by restricting to V the canonical isomorphism $V_C \to V_C/W$, so that the complex structure is induced by the resulting bijective map of V onto $\overline{W}$, and the almost complex structural operator J on V is the one whose C-linear extension to V_C has W and $\overline{W}$ as the eigenspaces corresponding to the eigenvalues $-\sqrt{-1}$ and $\sqrt{-1}$ respectively.

Consider the symmetric bilinear form obtained from g, as well as the alternating bilinear form ω, on V as bijective, R-linear maps of V onto the dual vector space V^*, and extend them by C-linearity to bijective operators of V_C onto $V_C^* = \mathrm{Hom}_C(V_C, C)$. Then J is compatible with g (respectively ω) if and only if $g(W) = W^*$ (respectively, $\omega(W) = \overline{W}^*$, $\overline{W}^* = \mathrm{Hom}_C(\overline{W}, C) \subset V_C^*$); thus the complex structure defined by W is compatible with g (respectively ω) if only if W—and hence also $\overline{W}$—is totally isotropic with respect to G (respectively ω). Consider now the automorphism $P = g^{-1}\omega : V_C \to V_C$; first of all, if we express P in terms of a real basis (that is to say, belonging to V) that is orthonormal with respect to g, the matrix of P is real and skew-symmetric; therefore P is diagonalizable and its eigenvalues are purely imaginary, nonzero, and conjugate in pairs, say $(\pm \lambda'_\alpha \sqrt{-1})$ $(1 \leqslant \alpha \leqslant n, 0 < \lambda'_1 \leqslant \lambda'_2 \leqslant \cdots \leqslant \lambda'_n)$. Let us say that, among the n positive constants λ'_α, there are r distinct ones, λ_ρ $(1 \leqslant \rho \leqslant r,$ $0 < \lambda_1 < \lambda_2 < \cdots < \lambda_r)$, respectively with multiplicity m_ρ, so that $\lambda_\rho \sqrt{-1}$ is the eigenvalue corresponding to an m_ρ-dimensional eigenspace $E_\rho \subset V_C$, and $\overline{E}_\rho$ is the eigenspace for the eigenvalue $-\lambda_\rho \sqrt{-1}$. For the sake of simplicity let us write $E_{-\rho}$ for $\overline{E}_\rho$ and denote by $\lambda_{-\rho}$ the negative quantity $-\lambda_\rho$, so that $\lambda_\rho \sqrt{-1}$ is the eigenvalue for E_ρ for $\rho = \pm 1, \ldots, \pm r$.

If W is totally isotropic with respect to both g and ω, then $g^{-1}\omega(W) \subseteq g^{-1}(\overline{W}^*) \subseteq W$, so that W is an invariant space of $P = g^{-1}\omega$ and hence W is spanned by n out of $2n$ possible linearly independent eigenvectors of P. Let

$\dim_C(W \cap E_{-\rho}) = j_\rho$ $(1 \leqslant \rho \leqslant r)$, where $0 \leqslant j_\rho \leqslant m_\rho$; then $j_\rho = \dim_C(\overline{W} \cap E_\rho)$ as well. We can choose a basis $(e'_1, \ldots, e'_n)$ for W and set $e'_{-\alpha} = \bar{e}'_\alpha$ $(1 \leqslant \alpha \leqslant n)$ as a basis for $\overline{W}$ with the following properties:

(a) Each e'_α is an eigenvalue for $P = g^{-1}\omega$, belonging, say to E_{ρ_α} $(\rho_\alpha = \pm 1, \ldots \pm r)$.
(b) For each pair of indices α, β $(1 \leqslant \alpha, \beta \leqslant n)$, $g(e'_\alpha, \bar{e}'_\beta) = \delta_{\alpha\beta}$.

Then $\omega(e'_\alpha, \bar{e}'_\beta) = \lambda_{\rho_\alpha}\sqrt{-1}\, g(e'_\alpha, \bar{e}'_\beta)$ and $\omega(e'_\beta, \bar{e}'_\alpha) = \lambda_{\rho_\alpha}\sqrt{-1}\, g(e'_\beta, \bar{e}'_\alpha)$. Thus the Hermitian form $\sqrt{-1}\, g(\bar{z}, \zeta)$ for $z, \zeta \in W$ is diagonal with respect to the basis $(e'_1, \ldots, e'_n)$, and we verify that

$$\sqrt{-1}\, \omega(\bar{e}'_\alpha, e'_\beta) = \lambda_{\rho_\alpha} g(e'_\beta, \bar{e}'_\alpha) = \lambda_{\rho_\alpha}\delta_{\alpha\beta},$$

showing that there are as many negative numbers in the sequence $(\lambda_{\rho_1}, \ldots, \lambda_{\rho_n})$ as there are dimensions in $W \cap (\sum_{\rho=1}^e E_{-\rho})$, that is to say $\sum_{\rho=1}^r j_\rho$.

The conclusion now follows by straightforward verification. We remark now that, in general, the eigenvalues of P are all distinct, so that the set of complex structures J compatible with both G and ω, and for which ω has index j, is limited to the choice of subspaces W spanned by j out of the n eigenspaces $\overline{E}_1, \ldots, \overline{E}_n$, and by the conjugates of the remaining ones. On the other hand the uniqueness of the choice $W = \sum_{\rho=1}^n E_\rho$ that achieves the index $j = 0$ is obvious. $\square$

Corollary. *The pseudo-Kählerian, complex symmetric spaces*

$$O(2n)/U(n; j) \qquad (0 \leqslant j \leqslant n)$$

admits a proper, branched (not holomorphic) covering over the Kählerian symmetric space $O(2n)/U(n)$ with $\binom{n}{j}$ sheets, with branch sets equivalent to products of real Grassmannians, $\prod_{\rho=1}^r \mathrm{Gr}(m_\rho; j_\rho)$ with $0 \leqslant j_\rho \leqslant m_\rho$, $\sum_{\rho=1}^r j_\rho = j$ for each partition $\sum_{\rho=1}^r m_\rho$ of n.

The statement of the above is simply a rewording of the previous theorem in a less precise form.

4. Hyper-Kähler Manifolds

The only known locally irreducible Kähler manifolds that admit a continuum of complex structures all admitting a common Kähler metric are the *hyper-Kähler manifolds* (also called by other authors Hamiltonian Kähler manifolds or Kähler manifolds with an intergrable quaternionic structure). These are manifolds that are $4n$-dimensional and whose holonomy group with respect to a given Riemannian connection is contained in (and in general equals) $\mathrm{Sp}(n)$, so that the tangent bundle has the structure of a quaternionic module, the quaternion scalar operations (for instance, represented as acting on the right) are preserved under Levi–Civita parallel transports.

Hyper-Kähler manifolds may be presented as Kähler manifolds (M_0, g)

admitting holomorphic 2-form φ_0, whose coefficient matrix $(P_{\alpha\beta})$ ($\varphi_0 = P_{\alpha\beta} dz^\alpha \wedge dz^\beta$ in local coordinates) is everywhere nonsingular (hence $\dim_C M_0$ is even) and with identically vanishing covariant derivative with respect to the given Kähler metric. Let $\dim_C M_0 = 2n$; then the nth exterior power of φ_0 is a holomorphic $2n$-form on M_0 whose squared absolute value is a positive multiple of the volume density μ of the Kähler metric. Hence the Ricci tensor of M_0, $R_{\alpha\bar\beta} = -\partial^2(\log \mu)/\partial z^\alpha \partial \bar z^\beta$, vanishes identically.

Let T' and $T'' = \bar T'$ denote the subspaces of the complexified tangent bundle $T(M) \otimes C$ that are locally spanned, respectively, by $(\partial/\partial z^1, \ldots, \partial/\partial z^n)$ and by $(\partial/\partial \bar z^1, \ldots, \partial/\partial \bar z^n)$, and denote by $(\ ,\ )$ the Hermitian inner produce of any two (real- or complex-valued) tangent vectors. Thus, if we denote by $u = u' + u''$ the decomposition of any complex tangent vector according to the decomposition $T(M) \otimes C = T' \oplus T''$, and by $\langle\ ,\ \rangle$ the Riemannian inner product in $T(M_0)$ extended by C-linearity to complex tangents, then for any two real or complex tangents u, v, we have

$$(u,v) = \langle u', v'' \rangle, \qquad \langle u, v \rangle = \langle u', v'' \rangle + \langle v', u'' \rangle.$$

Furthermore we have the C-linear operators K_1' and K_1'', mapping $T(M_0) \otimes C$ respectively onto T'' and T', defined as follows: for any tangent vector $u = u' + u''$, $K_1'u'' = K_1''u' = 0$ and $K_1'u'$, $K_1''u''$ are characterized, respectively, by the equations, for any $v = v' + v''$,

$$(K_1'u', v) = \langle v', K_1'u' \rangle = \varphi_0(u', v'),$$
$$(K_1''u'', v) = \langle K_1''u'', v'' \rangle = \bar\varphi_0(u'', v''). \qquad (4.1)$$

It follows that K_1' maps T' bijectively onto T'', K_1'' maps T'' bijectively onto T', K_1'' is the complex conjugate of K_1', and thus the map $K_1 = K_1' + K_1''$ is a real, bijective map of $T(M_0)$ onto itself; its matrix with respect to any (real) orthonormal basis with respect to the metric is antisymmetric, so that K_1 is diagonalizable over $T(M_0) \otimes C$, and its spectrum is on the purely imaginary axis and symmetric about 0. Furthermore K_1 has covariant derivative zero. In addition, $K_1^2 = K_1'K_1'' + K_1''K_1'$ is symmetric with respect to the underlying Riemannian structure; it is negative definite, and has T' and T'' as invariant subspaces. Replacing the 2-form φ_0 by $\sqrt{-1}\,\varphi_0$, we obtain likewise a real, bijective map

$$K_2 : T(M_0) \to T(M_0),$$
$$K_2 = \sqrt{-1}\,(K_1' - K_1''),$$

with the same properties described for K_1 and such that $K_2^2 = K_1^2$.

We construct now from K_1 and K_2 two analogous operators, denoted respectively by J_1 and J_2,

$$J_\nu = \frac{1}{\pi} K_\nu \int_{-\infty}^{\infty} (t^2 - K_\nu^2)^{-1} dt, \qquad \nu = 1, 2. \qquad (4.2)$$

From the spectral decomposition of K_ν, one deduces, for either value of ν, that the integral above converges, that J_ν commutes with K_ν, and that J_ν maps

T' and T'' each bijectively onto the other, so that

$$J_\nu = J'_\nu + J''_\nu, \qquad J''_\nu u' = J'_\nu u'' = 0.$$

Furthermore

$$J_\nu^2 = J'_\nu J''_\nu + J''_\nu J'_\nu = -1 \qquad (\nu = 1, 2),$$

$$J'_2 = \sqrt{-1}\, J'_1, \qquad J''_2 = -\sqrt{-1}\, J''_1,$$

and both J_1 and J_2 have vanishing covariant derivatives with respect to the metric, so that they define integrable complex analytic structures with respect to the metric, and the metric is Kählerian with respect to either of the two resulting new complex structures.

If we denote, in addition, by $J_0 = J'_0 + J''_0$ the almost complex structure subordinate to the given complex structure of M_0, where

$$J'_0 = \sqrt{-1}\, \big|_{T'}, \qquad J''_0 = -\sqrt{-1}\, \big|_{T''},$$

then J_0, J_1, J_2 represent three pairwise anticommutative, almost complex structures, with $J_1 J_0 = -J_0 J_1 = J_2$, so that the algebra generated by adjoining $J_0, J_1,$ J_2 to the real scalars (as operators on $T(M_0)$) is isomorphic to the division algebra of quaternions. For any $\alpha = (\alpha_0, \alpha_1, \alpha_2) \in S^2 \subset R^3$ (i.e. such that $\alpha_0^2 + \alpha_1^2 + \alpha_2^2 = 1$), the almost complex structural tensor

$$J(\alpha) = \sum_{\nu=0}^{2} \alpha_\nu J_\nu$$

is integrable to a complex analytic structure on M_0, with the same given metric on M_0 for a Kähler metric, thus describing a family of isometric (with the common metric being Kählerian), complex analytic structures. We restate this fact formally as follows:

Theorem 4.1. *Any hyper-Kählerian manifold* $(M_0;\ g_0, \varphi_0)$ *has, in a cononical way, a family of complex analytic structures* $J(\alpha)$ $(\alpha \in S^2)$ *including the given Kähler structure of* M_0 *and its conjugate, each of them admitting the same metric* g_0 *as a Kähler metric; for antipodal points,* $J(-\alpha) = -J(\alpha)$ *is the conjugate complex structure of* $J(\alpha)$. *For any* $\beta \in S^2$ *that is at right angles to* $\alpha \in S^2$, *the 2-form* $\psi_\beta(u, v)$ *defined by the equation*

$$2\psi_\beta(u, v) = \langle J(\beta)u, v \rangle - \sqrt{-1}\, \langle J(\beta)u, J(\alpha)v \rangle$$

is holomorphic with respect to the complex structure defined by J_α.

We shall examine the S^2-parameterized family of isometric Kähler structures thus constructed in two typical cases, one compact and one not.

A recent result announced by Bogomolov [6] states that there are no compact, $2n$-complex dimensional, locally irreducible hyper-Kähler manifolds for $n \geqslant 2$. For $n = 1$, using Kodaira's classification theorem for compact, complex surfaces, the only nonflat, compact, complex manifolds with first Chern class equal to zero (a necessary condition in a Kähler manifold to admit a nowhere vanishing, holomorphic form of maximum degree) that can arise are the so-called $K3$-surfaces, a family of compact, complex manifolds homologically equivalent to the nonsingular quartic surfaces in the complex projective 3-space, or to those quartics with isolated, ordinary quadratic singularities (up to 16 of them are possible), if removed by a simple "blowing up". The latter family, disregarding the biregular equivalence under projective transformations, has complex structures depending on 19 complex parameters or *moduli*. On the other hand Kodaira and Spencer [9] calculated that the number of complex moduli for small deformation of complex structure of the $K3$-surfaces is 20, revealing that there is a complex analytic family of nonalgebraic $K3$-surfaces, in which the algebraic ones constitute analytic subsets of codimension at least one, much in the same way that the 2-complex-dimensional abelian varieties form analytic subsets of codimension at least one in the 4-(complex)-parameter family of complex analytic structures in a 4-torus.

Let M_0 be a $K3$-surface with any given Kähler metric g'. This surface admits a holomorphic 2-form φ_0 which is nowhere zero; we normalize φ_0 so that $\int_{M_0} \varphi_0 \wedge \bar{\varphi}_0$ equals the total volume of the M_0 under the given Kähler metric. A fundamental theorem due to S. T. Yau [16], to which T. Aubin [3] has made earlier contributions, asserts that there exists another Kähler metric in the same Kähler cohomology class as g', whose volume element coincides with $\varphi_0 \wedge \bar{\varphi}_0$ (such a metric is then unique); we call such a metric a Yau metric on the $K3$-surface. Thus there are as many Yau metrics in M_0 as there are real, 2-dimensional cohomology classes representable by principal Kählerian (positive definite) forms of bidegree $(1, 1)$. Each such metric in M_0 induces in it a structure of a hyper-Kähler manifold.

The second real cohomology group of M_0 is 22-dimensional (since M_0 is simply connected and has Euler characteristic 24); within the space of all real-valued, harmonic 2-forms on M_0, we distinguish the 3-dimensional subspace spanned by the principal Kähler form ω_0 ($= \sqrt{-1}\, g_{\alpha\bar{\beta}}\, dz^\alpha \wedge dz^{\bar{\beta}}$ in terms of local holomorphic coordinates), and by

$$\omega_1 = \varphi_0 + \bar{\varphi}_0, \qquad \omega_2 = \sqrt{-1}\, (\varphi_0 - \bar{\varphi}_0),$$

where φ_0 is the holomorphic 2-form in M_0 used earlier.

These three harmonic 2-forms have equal cup squares $\frac{1}{2}\omega_\nu \wedge \omega_\nu$ identical to the volume element of the Yau metric (by definition). The orthogonal complementary space of harmonic forms (19-dimensional over R) is made up of the $(1, 1)$-forms η that are *primitive* with respect to the Kähler class, i.e. satisfy $\eta \wedge \omega_0 = 0$ identically (and, trivially, also $\eta \wedge \omega_1 = \eta \wedge \omega_2 = 0$). The primitive $(1, 1)$-forms are also characterized as being skew-self-adjoint, i.e. $*\eta = -\eta$; in other words $\eta \wedge \eta$ is a 4-form that is negative at each point where $\eta \neq 0$. Thus the signature of M_0 is $3 - 19 = -16$.

Consider now the 2-parameter family of complex structures in M_0, obtained by integration of the structure tensors $J(\alpha)$ ($\alpha = (\alpha_1, \alpha_2, \alpha_3) \in S^2$) arising from the hyper-Kähler or Yau metric structure of M_0. As in the case of complex tori, in order to pick a reasonable analytic subfamily of algebraic structures in the moduli space of M_0, we shall introduce a *polarization*. This is defined by an integral, 2-dimensional cohomology class u in M_0, represented by a real-valued, closed, exterior 2-form Φ in M_0 (of mixed bidegree, in general), whose exterior product with itself has a strictly positive integral over M_0; such a polarization is called *weakly positive*. If, in addition, the polarization class u can be represented by a real-valued, closed 2-form Φ such that $\Phi \wedge \Phi$ is strictly positive at each point (relative to the natural orientation of M_0), we say that the polarization is *strongly positive*. Finally, if the polarization class u can be represented by a closed, real-valued 2-form Φ which, in terms of the given complex structure in M_0, is purely of bidegree $(1, 1)$ and also satisfies everywhere the condition $\Phi \wedge \Phi > 0$, we say that u is a *strictly positive* polarization. Disregarding for the moment the assumption that u is assumed to be an integral cohomology class, each of the three sets consisting of, respectively, weakly positive, strongly positive, and strictly positive, real, 2-dimensional cohomology classes in M_0, is an open cone in $H^2(M_0, R)$, each contained in the previous one, and containing the nonzero elements in the 3-dimensional subspace spanned by the de Rham classes of ω_0, ω_1, and ω_2 (these three forms span a maximal vector subspace in the space of 2-forms Φ for which the quadratic form $\Phi \wedge \Phi$ is strictly positive definite everywhere). The boundaries of the last two of these three cones are unknown, as is also the answer to the question whether any two of the cones coincide.

Lemma 4.2. *Let Φ be any real-valued, harmonic 2-form in M_0 such that $\int_{M_0} \Phi \wedge \Phi > 0$. With reference to a given Ricci flat Kähler metric in M_0, let the Hodge decomposition of Φ be expressed by $\Phi = a\omega_0 + c\varphi_0 + \bar{c}\bar{\varphi}_0 + \eta$, where ω_0 is the principal Kähler $(1, 1)$-form, $a \in R$, φ_0 is a holomorphic 2-form in M_0, normalized so that $\int_{M_0} \varphi_0 \wedge \bar{\varphi}_0 = \frac{1}{2} \int_{M_0} \omega_0 \wedge \omega_0$, $c \in C$, and η is a primitive, harmonic $(1, 1)$-form, then the following conclusions hold:*

(a) *At least one of the coefficients a, c is different from zero.*
(b) *The corresponding Hodge decomposition of Φ with respect to each of the complex structures induced by each $J(\alpha)$ ($\alpha \in S^2$), takes the form*

$$\Phi = c_\alpha \varphi_{(\alpha)} + \bar{c}_\alpha \bar{\varphi}_{(\alpha)} + a_\alpha \omega_{(\alpha)} + \eta,$$

where η is the same as for the originally assumed complex structure.
(c) *There exists a unique $\alpha \in S^2$ for which $c_\alpha = 0$ and $a_\alpha > 0$; for the conjugate complex structure we have $a_\alpha < 0$.*

Proof. Since η is a primitive $(1, 1)$-form, this means that $\int_{M_0} \eta \wedge \eta < 0$ unless η is identically zero; since $\int_{M_0} \Phi \wedge \Phi > 0$, it follows that Φ cannot equal η. This proves assertion (a).

In order to prove assertion (b), we refer to the calculations leading to the statement of Theorem 4.1. Observe that the operators K_1 and K_2 already satisfy $K_\nu^2 = -1$ ($\nu = 1, 2$); hence from Equation (4.2) we have $J_\nu = K_\nu$ ($\nu = 1, 2$).

In terms of the complex structure defined in M_0 by $J(\alpha) = \sum_{\nu=0}^{2} \alpha_\nu J_\nu$, $\sum_{\nu=0}^{2} \alpha_\nu^2 = 1$, we observe that the principal Kähler form of the Yau metric is

$$\omega_0(\alpha) = \sum_{\nu=0}^{2} \alpha_\nu \omega_\nu,$$

where

$$\omega_1 = \varphi_0 + \bar{\varphi}_0, \qquad \omega_2 = \sqrt{-1}\,(\varphi_0 - \bar{\varphi}_0);$$

in fact we have, from the definition of J_ν ($\nu = 0, 1, 2$),

$$\omega_0(\alpha)(u, v) = \langle J(\alpha)u, v \rangle = \sum_{\nu=0}^{2} \alpha_\nu \langle J_\nu u, v \rangle = \sum_{\nu=0}^{2} \alpha_\nu \omega_\nu(u, v).$$

Consider the 22-dimensional vector space $H^2(M_0, g)$ over the reals, consisting of all the real-valued, harmonic 2-forms Φ in M_0: this space is defined only in terms of the underlying Riemannian structure of M_0, independently of any of the superimposed, complex structures determined by any $J(\alpha)$. For each $\alpha \in S^2$, the Hodge decomposition of Φ is again of the form $a(\alpha)\omega_0(\alpha) + c(\alpha)\varphi_0(\alpha) = \overline{c(\alpha)\varphi_0(\alpha)} + \eta(\alpha)$, where $\omega_0(\alpha)$ is the principal $(1, 1)$-form (4.2) and $\varphi_0(\alpha)$ is a holomorphic $(2, 0)$-form normalized by $\int_{M_0} \varphi_0(\alpha) \wedge \overline{\varphi_0(\alpha)} = \frac{1}{2} \int_{M_0} \omega_0(\alpha) \wedge \omega_0(\alpha)$. It is easy to verify that, since the orientation of M_0 is fixed, for each $\Phi \in H^2(M_0, g)$, the dual form $*\Phi$ has the Hodge decomposition, in terms of $J(\alpha)$, $*\Phi = a(\alpha)\omega_0(\alpha) + c(\alpha)\varphi_0(\alpha) + \overline{c(\alpha)\varphi_0(\alpha)} - \eta(\alpha)$. Hence the anti-self dual part $\eta(\alpha)$,

$$\eta(\alpha) = \tfrac{1}{2}(\Phi - {}^*\Phi),$$

is determined independently of the complex structure $J(\alpha)$, proving assertion (b).

For each $\alpha \in S^2$ the choice of the normalized, holomorphic 2-form $\varphi_0(\alpha)$ is unique up to a complex constant factor with unit absolute value. A real basis for the self-dual subspace of $H^2(M_0, g)$ can be made up of $\omega_0(\alpha)$, $\omega_1(\alpha) = \varphi_0(\alpha) + \overline{\varphi_0(\alpha)}$, $\omega_2(\alpha) = \sqrt{-1}\,(\varphi_0(\alpha) - \overline{\varphi_0(\alpha)})$; such a basis is orthornormal both globally on M_0 and pointwise, since the pointwise norm of any self-dual harmonic form in M_0 is constant (a consequence of the metric being a Yau metric). Consequently the set of all real bases $\omega_\nu(\alpha)$ ($\nu = 0, 1, 2$) is related to the initial basis ω_ν by a linear transformation belonging to the group $SO(3)$, with the first row consisting of the coordinates $(\alpha_0, \alpha_1, \alpha_2)$ of α.

Finally assertion (c) is proved as follows. At any one point where $\Phi \neq 0$ there are exactly two antipodal values of $\alpha \in S^2$ such that Φ at that point is purely of bidegree $(1, 1)$ with respect to the complex structure $J(\alpha)$. At that point therefore, and for the chosen values of α, the $(2, 0)$- and $(0, 2)$-components of Φ vanish; but, since in a compact, complex manifold with vanishing first Chern class, a harmonic $(2, 0)$-form (or its conjugate) cannot vanish at a point unless such a form is identically zero, there are exactly two antipodal values of $\alpha \in S^2$, where Φ is purely of bidegree $(1, 1)$. Since Φ cannot be a primitive form, there is in either case a part in the decomposition that is a nonzero multiple of the principal form, with positive coefficient for one of the values of α, and negative for the other. This completes the proof of the lemma. $\square$

We now consider a strongly positive polarization u in the manifold M_0 and repeat the argument of the lemma just proved. We recall that, according to Kodaira's imbedding theorem, a compact Kähler surface is a projective algebraic variety if and only if it admits a strictly positive polarization (with integer coefficients) representable by a closed 2-form Φ of bidegree purely $(1, 1)$ and of positive type at each point; the sense in which such a surface becomes a projective variety is that any sufficiently high, positive integer multiple of the polarization class becomes the divisor class of a projective hyperplane intersection with M_0, as a result of a holomorphic imbedding of the given surface into a complex projective space. We have thus proved the following theorem.

Theorem 4.3. *Let M_0 be any analytic $K3$-surface. Given any strictly positive polarization u in M_0 and any Yau metric (i.e. a Ricci flat Kähler metric) in $\dot{M}_0$, there is a unique complex structure $J(\alpha)$ on M_0 with respect to which given Yau metric is a Kahler metric and the polarization class representable by a harmonic form.*

The question whether the conclusion of this theorem still holds if the polarization class u is assumed to be only strongly positive appears still uncertain. A recently announced result by A. N. Todorov [14] seems to imply it, by making use of a strong formulation of a Torelli theorem (Horikawa [8], Kulikov [11]) for $K3$-surfaces.

A second class of examples of hyper-Kähler structures is furnished by the holomorphic cotangent vector bundle $T^{*\prime}(CP^n)$ of the n-dimensional complex projective space CP^n. Let $(u_0, \ldots, u_n) \in C^{n+1}\backslash\{0\}$ denote the complex homogeneous coordinates in CP^n; let $z^\alpha = u_\alpha/u_0$ $(1 \leqslant \alpha \leqslant n)$ denote the inhomogeneous coordinate functions in the domain $U_0 = \{u_0 \neq 0\} \subset CP^n$, and ζ_α $(1 \leqslant \alpha \leqslant n)$ the fibre coordinates of $T^{*\prime}(CP^n)|_{U_0}$ at the point represented by the holomorphic form $\sum_{\alpha=1}^n \zeta_\alpha \, dz^\alpha$; and let the Fubini–Study metric in CP^n be represented by the Hermitian form in U_0

$$ds^2 = 2g_{\alpha\bar{\beta}}(z, \bar{z}) \, dz^\alpha \, dz^{\bar{\beta}},$$

where

$$g_{\alpha\bar{\beta}} = \frac{1}{2} \frac{\partial^2}{\partial z^\alpha \partial z^{\bar{\beta}}} \log\left(1 + \sum_{\lambda=1}^n |z^\lambda|^2\right).$$

Then the transposed inverse matrix $(g^{\alpha\bar{\beta}})$ of $(g_{\alpha\bar{\beta}})$ is expressed by

$$g^{\alpha\bar{\beta}} = 2\left(1 + \sum_{\lambda=1}^n |z^\lambda|^2\right)(\delta^{\alpha\beta} + z^\alpha z^{\bar{\beta}}).$$

The manifold $M_0 = T^{*\prime}(CP^n)$ is covered by $n + 1$ subdomains $W_a = T^{*\prime}(U_a)$ $(0 \leqslant a \leqslant n)$, where $U_a = \{u_a \neq 0\}$, each subdomain being equivalent to $T^{*\prime}(U_0)$ under permutations of the homogeneous coordinates $(u_0, \ldots, u_n)$. The norm function of M_0 is expressed in $T^{*\prime}(U_0)$ by the real-valued function

$$t = g^{\alpha\bar{\beta}}\zeta_\alpha\zeta_{\bar{\beta}} = (1 + |z|^2)(|\zeta|^2 + |z \cdot \zeta|^2),$$

where

$$|z|^2 = \sum_{\alpha=1}^n |z^\alpha|^2, \qquad |\zeta|^2 = \sum_{\alpha=1}^n |\zeta_\alpha|^2, \qquad z \cdot \zeta = \sum_{\alpha=1}^n z^\alpha \zeta_\alpha.$$

Then the real-valued function $\Psi(z, \bar{z}; \zeta, \bar{\zeta})$ in $T^{*'}(U_0)$, expressed in terms of the coordinates $(z^1, \ldots, z^n; \zeta_1, \ldots, \zeta_n)$ by

$$\Psi(z, \bar{z}; \zeta, \bar{\zeta}) = \log(1 + |z|^2) + \sqrt{1 + 4t} - \log(1 + \sqrt{1 + 4t}), \tag{4.3}$$

$$t = (1 + |z|^2)(|\zeta|^2 + |z \cdot \zeta|^2), \tag{4.4}$$

(cf. Calabi [7, §5]) is the Kähler potential for a metric that is hyper-Kählerian with respect to the given complex structure and the Hamiltonian form

$$\varphi_0 = \sum_{\alpha=1}^{n} dz^\alpha \wedge d\zeta_\alpha.$$

We shall describe now the complex structures that occur in this open variety, induced by the structural tensors $J(\alpha)$ for $\alpha = (\alpha_0, \alpha_1, \alpha_2) \in S^2$ other than the points $(\pm 1, 0, 0)$ corresponding to the originally given complex structure and its conjugate.

Let $\mathcal{V}$ be the $(2n + 1)$-dimensional complex analytic set in $C^{(n+1)^2}$ consisting of the complex $(n + 1) \times (n + 1)$ matrices (x_{ij}) $(0 \leqslant i, j \leqslant n; n \geqslant 1)$ that are of rank one or zero,

$$x_{ik}x_{jl} - x_{il}x_{jk} = 0 \qquad (0 \leqslant i, j, k, l \leqslant n),$$

and the map $\mathrm{tr} : \mathcal{V} \to C$ defined by the trace; for each $s \in C$, $V_s = \mathrm{tr}^{-1}(s)$ is an irreducible, $2n$-dimensional complex space, nonsingular for $s \neq 0$ and with an isolated singularity at the origin for $s = 0$. The absence of singularities in V_s away from the origin stems from the fact that for each $s \neq 0$, V_s, as well as $V_0 \backslash \{0\}$ for $s = 0$, is homogeneous under the adjoint action of the matrix group $GL(n + 1, C)$ on $C^{(n+1)^2}$. Since the variety $\mathcal{V}$ is a cone, the inhomogeneous hyperplane sections V_s $(s \neq 0)$ are projectively equivalent to the infinite part of $\mathcal{V} \backslash V_0$ in the projective completion of $C^{(n+1)^2}$. Thus the infinite part of $\mathcal{V}$ has an alternative analytic realization as the product of CP^n with the dual projective space, i.e. as the space of pairs (p, q^*) consisting of points p and hyperplanes q^* in CP^n, while the infinite part of V_0 is the subset consisting of incident pairs $\{(p, q^*) \mid p \in q^*\}$. Thus V_s, for each $s \neq 0$, is a nonsingular affine algebraic variety, holomorphically equivalent to the set of pairs (p, q^*) with $p \notin q^*$.

The singularity of $\mathcal{V}$ at the origin can be resolved in two distinct ways, mutually equivalent under duality; to each matrix $X = (x_{ij}) \in \mathcal{V} \backslash \{0\}$ we associate the row-augmented matrix $(u; X)$, obtained by adjoining to X an extra column whose entries $(u_0, \ldots, u_n)$ are homogeneous coordinates, proportional to any one of the nonzero columns of X, say $(u_0, \ldots, u_n) = \rho(x_{0b}, \ldots, x_{nb})$ for some $\rho \neq 0$. Thus we obtain the subvariety of $CP^n \times (\mathcal{V} \backslash \{0\})$ that can be represented by $(n + 2) \times (n + 1)$ matrices of rank 1, where the first column $(u_0, \ldots, u_n)$ is never zero, and identified under proportional equivalence. This variety is obviously biregularly equivalent to $\mathcal{V} \backslash \{0\}$, but its closure in $CP^n \times \mathcal{V}$ includes $CP^n \times \{0\}$ in correspondence with the singularity of $\mathcal{V}$. The resulting space is therefore a complex analytic subvariety of $CP^n \times C^{(n+1)^2}$; we shall denote it by $\mathcal{V}_1$, and show presently that it is a desingularized model for $\mathcal{V}$ under the obvious holomorphic map $\pi : \mathcal{V}_1 \to \mathcal{V}$. An equivalent desingularization can be obtained by interchanging the roles of rows and columns in $\mathcal{V}$, while the composite of the two dilatations, replacing $\{0\}$ by $CP^n \times CP^n$, is equivalent to the result on $\mathcal{V}$ of the standard dilatation of $C^{(n+1)^2}$ with respect to the origin.

Lemma 4.4. *The variety $\mathcal{V}_1$ obtained from $\mathcal{V}$ by replacing the origin by an n-dimensional complex projective space is nonsingular; moreover the map $\pi_1 = \mathrm{tr} \circ \pi : \mathcal{V}_1 \to C$ is a surjection with no critical points. In particular the inverse image $\pi^{-1}(V_0) = \pi_1^{-1}(0)$ of V_0 is a desingularized model of V_0; moreover it is holomorphically equivalent to $T^*(CP^n)$ and diffeomorphic to $V_s = \mathrm{tr}^{-1}(s)$ for each $s \neq 0$.*

Proof. Consider the open cover of $\mathcal{V}_1$ made up of the $n + 1$ sets $\mathcal{U}_a$ $(0 \leqslant a \leqslant n)$, where each $\mathcal{U}_a$ is the subset such that the homogeneous coordinate u_a occurring in the augmented matrix $(u_i; x_{ij})$ is $\neq 0$. We show that each $\mathcal{U}_a$ can be mapped bijectively onto C^{2n+1} by the $2n + 1$ functions $(z_i; \zeta_j; s)$ $(0 \leqslant i \leqslant n, 0 \leqslant j \leqslant n;$ $i \neq a, j \neq a)$, where z_i and ζ_j are defined by

$$z_i = \frac{u_i}{u_a}, \qquad \zeta_j = x_{aj}.$$

In order to show that this is indeed a local coordinate system for $\mathcal{V}_1$, it is sufficient to show that all the extrinsic coordinates x_{ij} of the point corresponding to any assigned set of values $(z_i; \zeta_j; s)$ are uniquely determined and are holomorphic functions of $(z_i; \zeta_j; s)$ defined in C^{2n+1}; we omit the homogeneous coordinates (u_k), since their mutual ratios are described by the values of $z_i = u_i/u_a$ alone. In fact, since the augmented matrix $(u_i; x_{ij})$ is, by definition, of rank one, it follows that, for $j \neq a$,

$$x_{ij} = \frac{u_i}{u_a} x_{aj} = z_i \zeta_j,$$

$$x_{aj} = \zeta_j,$$

while for $j = a$,

$$x_{aa} = s - \sum_{i \neq a} x_{ii} = s - \sum_{i \neq a} z_i \zeta_i,$$

$$x_{ia} = z_i x_{aa} = z_i \left(s - \sum_{j \neq a} z_j \zeta_j \right) \qquad (i \neq a).$$

In addition this argument shows that, for each fixed value of $s \in C$ (whether equal to zero or not), the $2n$ functions $(z_i; \zeta_j)$ constitute a system of local coordinates in $\pi_1^{-1}(x) \cap \mathcal{U}_a$, so that the submanifolds $\pi_1^{-1}(s) = \pi^{-1}(V_s)$ (trivially isomorphic to V_s for $s \neq 0$) are diffeomorphic to $\pi_1^{-1}(0) = \pi^{-1}(V_0)$.

There remains to be shown that $\pi_1^{-1}(0)$ is isomorphic (holomorphically) to $T^*(CP^n)$. Indeed, we shall prove not only this fact, but also that, for each $s \neq 0$, $\pi_1^{-1}(s)$, and hence V_s, is isomorphic to a holomorphic bundle of affine complex n-spaces over CP^n, obtained by a deformation of, but inequivalent to, $T^*(CP^n)$. We define the holomorphic differential $\psi_{a,s}$ in $\mathcal{U}_a \cap \pi_1^{-1}(s)$ for each a $(0 \leqslant a \leqslant n)$ and each $s \in C$,

$$\psi_{a,s} = \sum_{i \neq a} \zeta_i \, dz_i.$$

In $\mathfrak{A}_a \cap \mathfrak{A}_b \cap \pi_1^{-1}(s)$ we calculate the discrepancy between $\psi_{a,s}$ and $\psi_{b,s}$ as follows: $\mathfrak{A}_a \cap \mathfrak{A}_b$ is characterized by the condition that $u_a u_b \neq 0$; under this condition we have

$$
\begin{aligned}
\psi_{b,s} &= \sum_{i \neq b} x_{bi}\, d\left(\frac{u_i}{u_b} \right) = \sum_{i \neq b} x_{bi}\left(\frac{u_a}{u_b}\, d\left(\frac{u_i}{u_a} \right) + \frac{u_i}{u_a}\, d\left(\frac{u_a}{u_b} \right) \right) \\
&= \sum_{i \neq a,b} x_{ai}\, d\left(\frac{u_i}{u_a} \right) + \left(\sum_{i \neq b} x_{ii} \right) d\log\left(\frac{u_a}{u_b} \right) \\
&= \sum_{i \neq a,b} x_{ai}\, d\left(\frac{u_i}{u_a} \right) + (s - x_{bb})\, d\log\frac{u_a}{u_b} \\
&= \sum_{i \neq a,b} x_{ai}\, d\left(\frac{u_i}{u_a} \right) + x_{bb}\frac{u_a}{u_b}\, d\left(\frac{u_b}{u_a} \right) - s\, d\log\left(\frac{u_b}{u_a} \right) \\
&= \psi_{a,s} - s\, d\log\left(\frac{u_b}{u_a} \right).
\end{aligned}
$$

Thus we have, in $\pi_1^{-1}(s) \cap \mathfrak{A}_a \cap \mathfrak{A}_b$,

$$
\psi_{b,s} - \psi_{a,s} = -s\, d\log\left(\frac{u_b}{u_a} \right). \tag{4.5}
$$

This shows that, for $s = 0$, the $n + 1$ differential forms, defined each in one of the covering coordinate domains, blend into a global 1-form ψ_0. The latter defines a contact structure in $\pi^{-1}(V_0)$, since, locally $d\psi_0 = \sum_{i \neq a} d\zeta_i \wedge dz^i$, which is a Hamiltonian 2-form. It is easy to verify that the canonical map of $\pi^{-1}(V_0)$ onto $T^{*\prime}(CP^n)$ defined locally by the evaluation of ψ in terms of local holomorphic coordinates is a global isomorphism.

For $s \neq 0$, the manifolds $\pi_1^{-1}(s) = \pi^{-1}(V_s)$, canonically isomorphic to V_s, are obviously holomorphic bundles over CP^n under the map π_s assigning to the augmented matrix $(u_i; x_{ij})$ the point in CP^n with homogeneous coordinates (u_i); in terms of the local coordinates (z_i, ζ_j) for any $\mathfrak{A}_a \cap \pi_1^{-1}(s)$, the first n parameters (z_i) represent the base point, the remaining ones (ζ_j) the fibre coordinates ranging over C^n, and (4.5) determines an affine transformation from one system of fibre coordinates to another. This bundle is clearly a deformation of the one for $s = 0$, which was shown to be isomorphic to $T^{*\prime}(CP^n)$, but is inequivalent to it for $s \neq 0$, since V_s is an affine variety in $C^{(n+1)^2}$, so that it has no compact complex subvarieties of positive dimension; in particular V_s, as a bundle over CP^n, has no holomorphic cross section for $s \neq 0$, unlike the cotangent bundle: which has the zero section. This completes the proof of the lemma. □

We shall conclude by showing that the family $(V_s)_{s \in C}$, where $V_s = \pi_1^{-1}(s) \subset \mathcal{V}_1$, describes *locally* a model of the deformation of complex structure of $T^{*\prime}(CP^n)$ defined by the S^2-parameterized family of integrable almost complex structures $J(\alpha)$ associated to the hyper-Kählerian structure on the cotangent bundle.

Theorem 4.5. *The two deformations of the holomorphic cotangent bundle $T^{*\prime}(CP^n)$ of the n-dimensional complex projective space defined, respectively, by the S^2-parameterized family of isometric complex structures connected with the hyper-Kählerian structure, and by the desingularization of the family defined by the complex $(n+1) \times (n+1)$-matrices of rank $\leqslant 1$, fibered by the trace mapping over C, are locally isomorphic.*

Proof. The proof consists, roughly speaking, of introducing in each manifold $V_s = \mathrm{tr}^{-1}(s)$ $(s \neq 0)$, consisting of the complex $(n+1) \times (n+1)$ matrices with rank 1 and trace s, a hyper-Kählerian structure invariant under a maximal compact group of automorphisms, and converging to the one quoted earlier from [7], as $s \to 0$. We shall only outline the proof, since the details are not essentially different from the ones in the case just quoted. The maximal compact group of automorphisms of V_s, as a bundle over CP^n, acts equivariantly with the elliptic group action on the base space: it can be described easily by the adjoint representation of $U(n+1)$ on the space $C^{(n+1)^2}$ of all $(n+1) \times (n+1)$ matrices. This group action leaves invariant not only the rank and the trace function, but also the Hermitian norm function, assigning to each matrix $X = (x_{ij})$ its Hermitian norm $|X|^2 = \mathrm{tr}({}^t\overline{X}X) = \sum_{i,j=0}^n |x_{ij}|^2$. Indeed, the orbits of the action of $U(n)$ in each V_s are simply the sets defined by the values that this Hermitian norm takes, ranging over all real values $\geqslant |s|^2$. Thus the space of orbits under the action of $U(n)$ in each V_s has one real dimension. In terms of the local coordinates (z^i, ζ_j) defined in each of the $n+1$ domains $\mathfrak{U}_a \cap \pi_1^{-1}(s)$ covering $\pi^{-1}(V_s)$ used in Lemma 4.4, the Hermitian norm can be expressed by the function

$$t = \sum_{i,j=0}^n |x_{ij}|^2 = \sum_{i,j \neq a} |z^i \zeta_j|^2 + \sum_{j \neq a} |\zeta_j|^2 + |x_{aa}|^2 \left(1 + \sum_{i \neq a} |z_i|^2 \right)$$

$$= (1 + |z|^2)(|\zeta|^2 + |s - z \cdot \zeta|^2), \tag{4.6}$$

where $|z|^2 = \sum_{i \neq a} |z^i|^2$, $|\zeta|^2 = \sum_{i \neq a} |\zeta_i|^2$, $z \cdot \zeta = \sum_{i \neq a} z^i \zeta_i$. An easy calculation reduces this expression to a form analogous to (4.4) in the case of $T^{*\prime}(CP^n)$. Consider the family, indexed by s, of analytic (nonholomorphic) cross section in each V_s considered as a complex affine space bundle over CP^n and converging as $s \to 0$ to the zero section in $T^{*\prime}(CP^n)$, defined in terms of the local coordinates by

$$\zeta_i = s(1 + |z|^2)^{-1} \overline{z^i} \qquad (0 \leqslant i \leqslant n, i \neq a).$$

Also introduce the auxiliary coordinates in V_s replacing the holomorphic fibre coordinates ζ_i by η_i, where

$$\eta_i = \zeta_i - s(1 + |z|^2)^{-1} \overline{z^i}. \tag{4.7}$$

Then the hermitian norm t, expressed in terms of (z^i, η_j), takes the form

$$t = |s|^2 + (1 + |z|^2)(|\eta|^2 + |z \cdot \eta|^2). \tag{4.8}$$

We now consider the holomorphic bundle Kähler metric [7] in V_s derived from the local Kähler potential function Ψ_a defined in any of the $n+1$ open domains $V_{s,a} = \pi(\mathcal{U}_a \cup \pi_1^{-1}(s))$ $(0 \leqslant a \leqslant n)$ covering V_s, with holomorphic coordinates (z^i, ζ_j), as follows:

$$\Psi_a = \log(1 + |z|^2) + u(t), \tag{4.9}$$

where t is the norm function (4.6), and $u(t)$ a suitable, smooth function of one real variable t $(t \geqslant |s|^2)$, and calculate the Kähler form $\partial\bar{\partial}\Psi_a$. The calculation is facilitated if one uses the local complex tangent frame in $V_{s,a}$ defined in terms of the local holomorphic coordinates as follows:

$$\nabla_{z^i} = \frac{\partial}{\partial z^i}\bigg|_{\eta=\text{const.}} + \left\{ \begin{matrix} k \\ j \ i \end{matrix} \right\} \eta_k \frac{\partial}{\partial \zeta_j} ; \quad \frac{\partial}{\partial \zeta_j} ; \quad \overline{\nabla_{z^i}} ; \quad \frac{\partial}{\partial \bar{\zeta}_j} ,$$

where all indices range from 0 to n, excluding the value a, the summation convention holds, and $\{ \begin{smallmatrix} k \\ j \ i \end{smallmatrix} \}$ is the Christoffel symbol for the connection of the Fubini–Study metric in CP^n with respect to the inhomogeneous coordinates (z^i), and where (η_k) are the auxiliary functions defined by (4.7). The dual coframe is then made up of the differentials

$$dz^i, \quad D\eta_j, \quad \overline{dz^i}, \quad \overline{D\eta_j} ,$$

where

$$D\eta_j = d\eta_j - \eta_k \left\{ \begin{matrix} k \\ j \ i \end{matrix} \right\} dz.$$

The differential operators ∇_{z^i} and the differentials $D\zeta_j$ are conveniently used here, because they arise naturally from a connection in the holomorphic bundle $V_s \xrightarrow[2m]{\pi} CP^n$. In terms of this basis, the Hamiltonian form φ_0 in V_s can be expressed equivalently locally in any of the following three ways:

$$\varphi_0 = d\zeta_i \wedge dz^i = D\zeta_i \wedge dz^i = \partial\eta_i \wedge dz^i.$$

With this formalism, one obtains an ordinary differential equation for the function $u(t)$ in (4.9) whose solutions generate a Kähler metric for which the covariant derivative of φ_0 is identically zero. The details are essentially the same as in the concluding pages of [7], and the solution is

$$u(t) = \sqrt{1 + 4(t - |s|^2)} - \log\left(1 + \sqrt{1 + 4(t - |s|^2)}\right).$$

The resulting Kähler metric is then a hyper-Kähler metric in V_s, converging to the one already described in the limiting case of $T^{*\prime}(CP^n)$, the desingularized model for V_0. Except for the few remaining routine computational details, the proof of Theorem 4.5 is thus complete. $\square$

References

[1] D. V. Alekseevskiĭ, Classification of quaternionic spaces with transitive, solvable groups of motions. *Izv. Akad. Nauk SSSR* **39** (1975), *A.M.S. Transl.* **9**, 297–339 (1975).

[2] M. F. Atiyah, *Some Examples of Complex Manifolds*. Bonner Mathematische Schriften, No. 6, Bonn, 1958.

[3] T. Aubin, Equations du type Monge–Ampère sur les variétés kählériennes compactes. *Bull. Sci. Math.* **102**, 63–95 (1978).

[4] M. Berger, Remarques sur les groupes d'holonomie des variétés riemanniennes. *C. R. Acad. Sci. Paris Sér. A* **262**, 1316–1318 (1966).

[5] R. B. Brown and A. Gray, Riemannian manifolds with holonomy group *Spin*(9), in *Differential Geometry (Symposium in Honor of Kentaro Yano)*. Konokuniya, Tokyo, 1972, pp. 41–59.

[6] F. A. Bogomolov, Hamiltonian Kähler manifolds. (in Russian), *Dokl. Akad. Nauk SSSR* **243** (5), 1101–1104 (1978).

[7] E. Calabi, Métriques kählériennes et fibrés holomorphes. *Ann. Sci. Ec. Norm. Sup. Paris, 4me Sér.* **12**, 269–294 (1979).

[8] E. Horikawa, Surjectivity of the period map of $K3$ surfaces of degree 2. *Math. Ann.* **228**, 113–146 (1977).

[9] K. Kodaira and D. C. Spencer, On deformations of complex structures, II. *Ann. Math.* **67** (1958), esp. pp. 403–408.

[10] V. Kulikov, Epimorphicity of the period mapping for $K3$ surfaces. (in Russian) *Usp. Mat. Nauk* **32**, (4; 196), 257–258 (1977).

[11] V. Kulikov, Degeneration of $K3$ surfaces and Enriques surfaces. *Izv. Akad. Nauk SSSR* **41**, 1008–1042 (1977); *AMS Transl.* **11**, 957–989 (1977).

[12] A. Lichnérowicz, *Théorie Globale des Connexions et des Groupes d'Holonomie*. Rome, Cremonese, 1955, esp. pp. 250–251.

[13] A. Newlander and L. Nirenberg, Complex analytic coordinates in almost complex manifolds. *Ann. Math.* **65**, 391–404 (1957).

[14] A. N. Todorov, *Moduli of kählerian K-3 surfaces*. Mathematische Arbeitstagung, Bonn, 1979.

[15] A. Weil, Théorèmes fondamentaux de la théorie des fonctions thêta (d'après des mémoires de Poincaré et Frobenius), in *Sém. Bourbaki*, No. 16, Mai 1949; *Oeuvres scientifiques—Collected papers*. Springer, New York, Heidelberg, Berlin, 1979, I, pp. 414–421.

[16] S. T. Yau, On the Ricci curvature of compact Kähler manifolds and complex Monge–Ampère equations, I. *Comm. Pure Appl. Math.* **31**, 339–411 (1978).

Two Applications of Algebraic Geometry to Entire Holomorphic Mappings

Mark Green[1]*
Phillip Griffiths[2]*

In this paper we shall prove two theorems concerning holomorphic mappings of large open sets of $\mathbb{C}^k$ into algebraic varieties. Both are in response to well-known outstanding problems, and we feel that the techniques introduced should in each case have further applications.

To state our first result, we recall that a holomorphic mapping into an algebraic variety is said to be *algebraically degenerate* in case the image lies in a proper algebraic subvariety.

Theorem I. *Let X be an algebraic variety whose irregularity satisfies*

$$q > \dim X.$$

Then any entire holomorphic curve

$$f : \mathbb{C} \to X$$

is algebraically degenerate.

We remark that the irregularity $q = h^{1,0}(X)$ is the dimension of the space of holomorphic 1-forms on any smooth model for the function field of X. Since such desingularizations exist by Hironaka's well-known theorem, and since q is a birational invariant, our definition of the irregularity makes sense.

When X is a curve, this theorem was proved by Picard [26] in a paper closely related to his proof of the usual Picard theorem. Nowadays this case is an obvious consequence of the uniformization theorem, but unfortunately this latter result does not generalize. Some 47 years later Theorem I was formulated by A. Bloch [1], who established several special cases and contributed the essential technical idea of using jets. To a reader trained in modern mathematics Bloch's paper is obscure to put it mildly, and interest in the subject was revived by Ochiai [25], who considerably clarified matters and who formulated a technical result that would yield what he termed Bloch's conjecture.

Our approach is different in that rather than establishing Ochiai's technical result (which is, in fact, true), we use the method of negative curvature. The

---

[1] Research partially supported by NSF Grant and the Alfred P. Sloan Foundation.

[2] Research partially supported by NSF Grant #MCS7707782.

* Department of Mathematics, Harvard University, Cambridge, MA 02138, USA.

difference is in a sense more apparent than real, in that the essential ingredient in both proofs is the use of higher-order jets to detect geometric consequences of the assumption $q > \dim X$, consequences of a somewhat subtle character that may not be evident from first-order considerations.

In addition to the work of Bloch and Ochiai, our proof was motivated by the recent paper [2] of Bogomolov, who used symmetric differentials to show the existence of finitely many rational and elliptic curves on a surface of general type with $c_1^2 > c_2$. In this paper we shall use jet differentials to construct a negatively curved jet psuedometric that leads to the proof of Theorem I. For any surface of general type these exist in abundance—even when $H^0(\text{Sym}^m \Omega_X^1) = 0$ for all $m \geq 1$—and it is our feeling that the systematic use of higher-order differentials presents an algebrogeometric technique that may be useful in other contexts. For this reason we have, in Section 1, attempted to clearly explain the basic concepts. We have also shown that, for any smooth n-dimensional variety X for which $c_1(\Omega_X^1)^n > 0$, the Euler characteristic of the sheaf of jet differentials grows at the maximum rate, and for general type surfaces these jet differentials give a birational embedding of a suitably prolonged projectivized jet bundle.

Our second main result is in response to the following well-known

Conjecture. *An n-dimensional algebraic variety X is measure-hyperbolic if, and only if, X is of general type.*

(Cf. Kobayashi [20]—the relevant definitions together with additional references are given in Section 4 below.) The implication

$$X \text{ general type} \quad \Rightarrow \quad X \text{ measure-hyperbolic}$$

was established some time ago, and so the conjecture pertains to the converse. When $n = 1$ the result is a simple consequence of the uniformization theorem and classification of curves according to their Kodaira dimension. Turning to surfaces, the conjecture would follow from showing that, for any surface X not of general type, there is a holomorphic mapping

$$f : \Delta \times C \to X$$

that takes the origin to a given general point on X and whose Jacobian is not identically zero. Using the classification of surfaces, one is easily reduced to constructing f when X is an algebraic $K3$ surface. These fall into an infinite number of 19-dimensional algebraic families $\mathcal{F}_n$. Our result is

Theorem II. *An algebraic $K3$ surface $X \in \mathcal{F}_n$ is not measure-hyperbolic when $n = 1, 2,$ or 3.*

The proof is by showing that on any such X there is a family of ∞^1 elliptic curves (all singular), which then leads to the desired mapping f. These curves are constructed by projective methods. In fact, the construction is valid for all n, but the proof of Theorem II does not go through due to a certain technical point (involving singularities) that we are unable to resolve. This point is one of those issues that are in some sense "geometrically obvious" but whose proof will require a deeper understanding of possible degeneracies than we are able to muster. Nailing it down seems to us a very worthwhile project, as it would have the following geometric consequence:

(*) On any algebraic K3 surface $X \in \mathcal{F}_n$ there are a positive finite number of rational curves that have $(n + 1)$ distinct nodes.

Our proposed proof of this assertion—which has the existence of ∞^1 elliptic curves as an easy consequence—is by induction on the degree $2n$ of $X \subset \mathbb{P}^{n+1}$, and furnishes a technique that may be useful in other contexts.

It is a pleasure to thank Joe Harris for several conversations pertaining to Section 4, and for helping us with several incomplete but enjoyable "proofs" of (*).

Part A. Jet Differentials and Bloch's Conjecture

1. Jets and Jet Differentials

(a) Definition and basic properties of jet spaces. We shall first explain jets for holomorphic mappings into a smooth complex manifold X.

Given $x \in X$, we denote by Δ a disc of any positive radius and consider germs of holomorphic mappings

$$f : \Delta \to X$$

that satisfy $f(0) = x$. In a local holomorphic coordinate system any such f is given by its convergent series

$$f(z) = f^{(0)} + f^{(1)}z + f^{(2)} \frac{z^2}{2!} + f^{(3)} \frac{z^3}{3!} + \cdots, \qquad (1.1)$$

where $f^{(k)} \in \mathbb{C}^n$ and $f^{(0)} = x$.

Two germs f and $\tilde{f}$ osculate to order k in case

$$f^{(0)} = \tilde{f}^{(0)}, \quad f^{(1)} = \tilde{f}^{(1)}, \ldots, f^{(k)} = \tilde{f}^{(k)}.$$

The equivalence classes of such germs will be called *jets of order* k at x and denoted $J_k(X)_x$. It is clear that

$$J_k(X) = \bigcup_{x \in X} J_k(X)_x$$

forms a complex manifold of dimension $n + kn$, and if $U \subset X$ is an open set on which we have holomorphic coordinates, then this choice of coordinates induces an isomorphism

$$J_k(U) \cong U \times \mathbb{C}^{kn}.$$

Given a holomorphic arc $f : \Delta \to X$ with $f(z) = x$, we denote by $j_k(f)_x \in J_k(X)_x$ the k-jet defined by the germ of f at x. The notation

$$j_k(f) : \Delta \to J_k(X)$$

will be used to denote the natural lifting of f to k-jets. Intuitively, $J_k(X)_x$

consists of kth-order infinitesimal arcs centered at x, and $j_k(f)$ describes the family of these arcs along the holomorphic curve $f(\Delta)$. In general, a holomorphic mapping

$$h : X \to Y$$

between complex manifolds induces a mapping

$$h_* : J_k(X) \to J_k(Y)$$

on k-jets. For $k = 1$, we have the usual notion of tangent vectors and induced map on tangent spaces.

The jet manifolds $J_k(X)$ are holomorphic fibre bundles over X, but for $k \geqslant 2$ they are not vector bundles. However, there are obvious maps

$$J_{k+1}(X) \to J_k(X) \tag{1.2}$$

whose fibres are affine linear spaces.

Using local coordinates on X so that jets may be expressed in the form (1.1), the fibre of (1.2) amounts to fixing $x = f^{(0)}, f^{(1)}, \ldots, f^{(k)}$ and having $f^{(k+1)}$ free to vary over $\mathbb{C}^n$. If, moreover, $f^{(1)} = \cdots = f^{(k)} = 0$, then $f^{(k+1)}$ transforms like a tensor in $T_x(X)$. In other words, the fibres of (1.2) are affine bundles whose associated vector bundle is $T(X)$.

Next, we will define an action of $\mathbb{C}^*$ on jets that amounts to reparametrization by a constant dilation or contraction. Recalling that Δ denotes a disc of unspecified positive radius, given $f : \Delta \to X$ and $t \in \mathbb{C}^*$, we set $f_t(z) = f(tz)$ and define

$$t \cdot j_k(f) = j_k(f_t).$$

In the coordinates (1.1),

$$t \cdot \{ f^{(0)}, f^{(1)}, \ldots, f^{(k)} \} = \{ f^{(0)}, tf^{(1)}, \ldots, t^k f^{(k)} \}. \tag{1.3}$$

If $J_k^*(X)$ denotes the nonconstant jets—i.e., those with some $f^{(j)} \neq 0$ for $1 \leqslant j \leqslant k$—then this $\mathbb{C}^*$ action preserves $J_k^*(X)$ and we define

$$P_k(X) = J_k^*(X)/\mathbb{C}^*.$$

For $k = 1$ we obtain the projectivized tangent bundle $P_1(X) = \mathbb{P}T(X)$, whose elements will be written (x, ξ), where $x \in X$ and $\xi \in \mathbb{P}T_x(X)$ is a tangent direction. It is clear that $P_1(X)$ is a complex manifold, and is in fact a $\mathbb{P}^{n-1}$-bundle over X.

For $k \geqslant 2$ the objects $P_k(X)$ are perhaps less familiar. The fibre $F_{k,n}$ of $P_k(X) \to X$ is a *weighted projective space* (cf. Dolgacev [8]); it is the quotient of $\mathbb{C}^{kn} - \{0\}$ by the $\mathbb{C}^*$-action

$$t \cdot \{ w^{(1)}, w^{(2)}, \ldots, w^{(k)} \} = \{ tw^{(1)}, t^2 w^{(2)}, \ldots, t^k w^{(k)} \}, \tag{1.4}$$

where $w^{(j)} \in \mathbb{C}^n$. For $k \geqslant 2$ this action has fixed points, and when also $n \geqslant 2$ the fibre $F_{k,n}$ is a projective algebraic variety having what are usually termed quotient singularities. For example, when $k = n = 2$ the fibre is a quotient of $\mathbb{C}^4 - \{0\}$ under the action

$$t \cdot \{ x, y, u, v \} = \{ tx, ty, t^2 x, t^2 y \}.$$

Taking $t = -1$, the plane $x = y = 0$ is left fixed and $F_{2,2}$ has a singular line. In fact it is biholomorphic to the cone $\{w_1^2 + w_2^2 + w_3^2 = 0\}$ in $\mathbb{C}^4$. The presence of singularities in $P_k(X)$ will not cause any difficulty, since in fact these weighted projective spaces are quite nice varieties that are well understood.

We shall be using jets to study holomorphic curves $f: \Delta \to X$ in general complex analytic varieties X that may have singularities. It is, of course, desirable to define intrinsically the jet spaces associated to X. However, for our purposes this would take us too far afield, and is not necessary, for the following reason: Given any resolution $\tilde{X} \to X$ of the singularities of X, there is a unique lifting $\tilde{f}: \Delta \to \tilde{X}$ of any holomorphic curve whose image does not lie entirely in the singular locus of X. Since we shall be using jets to prove statements like "any entire holomorphic curve $f: C \to X$ is analytically degenerate", this device of using resolution of singularities will suffice for our needs.

(b) Formalism of jet differentials. We will now define the sheaves of jet differentials on a complex manifold X. On the weighted projective space $F_{k,n}$ given by the $\mathbb{C}^*$ action (1.4) we consider polynomials $\phi(w)$ in the kn variables $w_i^{(1)}, \ldots, w_i^{(n)}$. Assigning to $w_i^{(l)}$ the weight l, we consider polynomials that are homogeneous of weight m. Equivalently, the polynomial ϕ should satisfy

$$\phi(t \cdot w) = t^m \phi(w).$$

By taking local coordinates on X and allowing the coefficients of $\phi(w)$ to be holomorphic functions, we may define the *sheaf* $\mathcal{J}_{k,m}$ *of k-jet differentials on X of weight m*.

For example, when $k = 1$ we have

$$\mathcal{J}_{1,m} = \operatorname{Sym}^m \Omega_X^1.$$

For another example, when $k = 3$, sections of $\mathcal{J}_{3,m}$ are locally

$$
\begin{aligned}
\phi &= \sum a_i f_i', & m &= 1, \\
\phi &= \sum a_{ij} f_i' f_j' + b_i f_i'', & m &= 2, \\
\phi &= \sum a_{ijk} f_i' f_j' f_k' + b_{ij} f_i' f_j'' + c_i f_i''', & m &= 3, \\
\phi &= \sum a_{ijkl} f_i' f_j' f_k' f_l' + b_{ijk} f_i' f_j' f_k'' \\
&\quad + c_{ij} f_i'' f_j'' + d_{ij} f_i' f_j''', & m &= 4,
\end{aligned}
\tag{1.5}
$$

etc. Here the coefficient functions are holomorphic functions on X, and the obvious symmetry conditions—such as $a_{ij} = a_{ji}$ in the second equation in (1.5)— are assumed satisfied when applicable. In summary:

(1.6) *Sections of* $\mathcal{J}_{k,m}$ *are locally given by homogeneous polynomials with holomorphic coefficients in the variables* $f_i', f_i'', \ldots, f_i^{(k)}$ *of total weight m, where* $f_i^{(l)}$ *is assigned weight l.*

By considering the highest expression in the $f_i^{(k)}$ that occurs, we see that there is a natural filtration on the weighted homogeneous polynomials of total weight m. This gives an intrinsic filtration

$$\mathcal{J}_{k-1,m} = S_0 \subset S_1 \subset \cdots \subset S_{[m/k]} = \mathcal{J}_{k,m},$$

where

$$S_i/S_{i-1} \cong \text{Sym}^i \Omega_X^1 \otimes \mathcal{J}_{k-1,\,m-ki}.$$

Thus, inductively we have:

$\mathcal{J}_{k,m}$ *has a composition series whose factors are*

$$\text{Sym}^{i_1} \Omega_X^1 \otimes \text{Sym}^{i_2} \Omega_X^1 \otimes \cdots \otimes \text{Sym}^{i_k} \Omega_X^1,$$

where each combination of nonnegative indices satisfying

$$i_1 + 2i_2 + \cdots + ki_k = m$$

occurs exactly once.

The simplest example of this is the exact sequence

$$0 \to \text{Sym}^2 \Omega_X^1 \to \mathcal{J}_{2,2} \to \Omega_X^1 \to 0.$$

We will now give an alternate definition of $\mathcal{J}_{k,m}$, one that will be useful in our study below. Given any complex analytic variety Y with a $\mathbb{C}^*$ action and analytic quotient space $Z = Y/\mathbb{C}^*$, we denote the projection $Y \to Z$ by $\tilde{\omega}$ and define a sheaf $\mathcal{L}^m$ of $\mathcal{O}_Z$-modules as follows: For an open set $U \subset Z$ we set

$$\mathcal{L}^m(U) = {}'\left\{ \phi \in \mathcal{O}\big(\tilde{\omega}^{-1}(U)\big) : \phi(ty) = t^m \phi(y) \right\}.$$

The presheaf $U \to \mathcal{L}^m(U)$ then leads to a sheaf $\mathcal{L}^m$ on Z. We note that, under the new $\mathbb{C}^*$ action given by the standard covering $t \to t^m$ of $\mathbb{C}^*$, $\mathcal{L}^m$ becomes the first power of the new sheaf $\mathcal{L}$.

Taking $Y = J_k(X)$ and $Z = P_k(X)$ we have now defined the sheaves $\mathcal{L}^m$ upstairs on $P_k(X)$. In general these sheaves are not invertible (Dolgacev [8]). However, for any multiple $m = l \cdot k!$ of $k!$, the sheaf $\mathcal{L}^m$ is associated to a line bundle. Essentially this is because the action

$$t \times j \to t^{k!}j$$

gives a free action of

$$\mathbb{C}^*/(k\text{th roots of unity}) \cong \mathbb{C}^*,$$

and under a free action the sheaf $\mathcal{L} \to Z$ in the preceding paragraph is the one associated to the line bundle $L = Y \times_{\mathbb{C}^*}\mathbb{C}$ over Z.

The restriction of $\mathcal{L}^m$ to each fibre of $P_k(R) \xrightarrow{\pi} \mathbb{C}$ is Dolgacev's (m) on the weighted projective space $F_{k,n}$. In particular, by the Theorem in §1.4 of [8] we have

$$R_\pi^q(\mathcal{L}^m) = 0 \qquad \text{for } m \geqslant 0, q > 0,$$
$$R_\pi^0(\mathcal{L}^m) = \mathcal{J}_{k,m} \text{ is locally free on } X. \tag{1.7}$$

As the second equation suggests, and as is clear from the definitions, the previously defined sheaf $\mathcal{J}_{k,m}$ of m-fold jet differentials is the 0th direct image of $\mathcal{L}^m \to P_k(X)$. In general, by (1.7) and the Leray spectral sequence,

$$H^q(X, \mathcal{J}_{k,m}) \cong H^q(P_k(X), \mathcal{L}^m) \tag{1.8}$$

for all $q \geqslant 0$, $m \geqslant 0$. In particular, for $m = l \cdot k!$ giving the space of global sections $H^0(X, \mathcal{J}_{k,m})$ is equivalent to giving the rational mapping

$$\phi_m : P_k(X) \to P^N$$

($N + 1 = h^0(X, \mathcal{J}_{k,m})$) in which each fibre is developed onto a rational image of $F_{k,n}$.

It is clear that a holomorphic mapping $f : X \to Y$ between complex manifolds induces a pullback

$$f^* : H^0(Y, \mathcal{J}_{k,m}) \to H^0(X, \mathcal{J}_{k,m}) \tag{1.9}$$

on jet differentials. Somewhat more interestingly, if f is only assumed to be meromorphic and therefore perhaps not an actual map in codimension two, the usual argument invoking Hartogs' extension theorem shows that the transformation (1.9) is still defined. In particular,

The spaces $H^0(X, \mathcal{J}_{k,n})$ are bimeromorphic invariants of complex manifolds.

As a consequence, the space $H^0(X, \mathcal{J}_{k,m})$ of global jet differentials on any analytic variety may be defined to be $H^0(\tilde{X}, \mathcal{J}_{k,m})$ for any resolution $\tilde{X}$ of X.

To conclude this section we will introduce two formal operations on jet differentials. These will not be used explicitly in our work, but help to clarify the nature of these objects. The first is simply multiplication. More precisely, the projection (1.2) induces inclusions

$$\mathcal{J}_{k,m} \subset \mathcal{J}_{k+1,m},$$

and we shall denote the limit $\bigcup_k \mathcal{J}_{k,m}$ by $\mathcal{J}_{\cdot,m}$. Then multiplication of weighted homogeneous polynomials gives a product

$$\mathcal{J}_{\cdot,m} \otimes \mathcal{J}_{\cdot,m'} \to \mathcal{J}_{\cdot,m+m'}$$

that satisfies obvious algebraic rules.

The second operation is that of differentiation, to be denoted by

$$' : \mathcal{J}_{k,m} \to \mathcal{J}_{k+1,m+1}.$$

It is defined as follows: Given a section ϕ of $\mathcal{J}_{k,m}$ and holomorphic arc $f : \Delta \to X$, we set

$$\phi'(j_{k+1}(f))(z) = \frac{d}{dz}(\phi(j_k(f))(z)).$$

For example, in the case $k = m = 2$ of (1.5),

$$\phi = \sum a_{ij} f_i' f_j' + b_i f_i'',$$

$$\phi' = \sum \frac{\partial a_{ij}}{\partial z_k} f_i' f_j' f_k' + \left(2a_{ij} + \frac{\partial b_j}{\partial z_i}\right) f_i' f_j'' + b_i f_i'''.$$

As usual the Leibniz rule

$$(\phi\psi)' = \phi'\psi + \phi\psi'$$

is valid.

A simple but fundamental observation, one that will be discussed in detail in the next section, is this: We ask whether there may be global sections in

$H^0(X, \mathcal{J}_{k,m})$ that do not ultimately come from ordinary symmetric differentials (i.e., sections in $H^0(X, \operatorname{Sym}^m \Omega^1_X)$). In the following section, the answer to this will turn out to be yes, and as an indication that this should be so we consider the example of a section

$$\phi = \sum a_{ij} f_i' f_j', \qquad a_{ij} = a_{ji},$$

of $\operatorname{Sym}^2 \Omega^1_X$. The derivative

$$\phi' = \sum \frac{1}{3} \left(\frac{\partial a_{ij}}{\partial z_k} + \frac{\partial a_{ki}}{\partial z_j} + \frac{\partial a_{jk}}{\partial z_i} \right) f_i' f_j' f_k' + 2 a_{ij} f_i' f_j''$$

has a coefficient of $f_i' f_j''$ that is symmetric in its indices, but by (1.5) this is not necessarily the case for a general section of $\mathcal{J}_{2,3}$. This led us to suspect that the algebra $\bigoplus_{k,m} H^0(\mathcal{J}_{k,m})$ may be larger than that generated by symmetric differentials and their derivatives.

(c) Existence of jet differentials on a surface of general type. In this section we shall prove the following two results:

(1.10) Proposition. *For any smooth projective variety X,*

$$\chi(X, \mathcal{J}_{k,m}) = \frac{m^{(k+1)n-1}}{(k!)^n((k+1)n-1)!}$$

$$\times \left(\frac{(-1)^n}{n!} c_1(X)^n (\log k)^n + O\big((\log k)^{n-1}\big) \right)$$

$$+ O(m^{(k+1)n-2}).$$

We note that

$$\dim P_k(X) = (k+1)n - 1,$$

so that if $c_1(\Omega^1_X)^n > 0$, then by (1.8) the Euler characteristic of $\mathcal{J}_{k,m}$ grows at the maximum rate.

(1.11) Proposition. *Let X be a surface of general type. Then for k, m sufficiently large, the rational mapping*

$$\phi_m : P_k(X) \to \mathbb{P}^N$$

is birational onto its image.

As we shall see in the next section, there are simple surfaces of general type for which $H^0(X, \operatorname{Sym}^m \Omega^1_X) = 0$ for all $m \geq 0$, so that jet differentials definitely give more information than ordinary symmetric ones.

We need to calculate the leading term of $\chi(\mathcal{J}_{k,m})$ for a variety of dimension n. First,

$$\operatorname{ch}(\mathcal{J}_{k,m}) = \sum_{i_1 + 2i_2 + \cdots + ki_k = m} \operatorname{ch}\big(\operatorname{Sym}^{i_1} \Omega^1_X \otimes \operatorname{Sym}^{i_2} \Omega^1_X \otimes \cdots \otimes \operatorname{Sym}^{i_k} \Omega^1_X\big)$$

(with all i's integral and ≥ 0), as $\mathcal{J}_{k,m}$ has a composition series involving exactly these sheaves. If

$$c(\Omega^1_X) = (1 + \lambda_1)(1 + \lambda_2) \cdots (1 + \lambda_n)$$

formally, then

$$\text{ch}(\text{Sym}^i \Omega^1_X) = \sum_{x_1 + \cdots + x_n = i} e^{x_1\lambda_1 + \cdots + x_n\lambda_n}$$

(with all x's integral and $\geqslant 0$). So

$$\text{ch}(\mathcal{J}_{k,m}) = \sum_{\substack{x_{11} + \cdots + x_{1n} + 2(x_{21} + \cdots + x_{2n}) \\ + \cdots + k(x_{k1} + \cdots + x_{kn}) = m}} \exp\{(x_{11} + \cdots + x_{k1})\lambda_1 + \cdots$$

$$+ (x_{1n} + \cdots + x_{kn})\lambda_n\}$$

$$= \sum_{\substack{\frac{x_{11}}{m} + \cdots + \frac{x_{1n}}{m} \\ + 2(\frac{x_{21}}{m} + \cdots + \frac{x_{2n}}{m}) \\ + \cdots + k(\frac{x_{k1}}{m} + \cdots + \frac{x_{kn}}{m}) \\ = 1}} \exp\left\{m\left(\left(\frac{x_{11}}{m} + \cdots + \frac{x_{k1}}{m}\right)\lambda_1 + \cdots \right. \right.$$

$$\left. \left. + \left(\frac{x_{1n}}{m} + \cdots + \frac{x_{2n}}{m}\right)\lambda_n\right)\right\}$$

(with all x's integral and $\geqslant 0$). This can be approximated by an integral modulo lower-order terms, so

$$\text{ch}(\mathcal{J}_{k,m}) = m^{kn-1} \int \cdots \int_{\substack{y_{11} + \cdots + y_{1n} + 2(y_{21} + \cdots + y_{2n}) \\ + \cdots + k(y_{k1} + \cdots + y_{kn}) = 1}} e^{m((y_{11} + \cdots + y_{k1})\lambda_1 + \cdots + (y_{1n} + +y_{kn}\lambda_k)}d\omega$$

$$+ O(m^{(k+1)n-2})$$

(with all y's $\geqslant 0$), where $d\omega$ is the element of area $dy_{12}\, dy_{13} \cdots dy_{kn}$ pulled back to the hyperplane $y_{11} + \cdots + y_{1n} + \cdots + k(y_{k1} + \cdots + y_{kn}) = 1$ by the projection map. Recalling that our exponential is purely formal and represents a polynomial of degree n, we have

$$\text{ch}(\mathcal{J}_{k,m}) = m^{kn-1} \int \cdots \int_{\substack{y_{11} + \cdots + y_{1n} + 2(y_{21} + \cdots + y_{2n}) \\ + \cdots + k(y_{k1} + \cdots + y_{kn}) = 1}} \frac{m^n}{n!}((y_{11} + \cdots + y_{k1})\lambda_1$$

$$+ \cdots + (y_{1n} + \cdots + y_{kn})\lambda_n)^n\, d\omega + O(m^{(k+1)n-2})$$

(with all y's $\geqslant 0$). By a substitution,

$$\text{ch}(\mathcal{J}_{k,m}) = \frac{m^{(k+1)n-1}}{n!\,(k!)^n} \times \int \cdots \int_{y_{11} + \cdots + y_{kn} = 1} \left(\left(y_{11} + \frac{y_{21}}{2} + \cdots + \frac{y_{k1}}{k}\right)\lambda_1 \right.$$

$$\left. + \cdots + \left(y_{1n} + \frac{y_{2n}}{2} + \cdots + \frac{y_{kn}}{k}\right)\lambda_n\right)^n d\mu$$

$$+ O(m^{(k+1)n-2})$$

(with all y's $\geqslant 0$), where $d\mu$ is $dy_{12} \cdots dy_{kn}$ pulled up by projection. Thus

$$
\text{ch}(\mathcal{I}_{k,m}) = \frac{m^{(k+1)n-1}}{(k!)^n} \times \int \cdots \int_{y_{11}+\cdots+y_{kn}=1} \sum_{q_1+\cdots+q_n=n} \frac{1}{q_1! \cdots q_n!}
$$

$$
\times \left(y_{11} + \frac{y_{21}}{2} + \cdots + \frac{y_{k1}}{k} \right)^{q_1} \cdots
$$

$$
\times \left(y_{1n} + \frac{y_{2n}}{2} + \cdots + \frac{y_{kn}}{k} \right)^{q_n} \lambda_1^{q_1} \cdots \lambda_n^{q_n} d\mu
$$

$$
+ O(m^{(k+1)n-2})
$$

(with all y's $\geqslant 0$, all q's integral and $\geqslant 0$). Setting

$$
F_k(q_1, \ldots, q_n) = \int \cdots \int_{y_{11}+\cdots+y_{kn}=1} \left(y_{11} + \frac{y_{21}}{2} + \cdots + \frac{y_{k1}}{k} \right)^{q_1} \cdots
$$

$$
\times \left(y_{1n} + \frac{y_{2n}}{2} + \cdots + \frac{y_{kn}}{k} \right)^{q_n} d\mu \qquad (1.12)
$$

(with all y's $\geqslant 0$), we have

$$
\text{ch}(\mathcal{I}_{k,m}) = \frac{m^{(k+1)n-1}}{(k!)^n} \sum_{q_1+\cdots+q_n=n} \frac{F_k(q_1 \cdots q_n)}{q_1! \cdots q_n!} \lambda_1^{q_1} \cdots \lambda_n^{q_n}
$$

$$
+ O(m^{(k+1)n-2}) \qquad (1.13)
$$

(with all q's integral and $\geqslant 0$).

To evaluate $F_k(q_1, \ldots, q_n)$, we introduce two notations. Let

$$
I_{j_1, \ldots, j_r}(p) = \int \cdots \int_{y_1+\cdots+y_p=1} y_1^{j_1} \cdots y_r^{j_r} dy_1 \cdots dy_p
$$

(with all y's $\geqslant 0$, $r, j_1, \cdots, j_r$ integers $\geqslant 0$, $r \leqslant p$). By calculus,

$$
I_{j_1, \ldots, j_r}(p) = \frac{j_1! \cdots j_r!}{(j_1 + \cdots + j_r + p)!} .
$$

The sums

$$
S_{j_1, \ldots, j_r}(k) = \sum \frac{1}{i_1^{j_1} i_2^{j_2} \cdots i_r^{j_r}} , \qquad 0 \leqslant j_1 \leqslant j_2 \cdots \leqslant j_r \text{ integers} \qquad (1.14)
$$

(where the summation is over $i_s < i_t$ if $j_s = j_t$ and $s < t$, with all i's integers $\geqslant 1$ and $\leqslant k$) grow asymptotically like a constant times $(\log k)^\nu$, where ν is the number of j_i's equal to 1. In particular,

$$
S_{11\ldots1}(k) \sim \frac{1}{r!} (\log k)^r .
$$

Returning to the expression (1.12) for $F_k(q_1, \ldots, q_n)$, we have

$$F_k(q_1, \ldots, q_n) = \sum \frac{q_1! \, q_2! \cdots q_n!}{j_{1,1}! \cdots j_{n,q_n}!} S_{j_{1,1}, \ldots, j_{1,q_1}}(k) \cdots$$

$$\times S_{j_{n,1}, \ldots, j_{n,q_n}}(k) I_{j_{1,1}, \ldots, j_{n,q_n}}(kn - 1) \qquad (1.15)$$

(where in the summation $j_{\nu,1} + \cdots + j_{\nu,q_\nu} = q_\nu$, $0 \leqslant j_{\nu,1} \leqslant j_{\nu,2} \leqslant \cdots \leqslant j_{\nu,q_\nu}$ for $\nu = 1, \ldots, n$, all j's integers). So

$$F_k(q_1, \ldots, q_n) = \sum \frac{q_1! \cdots q_n!}{((k+1)n - 1)!} S_{j_{1,1}, \ldots, j_{1,q_1}}(k) \cdots S_{j_{n,1}, \ldots, j_{n,q_n}}(k) \quad (1.16)$$

with the same conditions on the summation indices. Thus

$$F_k(q_1, \ldots, q_n) = \frac{1}{((k+1)n - 1)!} (\log k)^n + O\big((\log k)^{n-1}\big)$$

since the only way $(\log k)^n$ can occur is when all of the j's are 1. So

$$\mathrm{ch}(\mathscr{I}_{k,m}) = \frac{m^{(k+1)n-1}}{(k!)^n((k+1)n - 1)!} (\log k)^n \sum_{\substack{q_1 + \cdots + q_n = n \\ q\text{'s integers} \,\geqslant 0}} \frac{\lambda_1^{q_1} \cdots \lambda_n^{q_n}}{q_1! \cdots q_n!}$$

$$+ O\big((\log k)^{n-1}\big) + O(m^{(k+1)n-2})$$

$$= \frac{m^{(k+1)n-1}}{(k!)^n((k+1)n - 1)!}$$

$$\times \left(\frac{(-1)^n}{n!} c_1(X)^n (\log k)^n + O\big((\log k)^{n-1}\big) \right)$$

$$+ O(m^{(k+1)n-2}).$$

By the Hirzebruch Riemann–Roch theorem,

$$\chi(\mathscr{I}_{k,m}) = \frac{m^{(k+1)n-1}}{(k!)^n((k+1)n - 1)!}$$

$$\times \left(\frac{(-1)^n}{n!} c_1(x)^n (\log k)^n + O\big((\log k)^{n-1}\big) \right)$$

$$+ O(m^{(k+1)n-2}), \qquad (1.17)$$

which proves the desired result on the leading term of $\chi(\mathscr{I}_{k,m})$.

Returning to the explicit formula (1.16) for $F_k(q_1, \ldots, q_n)$, we can calculate the leading term explicitly for low dimensions. The leading terms are the following combinations of Chern numbers, where c_i denotes $c_i(\Omega_X^1)$:

$$n = 1, \qquad \frac{1}{(k!)^2} c_1,$$

$$n = 2, \qquad \frac{1}{(k!)^2(2k+1)!} \left((S_{1,1}(k) + S_2(k))c_1^2 - S_2(k)c_2 \right),$$

$$n = 3, \qquad \frac{1}{(k!)^3(3k+2)!} \left((S_{1,1,1}(k) + S_{2,1}(k) + S_3(k))c_1^3 \right.$$

$$\left. + (S_{2,1}(k) - 2S_3(k))c_1 c_2 + (S_3(k) - S_{2,1}(k))c_3 \right).$$

In particular, for surfaces the leading term is

$$k = 1, \qquad \frac{1}{3!}\left(c_1^2 - c_2\right),$$

$$k = 2, \qquad \frac{1}{4^3 \cdot 3!}\left(7c_1^2 - 5c_2\right),$$

$$k = 3, \qquad \frac{1}{6^5}\left(85c_1^2 - 49c_2\right).$$

For surfaces of general type, we have the result of Bogomolov [2]:

(1.18) *If a section of*

$$H^0\left(X, \mathrm{Sym}^{i_1}\Theta_X \otimes \mathrm{Sym}^{i_2}\Theta_X \otimes \cdots \otimes \mathrm{Sym}^{i_k}\Theta_X \otimes K^{(i_1 + \cdots + i_k)/2}\right),$$

$$i_1 + \cdots + i_k \text{ even},$$

vanishes at a point of X, it vanishes identically.

Thus if $i_1 + \cdots + i_k$ is even and $q < (i_1 + \cdots i_k)/2$,

$$H^0\left(X, \mathrm{Sym}^{i_1}\Theta_X \otimes \cdots \otimes \mathrm{Sym}^{i_k}\Theta_X \otimes K_X^q\right) = 0.$$

By squaring, we see we may drop the hypothesis that $i_1 + \cdots + i_k$ is even. Then using Serre duality,

$$H^2\left(X, \mathrm{Sym}^{i_1}\Omega_X^1 \otimes \cdots \otimes \mathrm{Sym}^{i_k}\Omega_X^1\right) = 0 \quad \text{for } i_1 + \cdots + i_k > 2. \quad (1.19)$$

As $\mathcal{J}_{k,m}$ has a composition series involving $\mathrm{Sym}^{i_1}\Omega_X^1 \otimes \cdots \otimes \mathrm{Sym}^{i_k}\Omega_X^1$ with $i_1 + 2i_2 + \cdots + ki_k = m$, we infer that

(1.20) $H^2(X, \mathcal{J}_{k,m}) = 0$ *for $m > 2k$, $k \geqslant 1$, and for X a surface of general type.*

Since

$$X(\mathcal{J}_{k,m}) = h^0(X, \mathcal{J}_{k,m}) - h^1(X, \mathcal{J}_{k,m}) + h^2(X, \mathcal{J}_{k,m}),$$

we conclude that for a minimal surface of general type

$$h^0(X, \mathcal{J}_{k,m}) \geqslant A m^{n(k+1)-1} + O(m^{n(k+1)-2}), \qquad A > 0,$$

for k sufficiently large.

From a result of Iitaka [18], it follows that

(1.21) *For X a surface of general type, if*

$$\left(\sum_{1 \leqslant i < j \leqslant k} \frac{1}{ij} + \sum_{1 \leqslant i \leqslant k} \frac{1}{i^2} \right) c_1^2 - \left(\sum_{1 \leqslant i \leqslant k} \frac{1}{i^2} \right) c_2 > 0,$$

then

$$\phi_m : P_k(X) \to \mathbb{P}_N$$

is birational to its image for m sufficiently large.

Less specifically, the hypothesis on the Chern classes always holds for k sufficiently large. This follows because $c_1^2(\Omega_X^1) > 0$ for a minimal surface of general type.

(d) Examples

(1) *Smooth hypersurfaces in $\mathbb{P}_n$.* Let X be a smooth hypersurface in $\mathbb{P}_n$ of degree d. The main facts are:

(1.22) $H^0(X, \operatorname{Sym}^k \Omega^1) = 0$ *for all $k \geqslant 1$ if $n \geqslant 3$.*

(1.23) $\phi_{\S2. m}$ *is a birational embedding for m sufficiently large, for X a surface and $d \geqslant 16$.*

To see (1.22), which is due to F. Sakai [28], begin with the exact sequence

$$0 \to \mathcal{O} \to \bigoplus_{n+1} \mathcal{O}(1) \to \Theta_{\mathbb{P}_n} \to 0$$

and its analogue

$$0 \to \bigoplus_{\binom{n+k-1}{k-1}} \mathcal{O}(k-1) \to \bigoplus_{\binom{n+k}{k}} \mathcal{O}(k) \to \operatorname{Sym}^k \Theta_{\mathbb{P}_n} \to 0.$$

Dualizing,

$$0 \to \operatorname{Sym}^k \Omega_{\mathbb{P}_n}^1 \to \bigoplus_{\binom{n+k}{k}} \mathcal{O}(-k) \to \bigoplus_{\binom{n+k-1}{k-1}} \mathcal{O}(1-k) \to 0.$$

Thus

$$H^i\left(X, \operatorname{Sym}^k \Omega_{\mathbb{P}_n}^1 \big|_X \otimes \mathcal{O}(l) \right) = 0$$

unless either

(1) $i = 0$, $l - k \geqslant 0$, or
(2) $i = n - 1$, $d - (n+1) + k - l \geqslant 0$.

From this and the sequence

$$0 \to \operatorname{Sym}^{k-1} \Omega_{\mathbb{P}_n}^1 \big|_X \otimes \mathcal{O}(-d) \to \operatorname{Sym}^k \Omega_{\mathbb{P}_n}^1 \big|_X \to \operatorname{Sym}^k \Omega_X^1 \to 0$$

we conclude

$$H^i\left(X, \operatorname{Sym}^k \Omega_X^1 \otimes \mathcal{O}(l) \right) = 0$$

unless either

(1) $i = 0$, $l - k \geqslant 0$,
(2) $i = n - 2$, $k - (n + 2) - l \geqslant 0$, or
(3) $i = n - 1$, $d - (n + 1) + k - l \geqslant 0$.

For $l = 0$ we conclude

$$H^0(X, \mathrm{Sym}^k \Omega_X^1) = 0.$$

To see (1.23), the exact sequence

$$0 \to \Theta_X \to \Theta_{P_n}|_X \to \mathcal{O}(d) \to 0$$

implies

$$(1 + c_1(\Theta_X) + c_2(\Theta_X))(1 + dH) = (1 + H)^4,$$

where H is the hyperplane class. Thus

$$c_1^2(\Theta_X) = (d - 4)^2 d,$$

$$c_2(\Theta_X) = (d^2 - 4d + 6)d.$$

Therefore

$$c_1^2(\Omega_X^1) - c_2(\Omega_y^1) = (10 - 4d)d,$$

$$7c_1^2(\Omega_X^1) - 5c_2(\Omega_X^1) = 2(d^2 - 18d + 41)d.$$

Thus $\phi_{\mathcal{J}_{1,m}}$ cannot be a birational embedding—indeed, we have seen there are no symmetric differentials—while $\phi_{\mathcal{J}_{2,m}}$ is a birational embedding for large m when $d \geqslant 16$.

(2) *Subvarieties of Abelian Varieties.* These will be discussed at length in Section 3 when we give the proof of Bloch's conjecture. Here we will merely assert without proof that:

(1.24) For $X_n \subset A_N$, if $k \geqslant n/(N - n)$ and X is not ruled by subtori, then $P_k(X) \xrightarrow{\phi_{L,m}} \mathbb{P}_M$ is a birational embedding for m sufficiently large.

For X a smooth surface, from the formula

$$(1 + c_1(X) + c_2(X))(1 + c_1(N_X) + c_2(N_X)) = 1$$

we conclude that

$$c_1^2(X) - c_2(X) = 0, \qquad N = 3,$$

and with a little geometry that

$$c_1^2(X) - c_2(X) > 0, \qquad N > 3.$$

Thus, for $X_2 \subset A_3$, 1-jets are not enough, while 2-jets are. For $X_2 \subset A_N$, $N > 3$, 1-jets are enough.

2. Metrics of Negative Curvature from Jet Differentials

(a) The Ahlfors lemma. The Ahlfors lemma is central to differential-geometric methods of studying holomorphic mappings, recurrently surviving all changes in viewpoint. We will use a variant of it here.

Definition. Let X be a complex space. A *jet pseudometric* is given by a function

$$| \; | : J_k(X) \to \mathbb{R}^+$$

that is continuous and smooth except when it is zero, and satisfies

$$|tj| = |t| \cdot |j| \qquad (j \in J_k(X), \, t \in \mathbb{C}^*).$$

Here the action of $\mathbb{C}^*$ on jets is by reparametrization as discussed in Section 1 (a). Intuitively, $| \; |$ assigns a length to kth-order infinitesimal arcs in X. In local coordinates in a neighborhood consisting of smooth points on X a jet pseudometric will be given by

$$|j_k(f)(z)| = F\big(f_1'(z), \ldots, f_n'(z), \ldots, f_1^{(k)}(z), \ldots, f_N^{(k)}(z)\big),$$

where $F(f_1', \ldots, f_n', \ldots, f_1^{(k)}, \ldots, f_N^{(k)})$ is a nonnegative function, smooth except when zero, that satisfies a suitable weighted homogeneity condition.

Definition. The jet pseudometric $| \; |$ *has holomorphic sectional curvatures* $\leqslant -A$ $(A > 0)$ *on discs* if for any holomorphic mapping $f : \Delta \to X$ and point $x \in f(\Delta)$ we have at x either $|j_k(f)| = 0$ or

$$\sqrt{-1} \, \partial\bar{\partial} \log|j_k(f)|^2 \geqslant A|j_k(f)|^2. \tag{2.1}$$

(Compare this definition with Wu [29].)

We remark that, multiplying $| \; |$ by A^{-1}, we may always make the constant in (2.1) to be -1.

EXAMPLES

(i) The standard example is the Poincaré metric

$$\rho(z)\,|dz| = \frac{|dz|}{1 - |z|^2} \; ;$$

it has constant holomorphic section curvature -1.

(ii) In [10] Grauert and Reckziegel introduced negatively curved Finsler metrics (cf. also Cowen [5]), given by a nonnegative function $F(x, \xi)$ on the tangent bundle satisfying

$$F(x, t\xi) = |t|F(x, \xi) \qquad (\xi \in T_x(X), \, t \in \mathbb{C}).$$

A useful remark they made is that the sum of two negatively curved Finsler metrics is again negatively curved; the same is true for jet pseudometrics having holomorphic sectional curvatures $\leqslant -A$.

(iii) The Kobayashi metric $| \; |_\kappa$ is the pseudometric on $J_1(X)$ defined by

$$|\xi|_\kappa = \inf_f \left| \xi/f_*\left(\frac{\partial}{\partial z}\right) \right|,$$

where the inf is taken over all holomorphic mappings $f : \Delta \to X$ that satisfy $f(0) = x$ and $f_*(\partial/\partial z)_0 = a\xi$, $a \in \mathbb{C}^*$ (cf. Kobayashi [20]).

If we have $f : \Delta_r \to X$ with $f(0) = x$ and $f_*(\partial/\partial z)_0 = \xi$, then setting $f_r(z) = f(z/r)$ gives $f_r : \Delta \to X$ with $f_r(0) = x$ and $(f_r)_*(\partial/\partial z_0) = r\xi$. It follows that

$|\xi|_\kappa = 0$ in case there is an entire holomorphic curve passing through x in the direction $\xi \in T_x(X)$. For X compact the converse is due to Brody [3].

(iv) Let $H_0, \ldots, H_{n+1}$ be a collection of $n + 2$ hyperplanes in general position in $\mathbb{P}^n$. There has been an extensive study of the position of a nondegenerate holomorphic curve in $\mathbb{P}^n$ relative to these hyperplanes (cf. the introductions to Cowen and Griffiths [6] and Green [11]). In particular, for $X = \mathbb{P}^n - \bigcup_{\mu=0}^{n+1} H_\mu$ a classical theorem of E. Borel says that an entire holomorphic curve $f : \mathbb{C} \to X$ must lie in a $\mathbb{P}^{n-1}$. The corresponding defect relations were established by H. Cartan and Ahlfors.

In Cowen and Griffiths [6] there is a proof of these defect relations and Borel's theorem using what amounts to a negatively curved jet pseudometric on X; cf. (6.3) and (6.4) on p. 132 of that paper. Comparing (6.3) and (5.13) one sees that for $n \geqslant 2$ higher derivatives enter in an essential way in this metric. In fact, the pseudometric vanishes at a point in case the curve osculates to high order to a hyperplane at that point, and it vanishes identically exactly when the image $f(\Delta)$ lies in a $\mathbb{P}^{n-1}$.

As with ordinary metrics, the basic fact concerning jet pseudometrics is the

Ahlfors lemma for jet pseudometrics. *On a complex space X we let $|\ |$ be a jet pseudometric that has holomorphic sectional curvatures $\leqslant -1$ on discs. Then any holomorphic mapping $f : \Delta \to X$ is distance decreasing relative to the Poincaré metric; i.e.,*

$$|j_k(f)(z)| \leqslant \rho(z) \tag{2.2}$$

for all $z \in \Delta$.

Proof. If not identically zero, the pseudometric $|j_k(f)(z)|^2 |dz|^2$ has Gaussian curvature $\leqslant -1$ at the points where it does not vanish. The result now follows from the usual Ahlfors lemma (Kobayashi [20]). $\square$

To state a corollary we let $(x, \xi) \in T(X)$ and denote by $J_k(X)_{(x,\xi)}$ the set of all jets $j \in J_k(X)_x$ that project onto ξ; i.e., the linear part of j is ξ.

(2.3) Corollary. *Let $|\ |$ be a jet pseudometric whose holomorphic sectional curvatures on discs are $\leqslant -1$. Then the Kobayashi length satisfies*

$$|\xi|_\kappa \geqslant \inf_{j \in J_k(X)_{(x,\xi)}} |j|.$$

The proof is immediate from the Ahlfors lemma and the definition of $|\ |_\kappa$.

(2.4) Corollary. *Let $|\ |$ be a jet pseudometric on $J_k(X)$ having holomorphic sectional curvatures $\leqslant -1$ on discs. Then if $f : \mathbb{C} \to X$ is an entire holomorphic curve, then*

$$|j_k(x)(z)| \equiv 0.$$

(b) Construction of negatively curved pseudometrics from jet differentials. We will now show how negatively curved jet pseudometrics may be constructed by having enough holomorphic sections of a suitable line bundle.

(2.5) Proposition. *Let X be a projective algebraic variety and $E \to P_k(X)$ a very ample line bundle. If $t_0, \ldots, t_M$ is any basis for $H^0(P_k(X), E)$ and $s_0, \ldots, s_N$ any basis for $H^0(P_k(X), L^m \otimes E^{-1})$, then for a suitable constant $A > 0$ the jet pseudometric*

$$|j|^2 = A \left(\sum_{i,\alpha} |s_i t_\alpha(j)|^2 \right)^{1/m}$$

has holomorphic sectional curvatures $\leqslant -1$ on discs.

Proof. Let $U \subset P_k(X)$ be an open set over which E and L are trivial, and suppose that $f : \Delta \to X$ is a holomorphic mapping such that $j_k(f)(z) \in U$ for all $z \in \Delta$. Then using these trivializations,

$$t_\alpha(j_k(f))(z) = u_\alpha(z),$$
$$s_i(j_k(f))(z) = v_i(z)$$

are holomorphic functions of z and

$$|j_k(f)(z)|^2 = A \left(\sum_{i,\alpha} |v_i(z)u_\alpha(z)|^2 \right)^{1/m}.$$

Assuming that $|j_k(f)|^2$ is not identically zero, the $(1,1)$ form

$$\sqrt{-1}\, \partial\bar{\partial} \log|j_k(f)(z)|^2 = \frac{\sqrt{-1}}{m}\, \partial\bar{\partial} \log\left(\sum|v_i(z)|^2\right) + \frac{\sqrt{-1}}{m}\, \partial\bar{\partial} \log\left(\sum|u_\alpha(z)|^2\right)$$

$$= \frac{\alpha}{m} + \frac{\beta}{m}$$

is intrinsically defined—i.e., does not depend on the trivializations used. Each of the forms α and β is nonnegative; and β has the following geometric interpretation: Let $\phi_E : P_k(X) \to \mathbb{P}^M$ be the projective embedding induced by the sections $t_0, \ldots, t_M$ and

$$\omega = \phi_E^* \qquad \text{(Fubini–Study metric on } \mathbb{P}^M\text{)}.$$

If we denote by

$$f_k : \Delta \to P_k(X)$$

the canonical lifting of $f : \Delta \to X$ given by $f_k(z) = j_k(f)(z)$, then

$$\beta = f_k^*(\omega). \qquad \square$$

Next we need to know that:

(2.6) $\beta(z) = 0 \Leftrightarrow j_{k+1}(f)(z)$ *is a constant jet.*

Proof. We shall prove that the right-hand side is equivalent to the differential of f_k vanishing at z. Taking $z = (0)$ and a local embedding of a neighborhood of

$f(0)$ in X as a subvariety in an open set in C^N, we write

$$f(z) = f^{(j)}(0)\,\frac{z^j}{j!} + f^{(j+1)}(0)\,\frac{z^{j+1}}{(j+1)!} + \cdots$$

where $f^{(j)}(0) \neq 0$. If $j \leqslant k+1$ and if the differential of $j_k(f) \in P_k(X)$ is zero at $z = 0$, then this means that

$$\frac{d}{dz}\left(j_k(f)(z)\right)_{z=0} = t(j_k(f)(0)).$$

The right-hand side is the reparametrization of the jet $j_k(f)(0)$, and all terms of order $\leqslant j$ are zero. But the left-hand side has a nonzero term of order $j-1$, which contradicts our assumption $j \leqslant k+1$.

Now both of the mappings

$$z \to \beta(z),$$

$$z \to |j_k(f)(z)|^2$$

are quadratic with respect to a reparametrization, and consequently the ratio

$$\frac{|j_k(f)(z)|^2}{\rho(z)}$$

is locally bounded from above on the projectivized tangent bundle of $P_k(X)$. Since it is intrinsic and X is compact, this ratio will everywhere be $\leqslant B$ for some constant B. This implies that

$$\sqrt{-1}\,\partial\bar{\partial}\log|j_k(f)|^2 \geqslant \frac{\beta}{m} \geqslant \frac{1}{Bm}\,|j_k(f)|^2$$

for any holomorphic mapping $f : \Delta \to X$. Adjusting constants yields the proposition. $\square$

(2.7) Corollary. *Consider the map*

$$\phi_{L^m} : P_k(X) \to \mathbb{P}^N$$

defined by the linear system $|L^m|$ on $P_k(X)$. Let $B_{k,m}$ be the union of the base locus of ϕ_{L^m} and the points $j \in P_k(X)$ such that $\dim(\phi_{L^m}^{-1}(\phi_{L^m}(j))) \geqslant 1$. Then there exists a jet pseudometric on $P_k(X)$ with holomorphic sectional curvatures $\leqslant -1$ on discs and vanishing at most on $B_{k,m}$.

Remark. Noguchi [24] has a similar observation in the case of symmetric differentials.

Proof. Let $E \to P_k(X)$ be a very ample line bundle. It will suffice to show that for sufficiently large l, the base of the linear system $|L^{lm} \otimes E^{-1}|$ is contained in $B_{k,m}$.

By blowing up we may assume that the base of $|L^m|$ is a divisor F on the blown-up variety $P_k(X)$, and we set $\tilde{L} = L^m \otimes F^{-1}$. Then

$$\phi_{\tilde{L}} : P_k(X) \to \mathbb{P}^N$$

is a holomorphic mapping that is finite-to-one outside the total transform $\tilde{B}_{k,m}$ of $B_{k,m}$.

Given $j \in P_k(X) - B_{k,m}$, we may choose a divisor $D \in |E|$ such that

$$\phi_{\tilde{L}}^{-1}(\phi_{\tilde{L}}(j)) \notin \tilde{D},$$

where $\tilde{D}$ is the total transform of D. If we choose a hypersurface of sufficiently high degree l in $\mathbb{P}^N$ that passes through $\phi_{\tilde{L}}(\tilde{D})$ but does not contain $\phi_{\tilde{L}}(j)$, then we obtain a section of $\tilde{L}^l - \tilde{D}$ on $P_k(X)$ that does not pass through j. Projecting its divisor down to $P_k(X)$ gives a divisor in $|L^{lm} - D|$ that does not pass through j. $\square$

Combining Corollary 2.4 with Corollary 2.7 gives the

(2.8) Corollary. *Let $f: \mathbb{C} \to X$ be an entire holomorphic curve with canonical lifting $f_k: \mathbb{C} \to P_k(X)$. Then, with the above notation,*

$$f_k(\mathbb{C}) \subset \bigcup_m B_{k,m}.$$

Remark. Observe that the right-hand side of this inclusion is defined purely in terms of the geometry of the linear systems $|L^m|$ on $P_k(X)$. These are in turn described by the jet differentials on X.

3. Proof of Bloch's Conjecture (Theorem I)

(a) Proof of Theorems I and I′. We begin by establishing

Theorem I′. *Let X be an analytic subvariety of a complex torus A. If X is not the translate of a subtorus of A, then any entire holomorphic curve $f: \mathbb{C} \to X$ lies in a proper analytic subvariety of X.*

Remark. By induction, then, the image curve $f(\mathbb{C})$ will lie in a proper subtorus. In this connection, when A is a simple abelian variety an elementary proof of Bloch's conjecture has been given by one of us (cf. Green [12]).

Proof. Writing $A = \mathbb{C}^N/\Lambda$ where Λ is a lattice in $\mathbb{C}^N$ and using the monodromy theorem, we may assume that any holomorphic mapping $f: \Delta_r \to X$ has been lifted to $\mathbb{C}^N$. We shall continue to denote this lifting by f, and remark that it is unique up to translation by a constant vector in Λ. Thus $f(z) = (f_1(z), \dots, f_N(z))$ where the $f_i(z)$ are holomorphic functions.

We shall also use the notation

$$u(j_k(f)) = (f_1', \dots, f_N', \dots, f_1^{(k)}, \dots, f_N^{(k)})$$

$$= (f_i', \dots, f_i^{(k)})$$

(here, the index i is thought of as running from 1 to n) for the indicated global coordinates on the jet spaces $J_k(X)$. Equivalently, u is the composite map in the

diagram

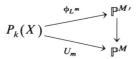

$$J_k(X) \to J_k(A) \cong A \times \mathbb{C}^{kN} \to \mathbb{C}^{kN}.$$

If we take any basis $P_0, \ldots, P_M$ for all polynomials with constant coefficients in the indeterminates $f_1', \ldots, f_N'; \ldots; f_1^{(k)}, \ldots, f_N^{(k)}$ that are homogeneous of total weight m when $f_i^{(l)}$ is assigned weight l, then the P_α form part of a basis for

$$H^0(P_k(X), L^m) = H^0(X, \mathcal{J}_{k, m}).$$

Using these to define a mapping U_m to projective space, we will have a diagram

$$
\begin{array}{ccc}
 & \overset{\phi_{L^m}}{\longrightarrow} & \mathbb{P}^{M'} \\
P_k(X) \diagup & & \big\downarrow \\
 & \underset{U_m}{\longrightarrow} & \mathbb{P}^M
\end{array}
$$

where ϕ_{L^m} is the mapping defined by the complete linear system $|L^m|$ and the vertical arrow is a linear projection. For $m = k!$ the mapping U_m involves $(f_i^{(l)})^{m/l}$ for all i and $l \leqslant k$, and consequently U_m has no base locus. Moreover, for this same m

$$U_m(j_1) = U_m(j_2) \quad \Leftrightarrow \quad u(j_1) = u(j_2)$$
$$\Leftrightarrow \quad j_1 = j_2 + a,$$

where $j_1, j_2 \in J_k(X)$ and $j_2 + a$ denotes the translation of j_2 by $a \in \mathbb{C}^N$. Summarizing: *If the jet j lies in the subvariety $B_{k, k!}$ (cf. Section 2), then*

$$\dim\{a \in A : j \in J_k(X) \cap J_k(X + a)\} \geqslant 1. \tag{*}$$

Now let $f: \mathbb{C} \to X$ be an entire holomorphic curve. By Corollaries 2.7 and 2.8 to Proposition 2.5, $j_k(f) \in B_{k, l}$ for all k and l. We define the sequence of complex-analytic varieties

$$V_k(f) = \{a \in A : j_k(f)(0) \in J_k(X) \cap J_k(X + a)\}.$$

These form a nested sequence

$$V_1(f) \subseteq V_2(f) \subseteq V_3(f) \subseteq \cdots$$

that eventually stabilizes at a variety V. By power series,

$$a \in V \quad \Leftrightarrow \quad f(\mathbb{C}) \subseteq X \cap (X + a).$$

On the other hand, by (*) above

$$\dim V = \dim \{a \in A : f(\mathbb{C}) \subseteq X \cap (X + a)\} \geqslant 1.$$

Now, either $X \cap (X + a)$ is a proper analytic subvariety of X for some $a \in A - \{0\}$—in which case we are done—or else

$$X = X + a \quad \text{for all } a \in V.$$

Assuming this alternative holds, we note that

$$\{a \in A : X = X + a\} = B$$

is a subgroup of A that must have positive dimension; in this case we shall say that X is *ruled by subtori*.

Letting $B^0 \subset A$ be the identity component of the group B and setting $\bar{A} =$

A/B^0, we have a diagram of entire holomorphic mappings

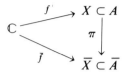

where $\overline{X}$ is not ruled by subtori. Applying the argument thus far, if $\overline{X}$ is not a point, then $\bar{f}(\mathbb{C})$ lies in a proper analytic subvariety $\overline{Z}$ of $\overline{X}$, $f(\mathbb{C})$ lies in $Z = \pi^{-1}(\overline{Z})$, and we are done. If $\overline{X}$ is a point, then X is a subtorus, and this contradicts our initial assumption. $\square$

Theorem I is an easy consequence of Theorem I'. If X is any algebraic variety with irregularity $q > \dim X$, then we denote by A the Albanese variety of X and by

$$\alpha : X \to A$$

the standard map. Setting $\alpha(X) = Y$, Y is not a subtorus of A, and consequently the image of

$$\alpha \circ f : \mathbb{C} \to Y$$

lies in a proper subvariety of Y. The same must be true of the image $f(\mathbb{C}) \subset X$.

Remark. The same argument applies whenever X is a compact Kähler manifold or a Moisezon space.

(b) Some remarks on analytic subvarieties of complex tori. We want to make some general observations about subvarieties of abelian varieties, and then in the following section shall give some related remarks on how our proof compares with the argument of Bloch and Ochiai.

Given an analytic subvariety $X \subset A$ of an abelian variety, the most obvious way to study its geometry is via its *Gauss mapping*

$$\gamma : X \to G(n, N). \tag{3.1}$$

Here, γ is defined at the smooth points of X by

$$\gamma(x) = T_x(X)$$

$$= \text{translate to the origin of the tangent space at } x \in X,$$

and is then extended to all of X as a natural mapping (see Griffiths [14] for further discussion). Actually, for the purposes of this discussion the singularities of X are not that essential, so the reader may either assume X is smooth or replace X by its Nash blowup on which γ is everywhere defined [14]. By definition

$$T^*(X) = \gamma^* E,$$

where $E \to G(n, N)$ is the *dual* of the universal subbundle.

The first step is to analyze the case when γ is degenerate in the sense that $\dim \gamma(X) < \dim X$. In this regard there is a classical structure theorem which may be found, e.g., in Section 4 of [15]:

(3.2) *Given $X \subset A$, we may pass to a finite unramified covering of A and make a translation to have*

$$A = A' \times A'',$$
$$X = A' \times X'',$$

where A', A'' are abelian subvarieties of A, $X'' \subset A''$ is an analytic subvariety whose Gauss mapping is nondegenerate, and the Gauss mapping of X has fibres $A' \times \{x''\}$.

Briefly, the fibres of γ are translates of abelian subvarieties that give a ruling of X.

We should like to make two further observations, of which the first is this:

(3.3) *For an n-dimensional subvariety $X \subset A$ of an abelian variety, the Kodaira number*

$$\kappa(X) = n$$

if, and only if, the Gauss mapping of X is nondegenerate.

In (3.3) X is assumed to be irreducible, but it may have singularities. We observe that we have a diagram

$$X \xrightarrow{\ \gamma\ } G(n, N) \xrightarrow{\ p\ } \mathbb{P}^{\binom{N}{n}-1}$$

with ϕ_K to $\mathbb{P}^{p_g-1}$ and π

where p is the Plücker embedding, ϕ_K is the canonical map of X ($p_g = h^{n,0}(X)$), and π is a linear projection. From this it follows that

$$\dim \gamma(X) = n \ \Rightarrow\ \kappa(X) = n,$$

and the converse is provided by the structure theorem (3.2).

Our second remark is that we have always taken A to be an abelian variety as opposed to just a complex torus. There is no particular reason for this, and there is also no essential loss of generality, because of the following:

(3.4) *Suppose $X \subset A$ is an analytic subvariety of a complex torus that is not contained in a subtorus, and assume that X is not ruled by subtori. Then A is an abelian variety.*

Proof. If X is not ruled by subtori, then the Gauss map of X is nondegenerate. Since there is an equidimensional mapping of X to the projective algebraic variety $\gamma(X)$, X is a Moisezon space. The Albanese variety $\mathrm{Alb}(X)$ is then an abelian variety, and there is a diagram of holomorphic mappings where the vertical arrow is surjective, since X is not contained in an abelian subvariety. Being a quotient of $\mathrm{Alb}(X)$, A must be an abelian variety. $\square$

Returning to our general discussion, we assume that the Gauss mapping (3.1) is equidimensional. Then one might think that $T^*(X)$, being generated by its global sections and pulled back from the universal bundle by a nondegenerate mapping, might be close to being ample. For example, we might hope that Sakai's λ-invariant (cf. [28])

$$\lambda = \mathrm{Tr} \deg \left\{ \bigoplus_{m \geqslant 0} H^0(X, \mathrm{Sym}^m \, \Omega_X^1) \right\} - n$$

would achieve its maximum possible value $\max(n, N - n)$. In this regard we first observe the

(3.5) Lemma. *The universal bundle* $E \to G(n, N)$ *is not ample if* $n \geqslant 2$. *The transcendence degree of* $\bigoplus_{m \geqslant 0} H^0(G(n, N), \mathrm{Sym}^m E)$ *is* N.

Proof. We denote by $P = P(E^*)$ the projective bundle of hyperplanes in E, by $\mathcal{O}_P(1)$ the tautological line bundle over P, and recall that

$$H^0(P, \mathcal{O}_P(m)) \cong H^0(G(n, N), \mathrm{Sym}^m E)$$

$$\cong \mathrm{Sym}^m(\mathbb{C}^{N*}). \tag{3.6}$$

Moreover, by definition E is ample if, and only if, $\mathcal{O}_P(1)$ is ample on P.

We may realize P as the subvariety of $G(n, N) \times \mathbb{P}^{N-1}$ defined by incidence:

$$P = \{(\Lambda, p) : p \in \Lambda\} \subset G(n, N) \times \mathbb{P}^{N-1}. \tag{3.7}$$

Since $H^0(\mathcal{O}_P(1)) = \mathbb{C}^{N*}$, the mapping given by the complete linear system $|\mathcal{O}_P(1)|$ is projection on the second factor in

$$
\begin{array}{ccc}
P & \xrightarrow{\pi_2} & \mathbb{P}^{N-1} \\
{\scriptstyle \pi_1} \downarrow & & \\
G(n, N) & &
\end{array}
$$

This mapping is everywhere defined and has fibres

$$\pi_2^{-1}(p) = \{\Lambda : p \in \Lambda\}$$

$$\cong G(n - 1, N - 1).$$

It follows that

$$\mathcal{O}_P(1) = \pi_2^* \, \mathcal{O}_{\mathbb{P}^{N-1}}(1)$$

cannot be ample if $n \geqslant 2$. By (3.6),

$$\bigoplus_{m \geqslant 0} H^0(\mathcal{O}_P(m)) \cong \bigoplus_{m \geqslant 0} \mathrm{Sym}^m(\mathbb{C}^N)^*$$

$$= C[z_1, \ldots, z_N]$$

has transcendence degree N, and for $n \geqslant 2$

$$N < n(N - n) + n = \dim P + 1. \quad \square$$

Because of the lemma we have the possibility that $T^*(X)$ may not be ample, even if γ is an embedding. To determine what λ is we consider the tangential variety

$$\tau(X) \subset \mathbb{P}^{N-1},$$

defined to be the union of the projectivized tangent spaces $\mathbb{P}T_x(X)$ as x varies over X. Alternatively, in the diagram

$$
\begin{array}{ccc}
P_1(X) \xrightarrow{\ \gamma^*\ } P \xrightarrow{\ \pi_2\ } \mathbb{P}^{N-1} \\
\downarrow \qquad\qquad \downarrow \\
X \xrightarrow{\ \ \gamma\ \ } G(n,N)
\end{array}
$$

the tangential variety is the image of $\tau = \pi_2 \circ \gamma^*$. It follows that

$$\lambda = \dim \tau(X) - n + 1,$$

so all we can easily say is that $\lambda \geqslant 1$. In general the behavior of $\tau(X)$ is not well understood, especially when $\dim X \geqslant 4$ (cf. Section 5 of Griffiths and Harris [15]), but in any case $T^*(X)$ cannot be ample when $\mathrm{codim}\, X < \dim X$.

(c) Some observations about jet differentials associated to subvarieties of abelian varieties. Our approach to Theorem I′ differs from Bloch and Ochiai's in two respects: we substitute negative-curvature arguments for Nevanlinna theory, and we make a different geometric computation.

In fact these are related. The most naive way to use negative curvature is via the observation that holomorphic sectional curvatures decrease on submanifolds. The flat Euclidean metric on an abelian variety $A = \mathbb{C}^N/\Lambda$ induces on any subvariety $X \subset A$ a metric whose holomorphic sectional curvatures $K(\xi)$ are $\leqslant 0$ (here $\xi \in T_x(X)$ is a tangent vector). The condition $K(\xi) < 0$ is closely related to the tangential variety having dimension $2n - 1$. More precisely, from Section 4 of Griffiths and Harris [15] we have:

(3.8) *If* $\mathrm{II}(\xi, \eta)$ *denotes the 2nd fundamental form of* $X \subset A$, *then for* $\xi \in T(X)$

$$\mathrm{II}(\xi, \xi) = 0 \quad \Leftrightarrow \quad K(\xi) = 0.$$

On the other hand, if $\mathrm{II}(\xi, \xi) \neq 0$ for every nonzero tangent vector, then the linear system $|\mathrm{II}|$ has no base points, and it follows from Section 5 of [15] that the tangential mapping

$$\tau : P_1(X) \to \mathbb{P}^{N-1} \tag{3.9}$$

is equidimensional.

However, even if τ is equidimensional, it may have a branch locus or, worse still, blow down a subvariety of $P_1(X)$. For example, if C_1, C_2 are curves in A and we consider the translation-type surface

$$X = C_1 + C_2 = \{ p_1 + p_2 \mid p_1 \in C_1, p_2 \in C_2 \},$$

then for each $(p, \xi) \in T(C_1)$, the curve $E = \{(p, \xi) \times (q, 0) \mid q \in C_2\}$ in $P(X)$ collapses to a point under τ. In fact, the union of such E's projects down onto all of X. The main hitch in completing Bloch's argument was to get these blown-down varieties under control.

Moreover, in general the tangential mapping (3.9) will not be equidimensional (e.g., if $\mathrm{codim}\, X < \dim X$), and apparently the conclusion to be drawn is that it is necessary to go to higher-order jets to detect the geometry necessary to force negative curvature.

As mentioned in the introduction, jets already appeared in the original paper of Bloch in 1926, as well as in the work of Ochiai [25]. Their main computation centered on determining the branch locus of the mappings

$$u_k : J_k(X) \to \mathbb{C}^{kN} \qquad (3.10)$$

that we encountered in the proof of Theorem I'. (We remark that we only needed the exceptional locus of u_k, and not the full branch locus.) We have found a geometric interpretation of their computation that may illuminate what is going on. Given a line L through the origin in $\mathbb{C}^N$, we define the Schubert cycle

$$\Sigma_L = \{\Lambda \in G(n, N) : L \subset \Lambda\}.$$

If we consider the Gauss mapping

$$\gamma : X \to G(n, N),$$

then we note that

$\gamma^{-1}(\Sigma_L)$ = projection to X of the fibre $\tau^{-1}(L)$ of the tangent mapping (3.9).

Now, rather than stop with the Gauss mapping alone, we extend to k-jets to obtain

$$\gamma_k : J_k(X) \to J_k(G(n, N)). \qquad (3.11)$$

The main computation in Ochiai [25] may be expressed by saying that the branch locus B_k of the mapping (3.11) satisfies

$$B_k \subseteq \bigcup_{L \in \mathbb{P}^{N-1}} \gamma_k^{-1}(J_k(\Sigma_L)) \cup \pi_k^{-1}(X_{\text{sing}}),$$

where $\pi_k : J_k(X) \to X$ is the projection. As a consequence we have:

(3.12) *For a holomorphic mapping* $f : \Delta \to X$ *that satisfies* $j_k(f)(z) \in B_k$ *for all* k *and* $z \in \Delta$*, one of the following alternatives must hold:*

$$f(\Delta) \subset X_{\text{sing}},$$

$$f(\Delta) \subset \gamma^{-1}(\Sigma_L) \quad \text{for some } L \in \mathbb{P}^{N-1}.$$

Using the interpretation (3.12), we may complete Ochiai's argument. Alternatively, we may use the jet forms to construct a negatively curved jet pseudometric on $J_k(X) - B_k$, which is the approach we have followed in this paper.

Part B. Measure Hyperbolic Algebraic Surfaces

4. Proof of Theorem II

(a) Reduction to the K3 case. We first recall the definition of the Kobayashi–Eisenman intrinsic volume form Ψ defined on any n-dimensional complex analytic variety X (cf. Kobayashi [20]). Let

$$\mu = \prod_{j=1}^{n} \frac{4\sqrt{-1} \, dz_j \, d\bar{z}_j}{\left(1 - |z_j|^2\right)^2}$$

denote the Poincaré volume form on the unit polycylinder Δ^n in $\mathbb{C}^n$. Given a smooth point $x \in X$, we consider all holomorphic mappings

$$f : \Delta^n \to X$$

that satisfy $f(0) = x$ and $J_f(0) \neq 0$, where $J_f = \Lambda^n f_*$ is the Jacobian determinant of f. Then by definition

$$\Psi(x) = \inf_f (f^{-1})^* \mu.$$

To better understand this definition, if $\Psi(x) = 0$, then we must have a sequence

$$f_k : \Delta^n \to X$$

of holomorphic mappings that satisfy

$$f_k(0) = x, \qquad |J_{f_k}(0)| \geq k.$$

If we let

$$\Delta^n(k, 1) = \{(z_1, \ldots, z_n) \in \mathbb{C}^n : |z_1| < k, |z_2| < 1, \ldots, |z_n| < 1\}$$

and replace z_1 by z_1/k, then we obtain a sequence of holomorphic mappings

$$g_k : \Delta^n(k, 1) \to X$$

satisfying

$$g_k(0) = x, \qquad |J_{g_k}(0)| \geq 1.$$

The analytic variety X is said to be *measure hyperbolic* in case Ψ is positive outside a proper subvariety E of X. In this case, for any point $x \in X - E$ there is an upper bound on the size of polydiscs $\Delta^n(k, 1)$ that can be mapped into X sending the origin to x and having Jacobian ≥ 1 there. In particular:

(4.1) *If there is a holomorphic mapping*

$$f : \Delta^n(\infty, 1) \to X$$

whose image contains a Zariski open subset (actually, any open set will do), then X fails to be measure hyperbolic.

Turning to the conjecture of the introduction, we may assume that X is a smooth projective variety and recall that X is said to be of *general type* in case, for some $m > 0$, the rational map

$$\phi_{mK} : X \to \mathbb{P}^N$$

defined by the pluricanonical system $|mK_X|$ is equidimensional—i.e, the image $\phi_{mK}(X)$ is an n-dimensional algebraic subvariety of $\mathbb{P}^N$. Equivalently, the canonical ring

$$\bigoplus_{m > 0} H^0(mK_X)$$

should have maximal transcendence degree $n + 1$. It is known that

$$X \text{ general type} \quad \Rightarrow \quad X \text{ measure hyperbolic}$$

(see Griffiths [13] for the case $m = 1$, and Kobayashi and Ochiai [21] for the extension of this idea to the general situation).

For any variety X the *Kodaira number* $\kappa = \kappa(X)$ is defined to be the maximal dimension of the pluricanonical images $\phi_{mK}(X)$ $(m > 0)$. For algebraic surfaces that contain no exceptional curve of the 1st kind the classification theorem (Griffiths and Harris [17, p. 590]) gives the following list:

(a) $\kappa = -1 \Rightarrow X$ is $\mathbb{P}^2$ or is ruled by $\mathbb{P}^1$'s.
(b) $\kappa = 0 \quad \Rightarrow \quad$ (i) X is a $K3$ surface if $q = 0$ and $p_g = 1$,
$\qquad\qquad\qquad$ (ii) X is an Enriques surface if $q = p_g = 0$,
$\qquad\qquad\qquad$ (iii) X is a hyperelliptic surface if $q = 1$,
$\qquad\qquad\qquad$ (iv) X is an abelian surface if $q = 2$.
(c) $\kappa = 1 \quad \Rightarrow X$ is an elliptic surface.
(d) $\kappa = 2 \quad \Rightarrow X$ is of general type.

The surfaces of class (a) and (b)(iv) clearly fail to be measure hyperbolic (cf. (4.1)—in these cases there is a nondegenerate mapping of all of $\mathbb{C}^2$ to X). To treat the remaining ones we shall utilize the

(4.2) Lemma. *If on a surface X there is an algebraic family consisting of ∞^1 algebraic curves whose general member is either rational or elliptic, then X fails to be measure hyperbolic.*

Remark. We do *not* require that the general curve E in our family should be smooth—to say that E is rational or elliptic means that the genus of its normalization should be zero or one.

Proof. We may describe these curves as being a family $\{E_t\}_{t \in B}$ where B is an algebraic parameter curve. If a generic E_t is rational, then there is a finite covering $\tilde{B}$ of B and a surjective rational mapping

$$\tilde{f} : \tilde{B} \times \mathbb{P}^1 \to X. \tag{4.3}$$

By deleting the finite set of points Z in $\tilde{B}$ over which $\tilde{f}$ may not be defined as a holomorphic mapping and setting

$$B^* = \tilde{B} - Z, \qquad f^* = \tilde{f}|_{B^*},$$

we arrive at a holomorphic mapping

$$f^* : B^* \times \mathbb{P}^1 \to X$$

whose image is Zariski open in X. Finally, passing to the universal covering of B^* gives a mapping of $C \times \mathbb{P}^1$ $(C = \Delta, \mathbb{C}, \text{ or } \mathbb{P}^1)$ to X with Zariski dense image, and we may apply (4.1).

If a generic E_t is elliptic, then removing a finite set Z from B we may assume that for each $t \in B^* = B - Z$ the curve E_t is irreducible and the normalization $\tilde{E}_t$ is a compact Riemann surface of genus one. We also assume that the universal covering of B^* is the disc Δ and the E_t have nonconstant j-invariant—otherwise the argument is similar but easier. The covering $\Delta \to B^*$ will be denoted by $z \to t(z)$, and we may then write

$$\tilde{E}_{t(z)} = \mathbb{C}/\Lambda_z$$

where Λ_z is a holomorphically varying lattice in $\mathbb{C}$. More precisely, enlarging Z

to include all t where E_t has automorphisms, we may lift the mapping

$$B^* \to \{\text{moduli of elliptic curves}\}$$

to a holomorphic mapping

$$\Delta \xrightarrow{\tau} \{\text{upper half plane}\}$$

such that

$$\Lambda_z = \{m + n\tau(z)\}_{m,\,n \in \mathbb{Z}}.$$

We then obtain a holomorphic mapping

$$f : \Delta \times \mathbb{C} \to X$$

defined by

$$f(z, w) = w \in \mathbb{C}/\Lambda_z;$$

by construction the image of f is Zariski dense in X and we may apply (4.1). $\square$

We again emphasize it is not required that the general E_t be smooth.

Also, if we observe that on $\Delta \times \mathbb{P}^1$ any holomorphic section of $\mathrm{Sym}^m(\Omega^2_{\Delta \times \mathbb{P}^2})$ must be identically zero, then from (4.3) we have the corollary:

(4.4) *If $H^0(mK_X) \neq 0$ for some $m > 0$, then any rational curve on X must be isolated.*

More precisely, for any holomorphic mapping

$$f : \Delta \times \mathbb{P}^1 \to X$$

the Jacobian J_f must be identically zero.

Using the lemma, we see that elliptic surfaces, hyperelliptic surfaces, and Enriques surfaces—these all have elliptic pencils—fail to be measure hyperbolic. To establish the conjecture for algebraic surfaces it will suffice to show that any algebraic $K3$ surface fails to be measure hyperbolic, and again using Lemma 4.2, this would follow from the assertion:

(4.5) *On any smooth algebraic $K3$ surface X there are ∞^1 elliptic curves.*

As mentioned in the introduction we shall give a construction of ∞^1 curves on any X that we can show to be elliptic for the first three families of $K3$ surfaces, and in general serve to reduce the conjecture to establishing a certain technical algebro-geometric point to be explained below.

(b) Informal discussion of the proof. In this subsection we shall discuss the idea behind the proposed construction of the ∞^1 elliptic curves on any algebraic $K3$ surface X. We recall (Mayer [22] and Saint-Donat [27]) that these surfaces fall into a sequence of irreducible families $\mathcal{F}_n$ $(n \geq 1)$ that may be described as follows:

(4.6)

(i) *The surfaces $X \in \mathcal{F}_1$ are 2-sheeted coverings $X \to \mathbb{P}^2$ branched over a smooth curve of degree six.*

(ii) *The surfaces $X \in \mathcal{F}_n$ ($n \geqslant 2$) are nondegenerate smooth surfaces $X \subset \mathbb{P}^{n+1}$ that have degree $2n$ and $p_g \neq 0$.*

(iii) *If $X \in \mathcal{F}_n$, then the general hyperplane section $C \in |\mathcal{O}_X(1)|$ is a smooth canonical curve of genus $n + 1$ (if we take $\mathcal{O}_X(1)$ to be $f^*\mathcal{O}_{\mathbb{P}^2}(1)$, then this is also valid for $n = 1$).*

We recall that for generic $X \in \mathcal{F}_n$ the Neron–Severi group $\mathrm{Pic}(X) \otimes Q$ is generated by the class of a hyperplane section. Since a smooth rational or elliptic curve E has respectively

$$E^2 = -2, \qquad E^2 = 0,$$

it follows that on a general algebraic $K3$ surface there are no smooth curves of genus 0 or 1 (actually, for generic $X \in \mathcal{F}_n$ there are not smooth curves of genus $\leqslant n$), so the curves we are seeking in order to establish (4.5) must be singular.

It is also the case that for generic $X \in \mathcal{F}_n$ there are no curves other than those cut out by hypersurfaces in $\mathbb{P}^{n+1}$ (cf. Saint-Donat [27]). If $V \subset \mathbb{P}^{n+1}$ is a hypersurface of degree d such that

$$C = V \cap X$$

is smooth, then the genus $g(C) = d^2 n$. It follows that if on any algebraic $K3$ surface $X \in \mathcal{F}_n$ we are to find a curve of genus $\leqslant n$, then we should look for those of the form

$$C = H \cap X$$

where $H \in \mathbb{P}^{n+1*}$ is a hyperplane that fails to meet X transversely—i.e., H should be a tangent hyperplane to X

Now the tangent hyperplanes constitute the *dual variety* $X^* \subset \mathbb{P}^{n+1*}$. At a smooth point of X^* the corresponding hyperplane H is simply tangent to X at one point, and consequently the section $C = H \cap X$ has one ordinary double point ($=$ node), and the genus of its normalization is $g(\tilde{C}) = n$. Suppose next that H is simply tangent at two points; then $C = H \cap X$ has two ordinary nodes and $g(\tilde{C}) = n - 1$. In general:

(4.7) *If a hyperplane H is tangent to X at k distinct points, then the corresponding section $C = H \cap X$ has normalization $\tilde{C}$ with genus $g(\tilde{C}) \leqslant n - k + 1$.*

Our main result is the following:

(4.8) Proposition

(i) *For any algebraic $K3$ surface $X \in \mathcal{F}_n$, there are ∞^{n-k+1} hyperplanes that are tangent at k points.*

(ii) *If $n = 1$ or 2 or $n = 3$ and X is generic, then these k points may be taken distinct.*

It is clear that Theorem II follows from Proposition (4.8), (4.7), and Lemma (4.2). Moreover, the full conjecture of the introduction would follow if part (ii) of (4.8) were established for all n. As we shall presently discuss, there are compelling reasons that this should be the case, but as will also be seen during the proof of (4.8) in the next section, there is one technical issue dealing with the precise meaning of "k-fold tangent point" that we are unable to overcome.

The idea behind the proof of Proposition (4.8) is simply the following count of constants:

(4.9) *The dimension of X^* is n, and it is "k conditions" that a hyperplane H is tangent to X at k points.*

Obviously, there are several matters that require extreme caution here, the most serious of which is that the dual X^* may not have any k-fold points when $k = n$. This particular difficulty will be overcome in the next section. The next most serious question—and the one on which we are stuck— is just how to control the genus of the section $H \cap X$ where H is a k-fold tangent hyperplane but where the points of tangency may not be distinct. This much, however, can be said:

(4.10) *If there are ∞^1 hyperplanes H such that a general one is tangent to X at n distinct points, then these are points of simple tangency.*

Proof. For the corresponding section $C = H \cap X$ we consider the normalization $\tilde{C}_0$ of any irreducible component C_0 of C. Then the genus $g(\tilde{C}_0) \leqslant 1$, with equality holding if, and only if, $C_0 = C$ and the points of tangency are simple. The result then follows from (4.4). □

The intuitive reason, then, why (ii) in Proposition (4.8) should hold for all n is that in any case by part (i) there are ∞^1 n-fold tangent hyperplanes, and if a general one of these were not simply tangent at n distinct points, then (4.4) would be violated.

(c) Existence of n-fold tangent hyperplanes to a $K3$ surface. In this subsection we shall establish (i) in Proposition (4.8). It is instructive to begin by discussing the pitfalls in trying to directly rigorize the naive dimension count (4.9).

For example, consider the statement: "It is one condition that a hyperplane is tangent to X." What this means is that the dual variety X^* is a hypersurface in $\mathbb{P}^{n+1*}$. Although this is generally true, there are certainly smooth nondegenerate varieties $V \subset \mathbb{P}^{n+1}$ for which V^* fails to be a hypersurface (cf. Section 3 of Griffiths and Harris [15]). However, for any smooth surface or any variety X whose Kodaira number $\kappa(X) \geqslant 0$, the dual X^* is a hypersurface [15, Section 3]. Both reasons are applicable in our present case.

A more serious objection concerns the singular locus of X^*. For example, the hyperplanes tangent at two distinct points occur on the double locus of X^*, and there are varieties for which X^* is a smooth hypersurface (e.g., nonsingular quadrics) or, even worse, X^* may be a hypersurface whose singularities occur in high codimension (e.g., according to Donagi [9] the dual of the Plücker image of the Grassmannian $G(3.6)$ in $\mathbb{P}^{19}$ is a hypersurface whose singularities occur in codimension five). About all that can be easily said is this:

(4.11) *If the dual $X^* \subset \mathbb{P}^{N*}$ of an algebraic variety $X \subset \mathbb{P}^N$ is a hypersurface, and if there is one hyperplane that is tangent at k distinct smooth points of X, then there are at least ∞^{N-k} such k-fold tangent hyperplanes.*

This is because if there is one point in $\mathbb{P}^{N*}$ that lies on k distinct local branches of a hypersurface X^*, then the k-fold locus of X^* has codimension $\leqslant k$.

We remark that in our problem dealing with algebraic $K3$ surfaces it is expected that all such inequalities should in fact be equalities. The reason is that if, e.g., there were ∞^2 n-fold tangent hyperplanes to $X \in \mathcal{F}_n$, then we would have on X either (i) ∞^1 elliptic curves with the same j-invariant, or (ii) ∞^1 rational curves. Both of these are impossible.

We also remark that, in general, the k-fold locus X_k^* of X^* may be defined by a condition on Fitting ideals (see the beautiful survey paper [19] of Kleiman). Since this is a determinantal condition, it follows that

$$X_k^* \text{ nonempty} \quad \Rightarrow \quad \operatorname{codim} X_k^* \leqslant k.$$

What our proof of (i) in Proposition (4.8) will give us is that

(4.12) $\operatorname{codim} X_k^* \leqslant k$ *for any algebraic $K3$ surface X.*

This is a fairly strong condition, but, as will be discussed in the next section, it does not yield the conjecture of the introduction, since it need not be the case that for any $H \in X_k^*$ and C_0 any irreducible component of $H \cap X$, we have $g(\tilde{C}_0) \leqslant n - k + 1$.

To establish (i) in Proposition (4.8) we shall use induction on n for the families $\mathcal{F}_n$, together with the following linkage between $\mathcal{F}_{n-1}$ and $\mathcal{F}_n$:

Let $X_0 \in \mathbb{P}^{n+1}$ be a $K3$ surface having one ordinary double point p_0. It is well known that such exist for all $n \geqslant 1$,[3] and projecting x_0 from p_0 gives a smooth $K3$ surface $X^1 \subset \mathbb{P}^n$. In fact, X^1 belongs to the family $\mathcal{F}_{n-1}$ and is biholomorphic to the standard desingularization $\tilde{X}_0$ of X_0.

It will suffice to prove (4.12) in the crucial case $k = n$. Suppose first that $H^1 \cap \mathbb{P}^n$ is a hyperplane that is tangent to X^1 at $n - 1$ distinct points. Then the inverse image of H^1 under the projection $\mathbb{P}^{n+1} - \{p_0\} \to \mathbb{P}^n$ gives a hyperplane $H_0 \subset \mathbb{P}^{n+1}$ that passes through the double point and is tangent to X_0 at $n - 1$ points corresponding to the tangencies of H^1 and X^1. If $E \subset \tilde{X}_0$ is the exceptional curve appearing in the resolution of p_0, then in general we may expect that none of these tangencies of H^1 and X^1 will occur along E. In this case H_0 is tangent to X_0 at $n - 1$ distinct points away from p_0 and passes through this double point.

Assuming that this is the situation, suppose that $X \in \mathcal{F}_n$ is a smooth $K3$ surface that is close to X_0, and let $U \subset X$ be the inverse image of a neighborhood U_0 of the double point under the collapsing map $X \to X_0$. Now the set of tangent hyperplanes to U forms an open piece $U^* \subset X^*$ of the hypersurface X^*, and the crucial observation is that under the specialization $U \to U_0$ we have

$$U^* \to U_0^* + 2p_0^*.$$

Here, U_0^* is the closure in $\mathbb{P}^{n+1*}$ of the set of tangent hyperplanes to the complex manifold $U_0 - \{p_0\}$, and p_0^* is the $\mathbb{P}^n$ of hyperplanes through p_0^*. In

---

[3] For example we may consider trigonal $K3$'s. These appear as hypersurfaces in a 3-dimensional scroll W (cf. Mayer [22] and Saint-Donat [27]), and X_0 may be taken to be a singular section of W.

the language of Section 5 of Griffiths [14], $2p_0^*$ is the *Plücker defect* associated to the degeneration $U \to U_0$. In particular, every hyperplane through p_0 is the specialization of two tangent hyperplanes to X.

Now by (4.11) where $k = 2$, there are ∞^2 hyperplanes that are tangent to X at $n - 1$ distinct points close to the $n - 1$ points where H_0 is tangent to $X_0 - \{p_0\}$, and by the above observation ∞^1 of these must also be tangent to U. This establishes (i) in Proposition (4.8) provided that there is a hyperplane H^1 that is tangent to X^1 at $n - 1$ distinct points none of which is on the exceptional curve E. In the general case the same argument goes through provided that we adopt the definition given by Kleiman [19] for the k-fold locus X_k^*. Rather than write all this out in detail, we shall examine the low cases $n = 1, 2, 3$ and discuss what is needed to establish (ii) in (4.8) for all n.

(d) Completion of the proof of (ii) in Proposition (4.8) When $n = 1$ we have that any smooth $K3$ surface $X \in \mathscr{F}_1$ is a 2-sheeted covering

$$X \xrightarrow{\pi} \mathbb{P}^2$$

branched along a smooth sextic curve B. The "hyperplane sections" are $\pi^{-1}(L)$ where $L \subset \mathbb{P}^2$ is a line, and the section is singular exactly when L is tangent to B. Consequently, the $\pi^{-1}(L)$ for $L \in B^*$ give the desired ∞^1 elliptic curves E_L on X. We note that E_L becomes rational when L is bitangent to B, and that such L always exist.

When $n = 2$ a smooth $K3$ surface $X \in \mathscr{F}_2$ is a quartic $X \subset \mathbb{P}^3$. This case illustrates the difficulty in the general situation. Namely, a "nice" bitangent plane H will be simply tangent to X at two distinct points, and the corresponding section $E = H \cap X$ will be a plane quartic curve having two ordinary nodes. It is then clear that $g(\tilde{E}) = 1$. However, in exceptional cases we may imagine that E has either one tacnode (= two infinitely near nodes) or one cusp, and both of these contribute to the locus X_2^*. In the first case we still have $g(\tilde{E}) = 1$, but in the second $g(\tilde{E}) = 2$.

Because (ii) of Proposition (4.8) is true when $n = 1$, we may use the induction argument above to infer that a generic $X \in \mathscr{F}_2$ has ∞^1 planes $H \in X_2^*$ that are tangent at two distinct points. Then, by specialization on any $X \in \mathscr{F}_2$, there are ∞^1 planes $H \in X_2^*$ for which $g(H \cap X) = 1$. Actually, in this case we can say more. For any smooth surface $X \subset \mathbb{P}^3$ we may take a net $\{H_t\}_{t \in \mathbb{P}^2}$ of hyperplane sections and plot the discriminant curve $B \subset \mathbb{P}^2$ where $E_t = H_t \cap X$ is singular. For a generic choice of net this curve B will have δ ordinary double points and κ cusps, and there are classical Plücker-type formulas for the numbers of each (see Castelnuovo and Enriques [4]). In particular, in the case at hand we have $\delta > 0$, and so there exists one—and hence ∞^1—planes that are tangent to X at two distinct points. This in turn yields (ii) of (4.8) when $n = 2$, and then the assertion about $n = 3$ follows as before from the induction argument.

It is pretty clear that for increasing n the possibilities for what an n-fold tangent hyperplane $H \in X_n^*$ may cut out on X quickly get out of hand, so that some more efficient method for dealing with the singularities must be devised in order to establish the second part of Proposition (4.8) in general.

(e) Concluding remarks. We will conclude with an algebro-geometric implication of the above argument, assuming that it can be pushed through in general. Namely, the same method would establish the following result:

(4.13) *On any algebraic K3 surface $X \subset \mathbb{P}^{n+1}$ there are a finite number of rational curves of degree $2n$.*

In fact, these will be sections $C = H \cap X$ where H is tangent to X at $n + 1$ distinct points. We note that C is a *Castelnuovo canonical curve* in the sense of [16]. Giving such a Castelnuovo canonical curve *in abstracto* is the same as giving $2n + 2$ marked points on $\mathbb{P}^1$; consequently there are ∞^{2n-1} such curves and they form a family that has codimension $n + 1$ in the Deligne–Mumford compactification [7] of curves of genus $n + 1$.

The above result is related to a special case of the recent beautiful theorem of Mori [23]:

Let V be a smooth algebraic variety of dimension n such that $-K_V$ is ample. Then V contains a rational curve C such that $C \cdot (-K_V) \leqslant n + 1$.

Mori's proof is in two steps: He first uses a characteristic p argument to produce a rational curve $C_1 \subset V$, and then he employs elementary deformation-theoretic techniques to reduce the degree of C_1 to $n + 1$.

When $\dim V = 3$ we may use Kodaira vanishing plus the Riemann–Roch theorem to find a surface $X \in |-K_V|$. In case X is smooth, it is a $K3$ surface and (4.13) yields a rational curve. When X is not smooth it should be even easier to find a rational curve, but we have not tried to do this.

We feel that it would be an instructive project to establish Mori's result by projective methods. In particular, a consequence of Mori's theorem is that X is not measure hyperbolic in case $-K_X$ is ample. According to the conjecture of the introduction, this should be true if we only assume that $-K_X \geqslant 0$, and a different argument for Mori's theorem might shed some light on this question.

References

[1] A. Bloch, Sur les systémes de fonctions uniformes satisfaisant á l'equation d'une variété algébrique dont l'irrégularité dépasse la dimension. *J. de Math.* **V**, 19–66 (1926).

[2] F. Bogomolov, Families of curves on a surface of general type. *Sov. Math. Dokl.* **18**, 1294–1297 (1977).

[3] R. Brody, Intrinsic metrics and measures on compact complex manifolds. Ph.D. Thesis, Harvard Univ., 1975.

[4] G. Castelnuovo and F. Enriques, Grundeigenschaften der Algebraischen Flächen. In *Encyklop. d. Math. Wissensch.*, Vol. 3, 1903, pp. 635–768.

[5] M. Cowen, Families of negatively curved Hermitian manifolds. *Proc. Amer. Math. Soc.* **39** 362–366 (1973).

[6] M. Cowen and P. Griffiths, Holomorphic curves and metrics of negative curvature. *J. Analyse Math.*

[7] P. Deligne and D. Mumford, The irreducibility of the space of curves of a given genus. *Publ. Math. IHES* **36**, 75–109 (1969).

[8]   I. Dolgacev, Weighted Projective Varieties, to appear.

[9]   R. Donagi, On the geometry of Grassmannians. *Duke Math. J.* **44**, 795–837 (1977).

[10] H. Grauert and H. Reckziegel, Hermitesche Metriken und normale familien holomorpher Abbildungen. *Math. Z.* **89**, 108–125 (1965).

[11] M. Green, Holomorphic maps into complex projective space omitting hyperplanes. *Trans. Amer. Math. Soc.* **169**, 89–103 (1972).

[12] M. Green, Holomorphic maps into complex tori. *Amer. J. Math.* **100**, 615–620 (1978).

[13] P. Griffiths, Complex differential and integral geometry and curvature integrals associated to singularities of complex analytic varieties. *Duke Math. J.* **45**, 427–512 (1978).

[14] P. Griffiths, Holomorphic mappings into canonical algebraic varieties. *Ann. of Math. (2)* **93**, 439–458 (1971).

[15] P. Griffiths and J. Harris, Local differential geometry and algebraic geometry. *Ann. Ec. Norm sup.* (Dec. 1979).

[16] P. Griffiths and J. Harris, On the variety of special linear systems on a general algebraic curve. *Duke Math. J.*, to appear.

[17] P. Griffiths and J. Harris, *Principles of Algebraic Geometry*. Wiley, New York, 1978.

[18] S. Iitaka, On D-dimensions of algebraic varieties. *J. Math. Soc. Japan* **23**, 356–373 (1971).

[19] S. Kleiman, Rigorous foundations of Schubert's enumerative calculus, in Mathematical developments arising from Hilbert's problems. In *Proc. of Symposia*, AMS, Vol. 27.

[20] S. Kobayashi, Intrinsic distances, measures and geometric function theory. *Bull. Amer. Math. Soc.* **82**, 357–416 (1976).

[21] S. Kobayashi and T. Ochiai, Mappings into compact complex manifolds with negative first Chern class. *J. Math. Soc. Japan* **23**, 137–148 (1971).

[22] A. Mayer, Families of K-3 surfaces.

[23] S. Mori, Projective manifolds with ample tangent bundles. To appear.

[24] J. Noguchi, Meromorphic mappings into a compact complex space. *Hiroshima Math. J.* **7** (2), 411–425 (1977).

[25] T. Ochiai, On holomorphic curves in algebraic varieties with ample irregularity. *Invent. Math.* **43**, S3–96 (1977).

[26] E. Picard, Sur une propriété des fonctions uniformes, liées par une relation algébrique. *Compt. Rend.* **91**, 724–726 (1880).

[27] B. Saint-Donat, Projective models of K-3 surfaces.

[28] F. Sakai, Symmetry powers of the cotangent bundle and classification of algebraic varieties. In *Proc. Copenhagen Summer Meeting in Algebraic Geometry*, 1978.

[29] H. Wu, A remark on holomorphic sectional curvature. *Indiana Math. J.* **22**, 1103–1108 (1972/73).

The Canonical Map for Certain Hilbert Modular Surfaces

F. Hirzebruch*

It was a great pleasure for me to participate in the symposium in honor of Shiing-shen Chern. In my lecture I intended to give a survey on Hilbert modular surfaces. But actually I discussed examples of such Hilbert modular surfaces for which specific information is available on their structure as algebraic surfaces. The paper presented here is an extended version of the talk.

Algebraic surfaces are often investigated by means of their pluricanonical maps (see for example Bombieri [2]). The properties of the canonical map itself (given by the sections of the canonical bundle or in the case of Hilbert modular surfaces by the cusp forms of weight 2) are relatively complicated (compare Beauville [1]). In a certain range, namely for minimal surfaces of general type with $2p_g - 4 \leqslant K^2 \leqslant 2p_g - 2$, Horikawa's results are available [21–25]. To apply them, one has to prove that the surface being studied is minimal. For Hilbert modular surfaces this is a difficult problem, which was attacked first by van der Geer and Van de Ven [10]. Van der Geer has obtained many results on the structure of special Hilbert modular surfaces [8, 9] including some of the surfaces studied here.

The rough classification of Hilbert modular surfaces according to rational, $K3$, elliptic, and general type was considered by Hirzebruch, Van de Ven, and Zagier [17, 19, 16]. The present paper tries to show that in some cases a finer classification of the surfaces of general type can be obtained.

1. Some Examples of Canonical Maps

Let X be a nonsingular n-dimensional compact algebraic manifold and $H^0(X, \Omega^n)$ the complex vector space of holomorphic n-forms. An element $\omega \in H^0(X, \Omega^n)$ can be written with respect to a local coordinate system in the form

$$\omega = a(u_1, u_2, \ldots, u_n) \, du_1 \, du_2 \cdots du_n,$$

where a is holomorphic. The dimension of $H^0(X, \Omega^n)$ is the geometric genus p_g. We shall sometimes write g instead of p_g. If $\omega_1, \omega_2, \ldots, \omega_g$ is a base of

*Mathematisches Institut der Universität, Wegelstrasse 10, 53 Bonn, Federal Republic of Germany.

$H^0(X, \Omega^n)$, then we have the canonical "map"

$$\iota_K : X \to P_{g-1}(C),$$

$$p \in X \mapsto \omega_1(p) : \omega_2(p) : \cdots : \omega_g(p) \in P_{g-1}(C),$$

which is not necessarily everywhere defined.

The vector space $H^0(X, \Omega^n)$ is the space of holomorphic sections of the canonical bundle $\mathcal{K}$ of X. The complete linear system $|\mathcal{K}|$ consists of all canonical divisors (i.e. divisors of elements $\omega \in H^0(X, \Omega^n)$). These are exactly the inverse images under ι_K of the hyperplanes of $P_{g-1}(C)$. (*Canonical divisors are always assumed to be nonnegative if nothing is mentioned to the contrary.*)

If $n = 1$, then X is a compact Riemann surface (algebraic curve) of genus g. If $g = 0$, then ι_K is not defined. For $g = 1$ the image $\iota_K(X)$ is a point. For $g = 2$, the map ι_K realizes X as double cover of $P_1(C)$ with 6 ramification points. For $g = 3$ the following holds (see Griffiths and Harris [12, p. 247]): The canonical map ι_K is a biholomorphic map of X onto a nonsingular curve of degree $2g - 2$ in $P_{g-1}(C)$ (generic case), or the curve X is hyperelliptic and ι_K realizes X as a double cover of the rational normal curve of degree $g - 1$ in $P_{g-1}(C)$ with $2g + 2$ ramification points. The normal curve mentioned is the image of $P_1(C)$ under the map given by the homogeneous polynomials of degree $g - 1$.

For $n = 2$ not much is known about the canonical map. For algebraic surfaces ($n = 2$) recent investigations are due to Beauville [1], but for specific surfaces (e.g. Hilbert modular surfaces) it is difficult to obtain information on ι_K. We now restrict to the case $n = 2$. Let us first recall some basic facts on algebraic surfaces. By $\mathcal{K}$ we denote the canonical bundle, and by K a canonical divisor. For a nonsingular curve S on X we have the adjunction formula

$$KS + SS = -e(S), \qquad (1)$$

where KS and SS are intersection numbers and $e(S)$ is the Euler number of S, which equals $2 - 2g(S)$ if S is irreducible. The formula (1) is true also if K is negative.

The irreducible nonsingular curve S is called exceptional (of the first kind) if $g(S) = 0$ and $SS = -1$ (or equivalently $g(S) = 0$ and $KS = -1$). The surface X is called minimal if it does not contain any exceptional curves.

An exceptional curve S is contained in every (nonnegative) canonical divisor (because $KS = -1$). Therefore ι_K is not defined on an exceptional curve. The exceptional curve S on X can be blown down to a point, the resulting surface Y being nonsingular again. The vector spaces $H^0(X, \Omega^n_X)$ and $H^0(Y, \Omega^n_Y)$ are isomorphic. Let $\pi : X \to Y$ be the natural map; then every canonical divisor on X is of the form $\pi^* K_Y + S$, where K_Y is a canonical divisor on Y. For a nonsingular rational curve S with $SS = -2$ we have $KS = 0$; therefore for every canonical divisor K, the curve S either is contained in K or does not meet K. If S is not contained in all canonical divisors, then ι_K maps S to a point.

Let us now assume that the nonsingular irreducible surface X is of *general type*. Then X contains finitely many exceptional curves mutually disjoint. They can be blown down. The resulting surface again may contain exceptional curves. They are mutually disjoint and can be blown down. After a finite number of such blowing-down processes we reach a minimal algebraic surface, the unique (nonsingular) minimal model in the birational equivalence class of X (see for example Griffiths and Harris [12, pp. 510, 573], Bombieri [2], and Hirzebruch and Van de Ven [17]).

The self-intersection number $K^2 = KK$ is an important invariant. Since the characteristic class of the canonical bundle equals $-c_1$, where $c_1 \in H^2(X, Z)$ is the first Chern class of X, we have

$$K^2 = c_1^2[X].$$

Let $e(X)$ denote the Euler number. M. Noether's formula gives the relation

$$e(X) + K^2 = 12\chi(X), \tag{2}$$

where

$$\chi(X) = 1 - q + p_g$$

(q = irregularity = half first Betti number of X) is the arithmetic genus of X (in the terminology of [14]). If an exceptional curve is blown down, then $e(X)$ decreases by one and K^2 increases by one, whereas $\chi(X)$ remains invariant. It is a birational invariant; in fact q and p_g are birational invariants.

For a minimal algebraic surface X of general type, K^2 and $\chi(X)$ are positive. The number K^2 of X is the maximal K^2 of all nonsingular surfaces in the birational equivalence class of X. By (1), the number $K^2 + 1$ equals the genus of a nonsingular irreducible curve C if C is a canonical divisor. We have the inequality due to M. Noether (compare Bombieri [2, p. 208]),

$$K^2 \geqslant 2p_g - 4, \tag{3}$$

and also the Bogomolov–Miyaoka inequality (Miyaoka [28])

$$K^2 \leqslant 3e(X),$$

for which, of course, minimality is not needed.

We shall now give a few classical examples of minimal algebraic surfaces of general type where the canonical map ι_K is well known from the nature of the example.

EXAMPLE 1. Let X be the double cover of $P_2(C)$ ramified along a nonsingular curve of degree 8. We have $p_g = 3$ and $K^2 = 2$. The natural map $X \to P_2(C)$ is the canonical map ι_K. The complete linear system $|\mathcal{K}|$ consists of the lines in $P_2(C)$ lifted to X. Observe that a (negative) canonical divisor of $P_2(C)$ is given by $-3L$ where L is a line and

$$-3L + \tfrac{1}{2} \cdot 8L = L.$$

The surface X is simply connected. The Euler number equals 46 by Noether's formula (2).

EXAMPLE 2. Consider two nonsingular quartic curves A and B in $P_2(C)$ intersecting transversally. Let X be the fourfold cover of the plane obtained by first taking the double cover Y of the plane ramified along A and then the double cover of Y ramified along the lift of B to Y. This construction is actually symmetric in A and B. The surface X admits an action of $Z/2 \times Z/2$ with the plane as orbit space. We may denote the nontrivial elements of $Z/2 \times Z/2$ by α, β, γ in such a way that X/α and X/β are the double planes ramified along A and B respectively, whereas X/γ is the double cover of the plane ramified along $A \cup B$ (with 16 ordinary rational double points as singularities). The surfaces X/α and X/β are rational (Euler number $= 10$, $K^2 = 2$; they are isomorphic to a plane with 7 points blown up). Each of them contains 56 exceptional curves coming in pairs which are the lifts of the 28 double tangents of A or B respectively. The surfaces X/α and X/β are double covers of the plane by their anticanonical maps (i.e., the lifts of the lines of the plane are exactly the elements of $|\mathfrak{K}^{-1}|$). The branching locus of X over X/α (the lift of B to X/α) is a fourfold anticanonical divisor of X/α. The lifts of the anticanonical divisor of X/α to X are the canonical divisors of X. Observe $K + \frac{1}{2}(-4K) = -K$ (on X/α). The map from X to $P_2(C)$ (degree 4) is the canonical map. For X we have $p_g = 3$ and $K^2 = 4$. It is simply connected and has Euler number 44. The elements α, β, γ operate on $H^0(X, \Omega^2)$ by multiplication with $-1, -1, 1$. Thus $H^0(X, \Omega^2)$ can be identified with the space of holomorphic 2-forms on the nonsingular model X_γ of X/γ (obtained by blowing up each of the 16 singularities in a nonsingular rational curve of self-intersection number -2). The natural map of X_γ to $P_2(C)$ is the canonical map (in fact X_γ belongs to the family of surfaces in Example 1; see later remarks).

EXAMPLE 3. Consider a nonsingular quadric Q in $P_3(C)$. The quadric is isomorphic to $P_1(C) \times P_1(C)$ by the two systems of lines on Q. Let X be the double cover of Q ramified along a nonsingular curve of bidegree $(6, 6)$. We have $p_g = 4$ and $K^2 = 4$. The natural map of X onto Q followed by the embedding of Q in $P_3(C)$ is the canonical map ι_K. The complete linear system $|\mathfrak{K}|$ consists of the planes in $P_3(C)$ intersected with Q and lifted to X. The planes intersected with Q are exactly the curves of bidegree $(1, 1)$. Observe that a (negative) canonical divisor on Q is given by $-2L_1 - 2L_2$, where L_1, L_2 are lines on Q in different systems and

$$-2L_1 - 2L_2 + \tfrac{1}{2}(6L_1 + 6L_2) = L_1 + L_2.$$

The surface X is simply connected. The Euler number equals 56.

EXAMPLE 4. Let X be a nonsingular quintic surface in $P_3(C)$. We have $p_g = 4$ and $K^2 = 5$. The embedding of X in $P_3(C)$ is the canonical map ι_K. The complete linear system $|\mathfrak{K}|$ consists of the planes in $P_3(C)$ intersected with X. The surface X is simply connected. The Euler number equals 55.

EXAMPLE 5. Consider a nonsingular cubic surface W in $P_3(C)$. For such a surface the complete linear system $|\mathfrak{K}^{-1}|$ consists of all hyperplane sections. (These hyperplane sections are the nonnegative anticanonical divisors.) A nonnegative divisor on W is anticanonical if and only if it has intersection number 1 with each of the 27 lines on W. The 27 lines are exactly the exceptional curves on W. Let C be a nonsingular curve on W with $C \in |\mathfrak{K}^{-4}|$, i.e., C has intersection number 4 with each of the 27 lines. If we realize W as $P_2(C)$ with six points $p_1, \ldots, p_6$ blown up (Griffiths and Harris [12, p. 489]), then C corresponds to a curve of degree 12 in the plane with $p_1, \ldots, p_6$ as quadruple points and no other singularities. Let X be the double cover of W ramified along C; then the complete linear system $|\mathcal{K}|$ of X consists of the *nonnegative anticanonical divisors of W lifted to X.* Observe $K + \frac{1}{2}(-4K) = -K$ (on W). Thus $|\mathcal{K}|$ consists of all lifted hyperplane sections of W. The canonical map of X is the map onto W followed by the embedding of W in the projective space $P_3(C)$. For X we have $p_g = 4$ and $K^2 = 6$. It is simply connected and has Euler number 54.

EXAMPLE 6. Let X be the double cover of $P_2(C)$ ramified along a nonsingular curve of degree 10. We have $p_g = 6$ and $K^2 = 8$. The natural map $X \to P_2(C)$ followed by the Veronese embedding of $P_2(C)$ in $P_5(C)$ is the canonical map. The canonical divisors of X are the quadrics in $P_2(C)$ (i.e. the hyperplane sections of the Veronese surface) lifted to X. The surface X is simply connected and has Euler number 76.

Remark. Examples 2 and 5 can be regarded as special cases of the following construction. A del Pezzo surface (see Manin [27]) of degree $g - 1$ in $P_{g-1}(C)$ is obtained as follows ($4 \leqslant g \leqslant 10$). In the plane $P_2(C)$ we blow up $10\text{-}g$ points. The dimension of the space of sections of the anticanonical bundle of this surface is g. The complete linear system $|\mathcal{K}^{-1}|$ consists of all cubics of $P_2(C)$ passing through the $10\text{-}g$ points. The image of $P_2(C)$ under the anticanonical map is a del Pezzo surface W in $P_{g-1}(C)$. In W we take a nonsingular curve representing a fourfold anticanonical divisor and the double cover of W ramified along this curve. This is an algebraic surface X with geometric genus g and $K^2 = 2g - 2$. The canonical map for X is the map of degree 2 to the del Pezzo surface W followed by the embedding of W in $P_{g-1}(C)$. If $g = 3$, the del Pezzo surface can still be introduced, but its anticanonical map realizes it as double cover of the plane; therefore the canonical map of X is of degree 4 (Example 2).

In Examples 1–6 certain "degenerations" may be admitted. In Examples 3, 4, 5 we admit that the quadric, quintic, or cubic surface has rational double points (sometimes called Kleinian singularities; see Brieskorn [3, 5]). These are the singularities which resolve minimally into a configuration of type A_k, D_k, E_6, E_7, E_8 of nonsingular rational curves of self-intersection number -2. For a quadric we can have only one singularity A_1 (quadric cone). Some examples of quintics with rational double points occur in van der Geer and Zagier [11]. For cubics the complete list of possibilities is given by Schäfli [29]; see also Griffiths and Harris [12, p. 640]. A report on singular cubics was given recently by Bruce and Wall [6]. They list the possible combinations of singularities as A_1, $2A_1$, A_2, $3A_1$, A_1A_2, A_3, $4A_1$, A_22A_1, A_3A_1, $2A_2$, A_4, D_4, A_32A_1, $2A_2A_1$, A_4A_1, A_5, D_5, $3A_2$, A_5A_1, E_6.

For the ramification curve on the desingularized quadric (cubic) surface in Examples 3 and 5 we require that it represent a threefold (fourfold) anticanonical divisor. The ramification curve in Examples 1, 3, 5, 6 may have singularities, but they are restricted to the condition that the double cover acquires only rational double points. The admissible curve singularities are, with respect to suitable local coordinates,

$$x^2 + y^{k+1} = 0 \quad (a_k),$$

$$x(y^2 + x^{k-2}) = 0 \quad (d_k; \; k \geqslant 4),$$

$$x^3 + y^4 = 0 \quad (e_6),$$

$$x(x^2 + y^3) = 0 \quad (e_7),$$

$$x^3 + y^5 = 0 \quad (e_8).$$

The double cover has singularities of type A_k, D_k, E_6, E_7, E_8 respectively. They are resolved to give our modified examples of algebraic surfaces and canonical maps. In Example 2 we admit that A has singularities a_k, d_k, e_6, e_7, e_8, and we then take on the desingularized model of the double cover branched along A a ramification curve B which represents a fourfold anticanonical divisor. Also B is required to have only singularities a_k, d_k, e_6, e_7, e_8.

By Brieskorn's theory [3–5], we know that the algebraic surfaces thus obtained still belong to the same family (up to deformation). They are, in particular, of the same diffeomorphism type.

Minimal algebraic surfaces with $p_g = 3$ and $K^2 = 2$ are called Moishezon surfaces. They are all of the type studied in Example 1. The surfaces of Examples 1, 3, 6 satisfy $K^2 = 2p_g - 4$, i.e., K^2 is for given p_g as small as possible (by Noether's inequality (3)). For $K^2 = 4$ such a surface belongs to Example 3 by a result of Horikawa [22], who has classified surfaces with $K^2 = 2p_g - 4$. For $K^2 = 8$ our Example 6 is only one of several possibilities. In Example 4 we have $K^2 = 2p_g - 3$. Surfaces satisfying this relation are studied by Horikawa in [21] and [23]. In Examples 2 and 5 we have $K^2 = 2p_g - 2$ with $p_g = 3$ or 4 respectively. For these surfaces see [24] and [25]. In [24] Horikawa considers the case $p_g = 4$. In Section 2 of [25] we find the case $p_g = 3$.

2. Minimality Criterion

How to decide whether a given algebraic surface is minimal? Essentially the following criterion was used by Hirzebruch and Van de Ven [18].

Proposition. *Let X be a nonsingular algebraic surface with $K^2 > 0$, on which there exists a nonnegative divisor D with*

$$D^2 = KD = K^2. \tag{1}$$

Then the homology class of $D - K$ is a torsion class. Every exceptional curve of X is contained in D.

Proof. Since $K(D - K) = (D - K)^2 = 0$ and $K^2 > 0$, we have by the Hodge index theorem for divisors (Griffiths and Harris [12, p. 472]) that $D - K$ is homologous to zero mod torsion. For an exceptional curve S we have $DS = KS = -1$; therefore S is contained in D. $\square$

If a nonnegative divisor D on the surface X satisfies (1), if $K^2 > 0$, and if no exceptional curve is contained in D, then X is minimal.

If a nonnegative divisor D on a simply connected surface satisfies (1), then D is a canonical divisor. It is often difficult to give a nonnegative canonical divisor explicitly, just as it is difficult to prove minimality. One could think that it would be easiest to get one's hands on a nonsingular irreducible curve C with $C^2 = KC = K^2$, which then would have genus $K^2 + 1$. A generic canonical divisor on a minimal surface of general type is such a curve C. However, one often does not succeed in finding such a C; rather one gets complicated configurations D of curves with $D^2 = KD = K^2$. In a way, D is a degenerate curve of genus $K^2 + 1$.

The proposition can be generalized. Let m be a positive integer. Instead of (1) we assume $DK = mK^2$, $D^2 = m^2K^2$. Then the homology class of $D - mK$ is a torsion class, and every exceptional curve of X is contained in D.

We shall now indicate configurations of curves which lead to a divisor D satisfying (1).

Assume that we have in Example 1 of Section 1 a line L in the plane which passes through three double points of types a_{k_1}, a_{k_2}, a_{k_3} of the ramification curve C of degree 8 and intersects C in two points transversally. Then the lift of L in the Moishezon surface is a rational curve $\tilde{L}$ with $K\tilde{L} = 2$ together with a configuration of nonsingular rational curves of self-intersection -2 arising from the resolution of the singularities of types A_{k_1}, A_{k_2}, A_{k_3}. The result looks like Figure 1. All curves are nonsingular and rational. All intersections are transversal. We have $\tilde{L}\tilde{L} = -4$. There are three chains of (-2)-curves of lengths k_1, k_2, k_3. This is a configuration D satisfying (1) with $K^2 = 2$ (each component S has multiplicity $m_S = 1$ in D).

To check (1) in this case and all further cases we look at each component S of the configuration $D = \sum m_S S$ and prove

$$DS = KS,$$

$$\sum_S m_S KS = K^2.$$

The number KS is known from the adjunction formula if the genus and self-intersection of S are given, whereas DS can be read off from the configuration.

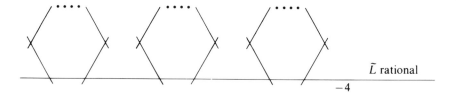

Figure 1. Configuration (I′). $D^2 = KD = 2$.

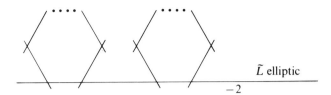

$\tilde{L}$ elliptic

-2

Figure 2. Configuration (I″). $D^2 = KD = 2$.

A second configuration D satisfying (1) with $K^2 = 2$ is the following. Assume that we have in Example 1 a line L in the plane which passes through two double points of types a_{k_1}, a_{k_2} of the ramification curve C of degree 8 and intersects C in four points transversally. Then the lift of L in the Moishezon surface is an elliptic curve $\tilde{L}$ with $K\tilde{L} = 2$ together with two chains of (-2)-curves. The result looks like Figure 2 (all multiplicities 1). In [18] the configuration (I″) and two other configurations were used.

Consider the configuration D in Figure 3 of nonsingular rational curves (four (-3)-curves and twelve (-2)-curves). The four (-3)-curves are joined by (-2)-curves. The divisor D is obtained by taking each curve with multiplicity 1, except the (-2)-curves drawn in boldface, which have multiplicity 2.

A configuration (II) occurs in Example 2 if each of the quartic curves A, B has a double point of type a_3 and if there is a line L tangent to A and B in these two double points (Figure 4). The divisor D is the lift of L to the fourfold cover

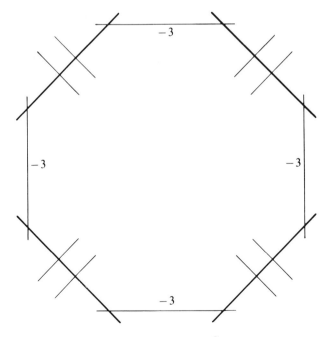

-3

-3

-3

-3

Figure 3. Configuration (II). $D^2 = KD = 4$.

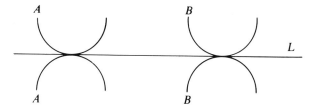

Figure 4

of the plane. The configuration (II) is a degeneration of the case when the two quartics have a common double tangent and we take the lift of the double tangent to the fourfold cover.

Many configurations can be constructed which are motivated by Example 3. Consider Figure 5. All curves are rational, except one which is elliptic. The

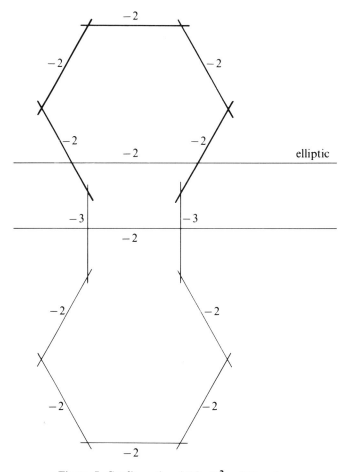

Figure 5. Configuration (III'). $D^2 = KD = 4.$

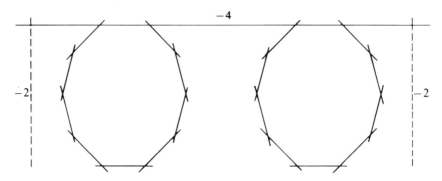

Figure 6. Configuration (III"). $D^2 = KD = 4$.

divisor D contains all curves with multiplicity 1, except the (-2)-curves drawn in boldface, which have multiplicity 2. Such a configuration arises if we take in Example 3 two lines L_1, L_2 on a nonsingular quadric belonging to different families of lines and assume that $L_1 \cap L_2$ is a double point of type a_5 of the branching curve C of bidegree $(6, 6)$, and that L_1 intersects C in 4 other points transversally, whereas L_2 passes through two other double points of C (of types a_1, a_5).

Let us now assume that in Example 3 the quadric is a cone and the ramification curve C does not pass through the vertex of the cone (Figure 6). Take a generating line L of the cone, and assume that it passes through two double points of type a_9 of C and intersects C transversally in two points. The lift of $2L$ leads to the above configuration (III") of rational curves, one with self-intersection -4, all others with self-intersection -2. All curves have multiplicity 2 except the two curves indicated by a broken line, which have multiplicity 1 and are mapped to the vertex of the cone. In (III') and (III") we could use other double points, i.e. the lengths of the chains of (-2)-curves could be changed.

Take the configuration shown in Figure 7, consisting of 4 nonsingular rational curves (each with multiplicity 1). The existence of such a configuration on a surface proves minimality if $K^2 = 5$. It is motivated by Example 4 if a hyperplane intersects the quintic in a conic and 3 lines. All the transversal

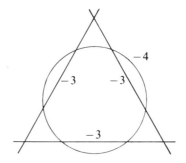

Figure 7. Configuration (IV). $D^2 = KD = 5$.

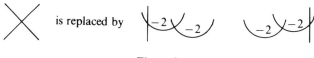

is replaced by

Figure 8

intersections may be replaced by chains of (-2)-curves, as shown in Figure 8. The modified configuration still satisfies $D^2 = KD = 5$.

There are cubic surfaces W for which the intersection of some plane with W consists of three lines L_1, L_2, L_3 which go through one point (Eckardt point). The three lines define an anticanonical divisor $L_1 + L_2 + L_3$ of W whose lift in the double cover (Example 5 of Section 1) is a configuration (∇) (Figure 9) of three elliptic curves (with multiplicity 1). On a surface with $K^2 = 6$ the existence of a configuration of type (V) proves minimality.

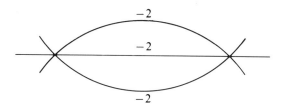

Figure 9. Configuration (V). $D^2 = KD = 6$.

Finally we consider Example 6 (Figure 10). The divisor D is of the form

$$D = 2C + \sum_{i=1}^{24} L_i,$$

where C is a nonsingular curve of genus 3 and there are twenty-four (-2)-curves intersecting C transversally. We have

$$D^2 = 2DK = 32.$$

The existence of a configuration (VI) on a surface with $K^2 = 8$ proves minimality. The divisor D and a double canonical divisor are homologous mod torsion. If the ramification curve in Example 6 is of the form $A + B$ where A and B are

Figure 10. Configuration (VI).

nonsingular curves of degrees 4 and 6 intersecting transversally, then the lift of
A gives a configuration (VI).

3. Hilbert Modular Surfaces

Let K be the real quadratic field of discriminant D over Q, and $\mathcal{O}$ its ring of
integers. The Hilbert modular group $G = SL_2(\mathcal{O})/\{\pm 1\}$ operates on $H \times H$,
where H is the upper half plane of C. The orbit space $H^2/SL_2(\mathcal{O})$ can be
compactified by finitely many cusps. The resulting surface $\overline{H^2/SL_2(\mathcal{O})}$ has
finitely many singularities, namely quotient singularities coming from elliptic
fixed points and the cusps. Minimal desingularization gives the simply connected algebraic surface $Y(D)$. (For an introduction to Hilbert modular surfaces
see [15] and [19].) The surface $Y(D)$ is rational for exactly 10 discriminants [19];
for all other discriminants it has a unique minimal model $Y_{min}(D)$. On a
nonrational $Y(D)$ certain curves can be blown down successively; they are
described in [19]. The resulting surface $Y^0(D)$ is conjectured to be equal to
$Y_{min}(D)$. This has been proved in many cases (van der Geer and Van de Ven
[10], Hirzebruch [16]). Freitag [7] and van der Geer [9] showed that for congruence subgroups Γ of G of sufficiently high level (Γ operates freely on H^2), the
minimal resolution of the cusps of $\overline{H^2/\Gamma}$ leads to a minimal algebraic surface.
The vector space of cusp forms of weight 2 for G is naturally isomorphic to
$H^0(X, \Omega^2)$ if X is any nonsingular model for the algebraic surface $\overline{H^2/G}$. The
same holds for a congruence subgroup Γ of G. (This result is due to Freitag;
compare Hirzebruch [15, Section 3.5, Lemma]). Thus the canonical map of X in
$P_{g-1}(C)$ (where $g = \dim H^0(X, \Omega^2)$) is induced by the "map" of $\overline{H^2/G}$ in
$P_{g-1}(C)$ given by g linearly independent cusp forms of weight 2. Therefore the
canonical map is especially interesting from the point of view of modular form
theory as well. A cusp form of weight $2m$ with $m > 1$, in general, cannot be
extended to the nonsingular model. The complete linear system of nonnegative
m-fold canonical divisors is a birational invariant, but for $m > 1$, in general, it is
smaller than the system of divisors of cusp forms of weight $2m$ (such a divisor
can have components with negative multiplicities on the nonsingular model).

If the Hilbert modular surface $Y^0(D)$ is of general type, then $K^2 > 0$. (See
Hirzebruch and Zagier [19].) Because it is simply connected (Švarčman [30]), a
divisor D satisfying (1) in the Proposition of Section 2 is a canonical divisor. It is
a nice program to write down such a divisor D in terms of explicitly known
curves, to prove minimality in this way, and to get information on the canonical
map by special properties of D. (Compare van der Geer and Zagier [8, 9, 11] for
very similar studies.) As mentioned before, the surface $Y(D)$ is rational for
exactly 10 discriminants; it is not rational and not of general type for 22
discriminants, the largest one being 165. In these 22 cases $Y^0(D)$ is minimal,
namely a $K3$-surface or an honestly elliptic surface [19]. In all other cases, the
surface is of general type. The calculations of [19] show that among those of
general type there are exactly five discriminants with geometric genus 3 and
seven discriminants with geometric genus 4. In these cases minimality can be

proved and the nature of the canonical map determined:

Theorem. *The Hilbert modular surfaces $Y^0(D)$ of general type with $p_g = 3, 4$ and their values of K^2 are given by the following lists.*

$p_g = 3$:

D	89	97	124	141	168
K^2	2	2	2	2	4

$p_g = 4$:

D	101	104	109	113	133	156	161
K^2	4	4	4	6	4	4	4

In these cases $Y^0(D)$ is minimal. The canonical map is everywhere defined. For $p_g = 3$ and $D \neq 168$ it is a map of degree 2 on the projective plane ramified along a curve of degree 8 having as singularities only double points of type a_k (Moishezon surface, Example 1 of Section 1). For $D = 168$ the canonical map is of degree 4 (Example 2 in Section 1: the quartic A is reducible and consists of two conics; the quartic B is irreducible). For $p_g = 4$ and $D \neq 113$ the canonical map is of degree 2 onto a quadric surface in $P_3(C)$ which is nonsingular for $D \neq 104$ and is a cone for $D = 104$ (Example 3 in Section 1). For $D = 113$ the surface $Y^0(D)$ is mapped with degree 2 onto a cubic surface (Example 5 in Section 1) which has one singularity (of type A_3). On this cubic surface there are three lines passing through one point (Eckardt point). The cubic surface is uniquely determined by these two properties: One singularity (of type A_3), one Eckardt point.

We cannot give complete proofs here. Every case has to be studied individually. The following remarks will make it possible for the reader to check the results.

The Hilbert modular group G admits the Hurwitz–Maass extension G_m (see [19]). We take the matrices $\begin{pmatrix} a & b \\ c & d \end{pmatrix}$ with entries in K such that $w = ad - bc$ is totally positive and $a/\sqrt{w}, b/\sqrt{w}, c/\sqrt{w}, d/\sqrt{w}$ are algebraic integers not necessarily in $\mathcal{O}$. The group G_m is the group of all these matrices divided by its center

$$\left\{ \begin{pmatrix} a & 0 \\ 0 & a \end{pmatrix} \mid a \in K^* \right\}.$$

The group G_m/G is abelian of type $(2, \ldots, 2)$ with $t - 1$ factors $Z/2$, where t is the number of primes dividing the discriminant D. The group G_m/G acts on $Y^0(D)$. It can be extended by the involution induced by the involution $\tau : (z_1, z_2) \to (z_2, z_1)$ of $H \times H$. This gives an abelian group M of type $(2, \ldots, 2)$ of order 2^t acting on $Y^0(D)$. Using results of Koll [26] and Hausmann [13] it is possible to determine the representation of M on $H^0(Y^0(D), \Omega^2)$, i.e. on the space of cusp forms of weight 2. Let M^0 be the subgroup of M consisting of the elements which operate on the cusps forms by $\pm$identity. Then the canonical map factors through $Y^0(D)/M^0$. For the discriminants in the theorem M^0 has order 2 except for $D = 168$, where it has order 4. The nontrivial element of M^0 ($D \neq 168$) is τ except for $D = 156$, where it is $\tau\alpha$. (Here α is the element of G_m/G represented by a matrix of determinant 13.) For $D = 168$ we have

$M^0 = \{1, \beta, \tau, \beta\tau\}$, where β is the element of G_m/G representable by a matrix of determinant 84.

The "explicitly known" curves which we use to construct canonical divisors are the curves coming from the resolution of the quotient singularities and the singularities at the cusps together with the modular curves F_N, also called skew-Hermitian curves. By a skew-Hermitian matrix we mean a matrix of the form

$$\begin{bmatrix} a_1\sqrt{D} & \lambda \\ -\lambda' & a_2\sqrt{D} \end{bmatrix} \qquad \text{where } \lambda \in \mathcal{O} \text{ and } a_1, a_2 \in Z.$$

The matrix is called primitive if there is no natural number >1 dividing a_1, a_2, λ. For given $N = a_1 a_2 D + \lambda\lambda'$ the curve F_n in H^2/G is defined to be the set of all points of H^2/G which have representatives $(z_1, z_2) \in H^2$ for which there exists a primitive skew-Hermitian matrix of determinant N such that

$$a_1\sqrt{D}\, z_1 z_2 - \lambda' z_1 + \lambda z_2 + a_2\sqrt{D} = 0.$$

It can be shown that F_N defines a curve in $\overline{H^2/G}$ and also in $Y(D)$ and $Y^0(D)$. The curve F_N is nonempty if and only if none of the character values (D_q/N) equal -1 where D is the product of the t prime discriminants D_q.

The group M leaves F_N invariant and permutes the connectedness components of F_N. The surface $Y^0(D)$ is obtained by blowing down all components of F_1, F_2, F_3, F_4, and F_9 (if $3 \mid D$), provided these curves are not empty, together with the curves into which the quotient singularities lying on F_1 and F_2 were resolved. The intersection behavior of the curves F_N is completely known (see Hirzebruch and Zagier [20] for D a prime, Hausmann [13] in general). It is also known how the F_N pass through the curves of the resolution of the cusps. The number of connectedness components of F_N was determined in [13]. Also the genus $g(C)$ and the value KC can be calculated for each connectedness component C of F_N (Hirzebruch and Zagier [15, 19]), at least if N satisfies certain number-theoretical conditions which are always fulfilled in the cases we need. Therefore all the information needed to construct canonical divisors is available.

If an element α of G_m/G can be represented by a primitive matrix

$$\begin{bmatrix} \lambda' & -a_2\sqrt{D} \\ a_1\sqrt{D} & \lambda \end{bmatrix}$$

with determinant N dividing D, then the curve F_N is pointwise fixed under $\alpha\tau$. Using this remark, one can determine the ramification curve C in the above cases. For $D \equiv 1 \bmod 4$ the curve C equals F_D. For $D \equiv 0 \bmod 4$ ($D \neq 156$, $D \neq 168$) the curve C is the union of F_D and $F_{D/4}$. For $D = 156$ the ramification curve is $F_{13} \cup F_{52}$. For $D = 168$ the curve $F_{168} \cup F_{42}$ is pointwise fixed under τ, and F_{84} pointwise fixed under $\beta\tau$.

We now give some examples. All canonical divisors constructed will be invariant under M with one exception, and all of them arise in the way indicated in Section 2. They not only help to prove minimality, but are also important to determine the nature of the canonical map.

$D = 124$

We construct a configuration of type (I′) involving F_{10} (as curve $\tilde{L}$), F_8, F_9, F_{16}, and curves coming from the singularities of H^2/G. We have $k_1 = k_2 = 5$ and $k_3 = 7$.

$D = 141$

We construct a configuration of type (I″) involving F_{21} as curve $\tilde{L}$, using F_7 and curves coming from the singularity at the unique cusp. We have $k_1 = k_2 = 3$.

$D = 168$

We get a configuration of type (II) involving F_7 and F_{18}. The two conics are the images of F_{42} and F_{168}; the quartic B is the image of of F_{84}. The two conics touch each other (double point a_3 of A); the quartic B touches each conic (double point a_3 of $A \cup B$) and has itself a double point a_3.

$D = 133$

We obtain a configuration of type (III′) involving F_7, F_9, F_{11} where one component of F_{11} is the elliptic curve. This divisor is not invariant under the nontrivial element of G_m/G.

$D = 104$

The configuration is of type (III″). The curve with self-intersection -4 is F_{10}. The chains of (-2)-curves come from the two cusp singularities. The two (-2)-curves which map to the vertex of the cone are the resolutions of the two quotient singularities of order 2 lying on F_{10}.

$D = 113$

The curves F_{15}, F_{16}, F_{18} are the three elliptic curves of configuration (V). The two intersection points are special points (Hirzebruch and Zagier [20]) with the quadratic form $15u^2 + 16uv + 13v^2$ of discriminant -7×113, which represents 15, 16, and 18. The A_3-singularity of the cubic comes from 2 chains of (-2)-curves of length 3 interchanged under the involution τ, which occur in the resolution of the unique cusp of H^2/G. As Wall pointed out to me, there is a one-parameter family of cubics with exactly one singular point (of type A_3), and in this family there is a unique cubic with an Eckardt point. In fact this cubic can be written down with respect to suitable homogeneous coordinates as

$$xyw + x^3 - xz^2 - yz^2 + y^3 = 0.$$

The cubic has 10 lines. They are the images of $F_{11}, F_{13}, F_{14}, F_{15}, F_{16}, F_{18}, F_{25}$, of two pairs of (-3)-curves of the cusp resolution, and of the pair of (-3)-curves which arise from the two quotient singularities of order 3 lying on F_7.

The singular point of the cubic surface is $(x, y, z, w) = (0, 0, 0, 1)$. The plane $w = 3(x + y)$ intersects the cubic surface in the three lines given by $x + y = 0$, $z = x + y$, $z = -(x + y)$, which pass through the Eckardt point $(1, -1, 0, 0)$.

The ramification curve C (corresponding to F_{113}) has on the cubic surface 5 singularities of type a_1, 3 of type a_2, 2 of type a_3, 1 of type a_8. The genus of C equals 3.

4. More Examples

It is interesting to study the canonical map for all Hilbert modular surfaces $Y^0(D)$ which fall into the range of Horikawa's classification ($2p_g - 4 \leqslant K^2 \leqslant 2p_g - 2$). These are only finitely many. For large D we have $K^2 \sim 8p_g$. Since K^2 is even for $Y^0(D)$, we should investigate the discriminants with $K^2 = 2p_g - 4$ or $K^2 = 2p_g - 2$. It turns out that $K^2 = 2p_g - 4$ happens only if $p_g = 3$ or 4. These discriminants were treated in Section 3. We have $K^2 = 2p_g - 2$ for $D = 168$ ($p_g = 3$) and $D = 113$ ($p_g = 4$). Also these cases were studied in Section 3. Otherwise $K^2 = 2p_g - 2$ if and only if $D = 129, 136, 184$ ($p_g = 5$), $D = 145$, 149 ($p_g = 6$), $D = 204$ ($p_g = 7$).

We only consider the prime discriminant $D = 149$. It turns out that $Y^0(149)$ is minimal. It is the double cover of a del Pezzo surface W of degree 5 in $P_5(C)$ ramified along a 4-fold anticanonical curve C of W corresponding to F_{149}. The curve C on W has 9 singularities of type a_1, 8 singularities of type a_2, 3 of type a_3, and 1 of type a_{10}. It has genus 3.

The surface W contains 10 lines (exceptional curves). They are the images of $F_{19}, F_{20}, F_{22}, F_{24}, F_{28}, F_{36}$, of two pairs of (-3)-curves coming from the resolution of the cusp, and of the two pairs of (-3)-curves coming from resolving the four quotient singularities of order 3 lying on F_6 and F_{16}.

The surface $Y^0(149)$ has exactly 44 nonsingular rational curves of self-intersection -2 which, together with the above-mentioned 14 irreducible curves (mapped to the 10 lines of the del Pezzo surface), generate a vector space of algebraic cycles of dimension 53. This vector space coincides with the space of algebraic cycles generated by the curves F_N on $Y^0(149)$ and the curves coming from resolving the singularities of H^2/G. In fact, the Picard number of $Y^0(149)$ is greater than or equal to 54. We have $h^{1,1} = 60$.

We now consider two examples of congruence subgroups.

Let G be the Hilbert modular group for $D = 13$ and Γ the principal congruence subgroup for the ideal (2) of $K = Q(\sqrt{13})$. Let Y be the surface obtained by resolving the 5 cusps of H^2/Γ. For Y we have $p_g = 4$ and $K^2 = -5$. However, on Y the curve F_1 has 10 components, all of which are exceptional curves. Blowing down these ten curves, we obtain a new surface with $K^2 = 5$. For each of the five cusps van der Geer and Zagier construct in [11] a configuration (IV) (with the three transversal intersections of the three lines in (IV) replaced by a chain of two (-2)-curves in each case). The 3 lines and the (-2)-curves correspond to the resolution of the cusp; the conic is one of the five components of F_4. In this way the authors of [11] obtain 5 cusp forms $s_0, s_1, \ldots, s_4$ of weight 2 satisfying $s_0 + s_1 + \cdots + s_4 = 0$ and prove that the canonical map ι_K is a holomorphic mapping of degree 1 of Y onto the quintic surface in $P_4(C)$ defined by the equations $\sigma_1 = 0$, $2\sigma_5 - \sigma_2\sigma_3 = 0$, where the 15 singularities of the quintic are of type A_2. They are images of the 15 chains of (-2)-curves of length 2 mentioned before. Otherwise ι_K is biholomorphic. (σ_i is the ith elementary symmetric function in the s_j.)

Let G be the Hilbert modular group for $D = 8$. The prime 7 splits in $K = Q(\sqrt{2})$. We have $(7) = \mathfrak{g}\mathfrak{g}'$. Consider the principal congruence subgroup Γ of G for the ideal $\mathfrak{g} = (3 + \sqrt{7})$. The group G/Γ is isomorphic to $PSL_2(F_7)$, the famous simple group G_{168} of order 168. The group Γ acts freely on H^2. The surface H^2/Γ has to be compactified by eight cusps corresponding to the points of $P_1(F_7)$. The resolution of each of the eight cusps of H^2/Γ consists of a cycle of six rational nonsingular curves with self-intersection number -4, -2, -4, -2, -4, -2. We denote the algebraic surface obtained by resolving the 5 cusps by Y. The normalized Euler volume of H^2/G equals $2\zeta_K(-1) = \frac{1}{6}$. Therefore the Euler number of H^2/Γ is $168/6 = 28$. For the Euler number of Y we get

$$e(Y) = 28 + 8 \times 6 = 76.$$

Since $Q(\sqrt{2})$ has a unit of negative norm, the arithmetic genus of Y equals $\frac{1}{4}e(H^2/\Gamma) = 7$. (See Hirzebruch [15, Section 3.6].)

The curve F_1 of Y is irreducible, nonsingular, and isomorphic to $\overline{H/\Gamma(7)}$ where $\Gamma(7)$ is the principal congruence subgroup of $SL_2(Z)/\{\pm 1\}$ of level 7. It has 24 cusps. The Euler number is given by the formula

$$e(F_1) = -\tfrac{1}{6} \cdot 168 + 24 = -4.$$

Here $-\frac{1}{6}$ is the normalized Euler volume of $H/SL_2(Z)$. Hence F_1 has genus 3. We also consider the curve F_2 of Y. It is irreducible and nonsingular. The quotient of F_2 by $G_{168} = G/\Gamma$ is the curve F_2 in H^2/G whose nonsingular model is $\overline{H/\Gamma_0^*(2)}$; see [15, Section 4.1]. Therefore

$$e(F_2) = -\tfrac{1}{6} \cdot \tfrac{3}{2} \cdot 168 + 24 = -18.$$

Thus F_2 has genus 10. It can be shown that F_1 and F_2 pass through each of the eight resolved cusps of Y as shown in Figure 11. The intersection of a canonical divisor K of Y with F_1 or F_2 can be calculated [15, Section 4.3, (19)]. This intersection number equals the normalized Euler volume of the curve multiplied by -2 decreased by the sum of the intersection numbers with the curves in the resolutions of the cusps.

We have

$$KF_1 = +\tfrac{1}{3} \cdot 168 - 48 = 8,$$

$$KF_2 = +\tfrac{1}{2} \cdot 168 - 72 = 12,$$

$$F_1F_1 = -4 \quad \text{and} \quad F_2F_2 = 6.$$

The curve F_1 (multiplicity 2) and the twenty-four (-2)-curves of the eight cusps of Y constitute a configuration (VI) of Section 2.

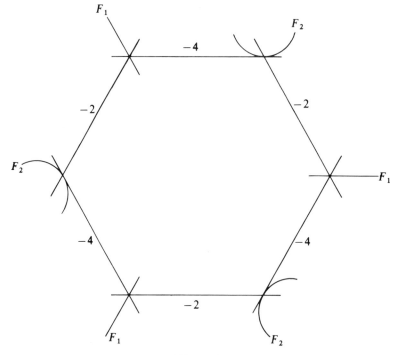

Figure 11

We have proved that Y is a minimal algebraic surface with $\chi = 7$ and Euler number 76. By Noether's formula $K^2 = 8$. We know that Y is simply connected. Hence $p_g = 6$. According to Horikawa's result [22] on minimal algebraic surfaces with $K^2 = 2p_g - 4$, the canonical map is of degree 2 onto a surface of degree 4 in $P_5(C)$. This is either the Veronese surface (projective plane) or a ruled surface. Since G_{168} acts on Y, it also acts on the image surface in $P_5(C)$. But a ruled surface does not admit such an action. Therefore Y is a double cover of the projective plane branched along a curve of degree 10 (Example 6 in Section 1).

We have seen that

$$D_1 = 2F_1 + \sum_{i=1}^{24} L_i$$

is a twofold canonical divisor of Y (where the L_i are the twenty-four (-2)-curves occuring in the cusps). We can show in the same way that

$$D_2 = 2F_2 + \sum_{i=1}^{24} L_i$$

is a threefold canonical divisor. Under the canonical map $\iota_K : Y \to P_5(C)$, where $\iota_K(Y) \simeq P_2(C)$, the image of $D_1 \cup D_2$ must be contained in the curve of ramification, because D_1, D_2 are not divisible by 2, but have F_1 and F_2 as components of multiplicity 2.

The group G_{168} acts on Y and also on $\iota_K(Y) \simeq P_2(C)$. This is the well-known action of G_{168} on $P_2(C)$, because this action, which is described by Weber [31, 15. Abschnitt] is essentially unique.

The action of G_{168} on the projective plane has a unique invariant curve A of degree 4 and a unique invariant curve B of degree 6. These must be the images of F_1 and F_2. The curve A is the famous curve of genus 3 studied by Felix Klein, which has G_{168} as automorphism group. The curves A and B are nonsingular. They intersect transversally in 24 points which are the flexes of A. The 24 tangents of inflexion of A are arranged in 8 "triangles": The tangent T_1 of A in a flex p_1 of A intersects A in a flex p_2 transversally; the tangent T_2 of A in p_2 intersects A in a flex p_3; the tangent T_3 of A in p_3 intersects A in p_1. The line T_1 is tangent to B in p_2 with intersection multiplicity 5; the same holds for T_2 in p_3 and for T_3 in p_1.

Under the canonical map the (-4)-curves of a resolution cycle of a cusp go to such a triangle T_1, T_2, T_3. The (-2)-curves are mapped to the flexes p_1, p_2, p_3.

The canonical involution σ on Y carries a (-4)-curve to another (-4)-curve which does not belong to any cuspidal resolution. This shows that σ is not modular, i.e., it is not induced by an automorphism of H^2. Because G_{168} is a maximal finite automorphism group of $P_2(C)$, the automorphism group of Y is the direct product $G_{168} \times Z/2$, where σ is the nontrivial element of $Z/2$. We collect the above information in the following theorem.

Theorem. *The* (*desingularized*) *Hilbert modular surface* Y *for the field* $Q(\sqrt{2})$ *and the principal congruence subgroup of the Hilbert modular group with respect to an ideal of norm* 7 *is a minimal algebraic surface with* $p_g = 6$ *and* $K^2 = 8$. *Under the canonical map it can be realized as the desingularized double cover of the Veronese surface* (*identified with* $P_2(C)$) *ramified along the curve* $A \cup B$ *of degree* 10, *where* A *and* B *are the unique invariant curves of degree* 4 *and* 6 *respectively for the action of* G_{168} *on* $P_2(C)$. *The full automorphism group of* Y *is* $G_{168} \times Z/2$, *where the canonical involution of* Y *is the nontrivial element of* $Z/2$. *This involution is not modular.*

5. Remarks on the Symmetric Hilbert Modular Group and on "Modular" Modular Forms

As mentioned in Section 3, the surface $Y^0(D)$ admits an action of the group M of order 2^t. If D is a prime p ($p \equiv 1$ mod 4), then M consists only of the involution τ induced by $(z_1, z_2) \to (z_2, z_1)$. It would be interesting to study the canonical map for the surfaces $Y^0(D)/M$. If, for example, the minimal model of such a surface is a Moishezon surface, then the involution on it is nonmodular. However, only some results on $Y^0(p)/\tau$ (p prime) are known (Hirzebruch and Van de Ven [18]). The surface $Y^0(p)/\tau$ is rational for exactly 24 primes. If it is not rational, then the surface $Y_\tau^0(p)$ is defined by blowing down on $Y^0(p)/\tau$

certain curves F_N and curves coming from the resolution of cusp and quotient singularities (Hirzebruch [16]).

It is conjectured that $Y_\tau^0(p)$ is minimal.

The geometric genus of $Y_\tau^0(p)$ equals 3 if and only if $p = 313, 653, 677, 773$. In these cases $K^2 = 2$, the surface is minimal and is a Moishezon surface [18].

The geometric genus of $Y_\tau^0(p)$ equals 4 if and only if $p = 337, 401, 541, 797$. In these cases $K^2 = 5, 6, 6, 5$. Are these surfaces minimal? If yes, is the canonical map holomorphic of degree 1 onto a quintic surface for $p = 337, 797$ and a double cover of a cubic surface for $p = 401, 541$?

The construction of a canonical divisor on $Y^0(D)$ by curves F_N and curves coming from the singularities gives a cusp form f of weight 2 for the Hilbert modular group G of $Q(\sqrt{D})$ whose zero divisor (f) on H^2/G consists only of modular curves F_N. Such a cusp form could be called "modular". The same remark applies to $Y_\tau^0(p)$, where such canonical divisors give "modular" cusps forms of weight 2 satisfying $f(z_1, z_2) = -f(z_2, z_1)$. (See [18, p. 147, Remark 2].) If one had a general theorem which guarantees the existence of a "modular" cusp form f of weight 2 (which is supposed to be skew-symmetric if one considers $Y_\tau^0(p)$), then the problem of minimality would be solved, because then all exceptional curves of a desingularization Y of $\overline{H^2/G}$ or $\overline{(H^2/G)}/\tau$ must be contained in the canonical divisor defined by f on Y, and this canonical divisor consists only of curves F_N and curves coming from singularities.

REFERENCES

[1]   A. Beauville, L'application canonique pour les surfaces de type général. *Inventiones Math.* **55**, 121–140 (1979).

[2]   E. Bombieri, Canonical maps of surfaces of general type. *Publ. Math. IHES* **42**, 171–219 (1973).

[3]   E. Brieskorn, Über die Auflösung gewisser Singularitäten von holomorphen Abbildungen. *Math. Ann.* **166**, 76–102 (1966).

[4]   E. Brieskorn, Die Auflösung der rationalen Singuliaritäten holomorpher Abbildungen. *Math. Ann.* **178**, 255–270 (1968).

[5]   E. Brieskorn, Singular elements of semi-simple algebraic groups. In *Actes, Congrès Intern. Math.*, 1970, Tome 2, pp. 279–284.

[6]   J. W. Bruce and C. T. C. Wall, On the classification of cubic surfaces. Preprint Liverpool 1978.

[7]   E. Freitag, Eine Bemerkung zur Theorie der Hilbertschen Modulmannigfaltigkeiten hoher Stufe. *Math. Z.* **171**, 27–35 (1980).

[8]   G. van der Geer, Hilbert modular forms for the field $Q(\sqrt{6})$. *Math. Ann.* **233**, 163–179 (1978).

[9]   G. van der Geer, Minimal models for Hilbert modular surfaces of principal congruence subgroups. *Topology* **18**, 29–39 (1979).

[10] G. van der Geer and A. Van de Ven, On the minimality of certain Hilbert modular surfaces. In *Complex Analysis and Algebraic Geometry.* Iwanami Shoten and Cambridge U. P., 1977, pp. 137–150.

[11] G. van der Geer and D. Zagier, The Hilbert modular group for the field $Q(\sqrt{13})$. *Inventiones Math.* **42**, 93–133 (1977).

[12] P. Griffiths and J. Harris, *Principles of Algebraic Geometry.* Wiley, New York, 1978.

[13] W. Hausmann, Kurven auf Hilbertschen Modulflächen. Dissertation, Bonn, 1979.

[14] F. Hirzebruch, *Topological Methods in Algebraic Geomery*, 3rd ed. Springer, New York, 1966.

[15] F. Hirzebruch, Hilbert modular surfaces. *Ens. Math.* **19**, 183–281 (1973).

[16] F. Hirzebruch, Modulflächen und Modulkurven zur symmetrischen Hilbertschen Modulgruppe. *Ann. Scient. Ec. Norm. Sup.* **11**, 101–166 (1978).

[17] F. Hirzebruch and A. Van de Ven, Hilbert modular surfaces and the classification of algebraic surfaces. *Inventiones Math.* **23**, 1–29 (1974).

[18] F. Hirzebruch and A. Van de Ven, Minimal Hilbert modular surfaces with $p_g = 3$ and $K^2 = 2$. *Amer. J. Math.* **101**, 132–148 (1979).

[19] F. Hirzebruch and D. Zagier, Classification of Hilbert modular surfaces. In *Complex Analysis and Algebraic Geometry*. Iwanami Shoten and Cambridge U. P., 1977, pp. 43–77.

[20] F. Hirzebruch and D. Zagier, Intersection numbers of curves on Hilbert modular surfaces and modular forms of Nebentypus. *Inventiones Math.* **36**, 57–113 (1976).

[21] E. Horikawa, On deformations of quintic surfaces. *Inventiones Math.* **31**, 43–85 (1975).

[22] E. Horikawa, Algebraic surfaces of general type with small c_1^2. I. *Annals Math.* **104**, 357–387 (1976).

[23] E. Horikawa, Algebraic surfaces of general type with small c_1^2. II. *Inventiones Math.* **37**, 121–155 (1976).

[24] E. Horikawa, Algebraic surfaces of general type with small c_1^2. III. *Inventiones Math.* **47**, 209–248 (1978).

[25] E. Horikawa, Algebraic surfaces of general type with small c_1^2. IV. *Inventiones Math.* **50**, 103–128 (1979).

[26] F. J. Koll, *Die elliptischen Fixpunkte und die Spitzen der diskreten Erweiterungen der Hilbertschen Modulgruppe.* Bonner Math. Schriften 84, 1976.

[27] Yu. I. Manin, *Cubic forms. Algebra, Geometry, Arithmetic.* North-Holland, Amsterdam, 1974.

[28] Y. Miyaoka, On the Chern numbers of surfaces of general type. *Inventiones Math.* **42**, 225–237 (1977).

[29] L. Schläfli, On the distribution of surfaces of the third order into species. *Phil. Trans. Roy. Soc.* **153**, 193–247 (1864).

[30] O. V. Švarcman, Simply-connectedness of the factor space of the Hilbert modular group (in Russian). *Funct. Anal. and Appl.* **8** (2) 99–100 (1974).

[31] H. Weber, *Lehrbuch der Algebra*, 2. Band, 2. Auflage. Vieweg, Braunschweig, 1899.

Tight Embeddings and Maps.
Submanifolds of Geometrical Class Three in E^N [1]

Nicolaas H. Kuiper*

Part I. Tight Embeddings and Maps

1. Introduction and Survey

Differential geometry is a field in which geometry is expressed in analysis, algebra, and calculations, and in which analysis and calculations are sometimes understood in intuitive steps that could be called geometric.

Professor Chern is a master in this interplay. I myself however have always been interested in geometrical phenomena with a minimal amount of calculation. The subject matter of this paper will illustrate this.

The paper is to a large extent a survey of known results, most of them obtained by various geometers in the last twenty years. There is a new flavor in the presentation, and there are some new results at the end (Sections 10–14).

The notion of *tightness* of a set or a map in Euclidean space E^N is defined in terms of intersections with half spaces in Section 2. By this definition it is a projective notion. It generalizes *convexity*, and the synthetic methods it leads to have a great deal in common with convexity theory. In Section 2 we prove an easy theorem concerning "top sets" which plays a fundamental role in all proofs dealing with the question how tight submanifolds or sets are placed in E^N. Tightness can be a condition that forces submanifolds occasionally into much more special shapes than convexity does. We will mention several examples.

The origin of the notion lies in the differential geometry of curves and smooth submanifolds of E^N. However, proofs of theorems concerning *total absolute curvature* (Section 4) and minima of these integrals (Section 5) (theorems of Fenchel, Chern, Lashof) can be simplified and generalized at the same time to the categories of piecewise linear and topological embeddings or maps as in Section 10.

[1] The author acknowledges with gratitude that part of this paper was written while he was a guest of IMPA in Rio de Janeiro in July 1979.

*Institut des Hautes Etudes Scientifiques, 35 Route de Chartres, 91 Bures-sur-Yvette, France.

Tightness of smooth submanifolds has also a relation with the "Ordnungs-geometrie" of Juel, Marchaud, and Haupt and Künneth [26]. Recall that the geometrical degree k of an embedded or immersed curve without straight pieces in the real projective plane P is the maximal number of intersection points with lines. This degree has been studied, and properties analogous to those of real algebraic curves of degree k were obtained. Analogous definitions hold for n-manifolds. The dual notion of *class* is explained as follows. Let $M \subset E^N \subset P^N$ be a smooth n-manifold in real projective N-space. The hyperplanes tangent to M in P^N form a subset M^+ of the dual projective space P^{N+} of all hyperplanes in P^N. *The geometrical class* $c(M) \leq \infty$ of M is the maximal number of intersection points of lines in P^{N+} with M^+, viz., the degree of M^+. It is the maximal number of tangent hyperplanes of M in P^N, in any pencil of hyper-planes. (For example, an algebraic curve of degree three without double points or cusps in the projective plane has algebraic class six.)

If M has an embedding in Euclidean space, $f : M \to E^N \subset P^N$, of *minimal possible geometrical class* $c(M) = \beta$ (to be explained), then f is tight. For example the Veronese surface in Section 6 is an embedding of a projective plane in $E^5 \subset P^5$ of class three, as we will see.

We now continue the description of the contents of this paper. We recall known results for smooth surfaces in Section 6, give examples of tight smooth n-manifolds in Section 7, and show how tightness can be very restrictive, possibly leading to nonexistence, in Section 8 in the smooth category. In Section 9 we report on Banchoff's examples and the theory of tightness for polyhedral sur-faces, and get to the latest results on tightness for topologically embedded or immersed surfaces, where high-codimension conditions completely fix the shape of the surface in some cases.

In Section 10, we return to the definition of tightness and deduce results on tight maps which generalize certain theorems on convex sets, and those of Fenchel, Chern, and Lashof concerning spheres: A tight "substantial" map of an n-sphere in E^{n+1} is onto the boundary of a convex body. The map is cellular for $n \leq 2$, but not necessarily so for $n \geq 4$. More general conclusions for tight maps of surfaces in E^N, rather near to those concerning tight imbeddings of surfaces, are obtained in Section 11.

In Part II, we study smooth submanifolds of geometrical class $c(M) = \beta = 3$, of high substantial codimension in E^N. The standard models for projective planes are examples (Section 12). Tightness makes these examples almost unique (Section 14, Theorem 15).

Finally we observe that the examples of tight n-manifolds due to Kobayashi and others (Section 17) are not only tight but taut as well. Tautness is a more special and conformal property, as it is defined in terms of intersections with round balls. Banchoff, Carter, and West opened that field some years ago. In another paper we determine all taut compact sets with finitely generated homol-ogy in three-space S^3. They are the Dupin cyclides, the i-spheres $i = 1, 2, 3$, and the point.

Throughout the paper we have indicated interesting open problems. For older surveys see Ferus [21], Willmore [69], and Kuiper [43].

2. The Definition of Tightness

Let $E^N = R^n$ be the Euclidean N-space, with the compatible structure of a vector space with an origin 0 and inner product. Then a *half space h*, bounded by a *hyperplane* $\partial h = H$, is a set given by an inequality

$$h : z \geqslant c$$

for some linear function z, and $c \in R$. We are interested in *embeddings* or *maps* $f: X \to E^N$ of a *connected compact space* X in E^N, such that half-space intersections $f^{-1}(h) \subset X$ are "not more complicated than strictly necessary". As a first approximation we recall and mention *Banchoff's two piece property* (TPP): *For every half space h, the part $f^{-1}(h)$ is connected.* Equivalently, every hyperplane H cuts X in at most two pieces, whence the name. In terms of homology (we use Čech homology with Z_2 as *coefficients*) this is expressed by the injectivity of

$$H_0\big(f^{-1}(h)\big) \to H_0(X) \qquad \forall h. \qquad (2.1)$$

Examples of TPP-sets are given in Figure 1. Only the sets in Figure 1(b) are not TPP. It is easy to decide, looking at Figure 1 (a), (b), and (c), that a compact plane set is TPP if and only if it is a point, a line segment, or a two-dimensional compact convex set, from which zero or more disjoint open convex sets are deleted. An example is the convex curve on the right in Figure 1(a). Another example, to which we come back in another paper, is the *"taut"* Swiss cheese, obtained from a round disk by deleting an everywhere dense set of mutually disjoint open discs. Observe that the homology of this last space is not finitely generated: each disc boundary carries a generator for $H_1(X)$.[2]

In figure 1(d) are shown TPP imbeddings of the 2-sphere, as the boundary of a convex set, the torus T, and a TPP immersion of the connected sum of the Klein bottle and the torus $K \# T$ (Klein bottle with one handle). We can explain their essential TPP properties after some definitions.

For a given $f: X \to E^N$, the half space h defined by $z \leqslant c$ and the hyperplane H with equation $z = c$ are called *supporting* in case the boundary, but not the interior of h, has points in common with $f(X)$. The intersection,

$$X_z = f^{-1}(h) = f^{-1}(H) \neq \text{void},$$

is then denoted by X_z, and it is called a *top set* in case it differs from X. Observe that the linear function z takes its maximal value (of zf) on X in the points of X_z. The restriction of f to X_z, $f|X_z$, is a *top map* of f. A top set X_{z,z_1} of a top map is called a *top^2 set*, etc. A *top* set* is a top^i set for some $i \geqslant 1$. A top* set is called an E^p-*top*-set* in case its image in E^N spans an affine p-plane E^p in that Euclidean space.

For later purposes we mention also that $f: X \to E^N$ is called *substantial* in case $f(X)$ does not lie in any hyperplane. For any convex set $Y \subset E^N$ the *convex hull*, denoted by $\mathfrak{K}Y$, is the smallest convex set containing Y. Its *boundary* $\partial \mathfrak{K}Y$ is the boundary in the p-plane that Y spans, $p \leqslant N$.

---

[2] If no disc boundaries have common points, then we get a unique topological space (always the same), as R. Edwards proved.

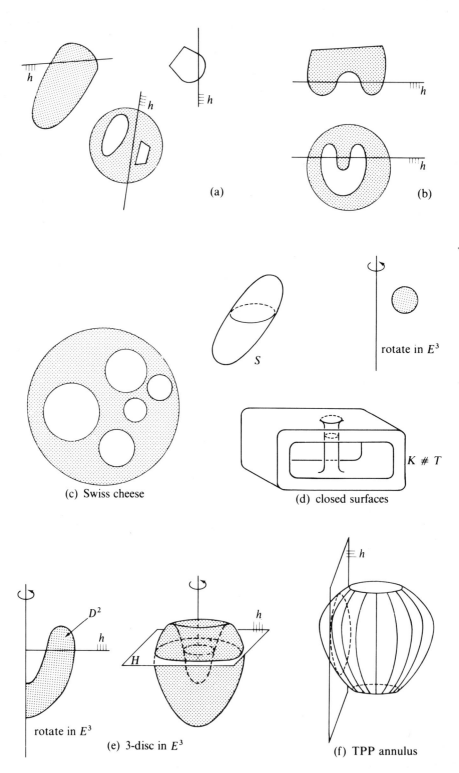

(a)

(b)

(c) Swiss cheese

S

rotate in E^3

$K \# T$

(d) closed surfaces

D^2

rotate in E^3

(e) 3-disc in E^3

H

(f) TPP annulus

Figure 1.

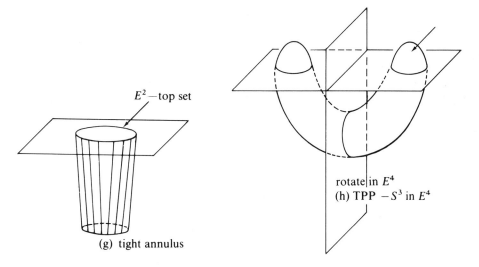

E^2—top set

rotate in E^4

(h) TPP $-S^3$ in E^4

(g) tight annulus

Figure 1.(cont.)

For embeddings $f: X \subset E^N$, top* sets are of course subsets of E^N; for maps the top* sets are determined, for given f, by their images (themselves top sets of $f(X)$ in E^N).

The 3-disc D^3 in E^3 in Figure 1(e) with a vertical axis of rotational symmetry is TPP. But as we see, there exists a half space h, bounded below by a horizontal hyperplane H, whose intersection $f^{-1}(h) = S^1 \times D^2$ has the nontrivial one-dimensional homology class of a circle in $f^{-1}(h) \subset D^3$. This class bounds in D^3, and $H_1(f^{-1}(h)) = Z^2 \to H_1(D^3) = 0$ is therefore not injective. A closed 3-manifold, namely a 3-sphere $M = S^3$, with analogous properties is illustrated in Figure 1(h). We see an embedded S^2 in $E^3 \subset E^4$, whose image is symmetric with respect to a vertical plane E^2. Rotation in E^4 about that E^2 gives an embedded S^3 in E^4 which is clearly TPP. But the half space $h \subset E^4$ intersects that S^4 in $f^{-1}(h) = S^1 \times D^2$, and as above,

$$H_1(S^1 \times D^2) = Z_2 \to H_1(S^3) = 0$$

is not injective. We reject such situations with a stronger condition in the following

Definition. The embedding or map $f: X \to E^N$ is called *tight* in case the induced homorphism in Čech homology with coefficients in Z_2,

$$H_*(f^{-1}(h)) \to H_*(X) \qquad (2.2)$$

is injective for every half space $h \subset E^N$.

Clearly tightness (2.2) implies TPP (2.1). *For compact sets in the plane and for maps of closed surfaces M^2 into E^N, TPP and tightness are clearly equivalent.*

For the surfaces in Figure 1(d) a necessary property for tightness or TPP is that every nonconvex top set X_z carries a cycle (see Section 3), which then must be an essential 1-cycle of X by the definition (2.2). Another necessary property is that all points of positive Gauss curvature are contained in $\partial \mathcal{H} f(X)$, the boundary of the convex hull.

The 3-disc in Figure 1(e) and the S^3 in Figure 1(h) are TPP but not tight. Also the set in Figure 1(f) is TPP but not tight, because for h as in the figure

$$H_1\big(f^{-1}(h)\big) = Z_2 \times Z_2 \quad \text{and} \quad H_1(X) = Z_2,$$

so that injectivity cannot hold.

3. First Consequences of the Definition

Tightness (as well as TPP) is a *projective property* in the following sense. Let $E^N = P^N \backslash P^{N-1}$ be the natural complement of a hyperplane P^{N-1} in real projective N-space P^N. Suppose $f : X \to E^N$ is tight and $\sigma : P^N \to P^N$ is a projective transformation such that the image $\sigma(fX) \subset E^N = P^N \backslash P^{N-1}$ is in E^N. Then $\sigma f : X \to E^N$ is tight as well. This follows from the definition being in terms of half spaces.

If $i : E^N = R^N \times 0 \to R^N \times R^r = R^{N+r}$ is the *natural embedding* and $p : E^N = R^{N-r} \times R^r \to R^{N-r}$ is a *natural projection*, tightness of f entails, by definition, tightness of $i \circ f$ and of $p \circ f$. So tightness is also invariant under these operations. We now prove the

Fundamental theorem 1. *A top set of a tight set $X \subset E^N$ is tight. A top map of a tight map $f : X \to E^N$ is tight, and the same holds for top* maps and top* sets.*

A more specific formulation is as follows: If $X_z = f^{-1}(h_z) = f^{-1}(H_z) \subset X$ is a top set for a tight map $f : X \to E^N$ of a compact space X, then

$$H_*(X_z) \to H_*(X) \tag{3.1}$$

is injective, and the top map

$$f' = (f|X_z), \qquad X_z \to E^N \tag{3.2}$$

is tight as well.

Proof. The first conclusion (3.1) is the definition of tightness applied to $h_z \cap X$. In order to simplify, mainly in notation, we write the proof of the remaining part for embeddings only. The generalization to maps is straightforward. So we assume a tight set $f(X) = X \subset E^N$ with top set X_z in the hyperplane H_z, bounding the half space h_z. Take any half space h which meets H_z in a half space of H_z. We need to prove that

$$H_*(X_z \cap h) \to H_*(X_z)$$

is *injective* for every h. Consider the orthogonal projection in a 2-plane (see Figure 2) orthogonal to h_z and h (and H_z and H). Then we see the existence of a sequence of half spaces h_i, $i = 1, 2, 3, \ldots$, giving a convergent nested sequence of subspaces of X that converge to $X_z \cap h$:

$$h_i \cap X \supset h_{i+1} \cap X \supset \bigcap_{j=1}^{\infty} h_j \cap X = h \cap X_z.$$

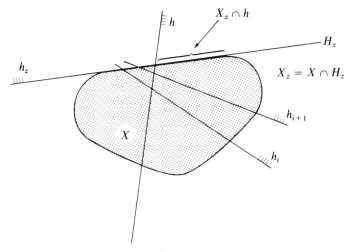

Figure 2.

For Čech homology the continuity of H_* is known (see Eilenberg-Steenrod [17], ch X) so that

$$\lim_{j=\infty} H_*(h_j \cap X) = H_*(h \cap X_z).$$

We now obtain a commutative diagram of morphisms in homology, with injective morphisms denoted $\overset{c}{\rightarrow}$, from a corresponding diagram of inclusions, as follows:

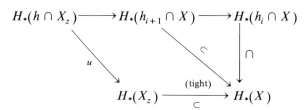

Then the morphism u is also injective and the fundamental theorem is proved.

Easy corollary 1a. *If $f : X \to E^N$ is a tight (hence TPP) embedding of a compact connected space X, then an E^0-top*-set is a point, an E^1-top*-set is a line segment, and an E^2-top*-set is a convex 2-dimensional plane disc, from which $k \geqslant 0$ disjoint convex open discs are deleted. For $k < \infty$, $k = \dim H_1(top^* set) \leqslant \dim H_1(X)$.*

Compare all top sets of the tight sets in Figure 1.

4. Total Absolute Curvature in Differential Geometry

Let $f : M^n \to E^N$ be a smooth (C^∞) immersion of a closed manifold in E^N—for example, a plane curve in E^2 or a torus in E^3 as in Figure 3. Then local and global invariants called the curvature of index k, the curvature, and the absolute curvature can be defined as in Kuiper [38, 44]. Here we only recall the definition

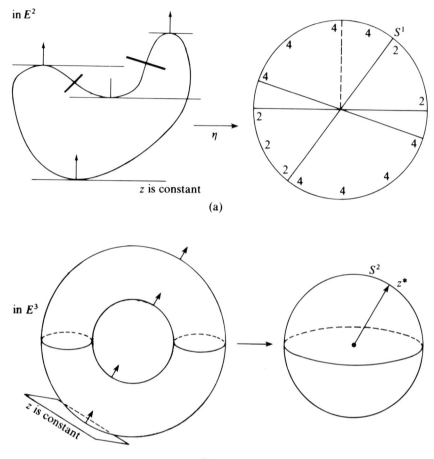

Figure 3

of the absolute curvature as follows. Let $\nu(M)$ be the abstract smooth manifold of dimension $N-1$ of unit normals in E^N of M. There is the natural map

$$\nu(M) \xrightarrow{\eta} S^{N-1}$$

which transports every unit vector parallel to the origin $0 \in E^N$. The unit sphere S^{N-1} in E^N has the origin as center, and has volume element $d\operatorname{vol} S^{N-1}$.

Definition. The total *absolute curvature* $\tau(f)$ of f is the integral of the pulled-back $(N-1)$-form $d\operatorname{vol} S^{N-1}$, with a factor $c_{N-1} > 0$:

$$\tau(f) = c_{N-1}\int_{\gamma(M)} |\eta^*(d\operatorname{vol} S^{N-1})| < \infty, \qquad (4.1)$$

where

$$c_i^{-1} = \int_{S^i} d\operatorname{vol} S^i.$$

Integration over the fibres of $\nu(M) \to M$ yields an integral over M:

$$\tau(f) = \int_M \tau_x |d \operatorname{vol} M|, \tag{4.2}$$

where $d \operatorname{vol} M$ is the integrand for the volume of the Riemannian metric of M, and $\tau_x \geqslant 0$ is the *absolute curvature* at $x \in M$. For the relation to the Lipschitz–Killing curvature see Willmore [69]. For hypersurfaces, $\nu(M)$ covers M twice, and η is the normal *Gauss map*.

For *closed curves* in E^N, $n = 1$, one finds (Kuiper [39, 40])

$$\tau(f) = \oint \frac{|\rho \, ds|}{\pi}, \tag{4.3}$$

where ds is arc length and ρ is the usual curvature. For *closed surfaces* in E^3, $n = 2$, one finds

$$\tau(f) = \int_M \frac{|K \, d\sigma|}{2\pi}, $$

where $d\sigma = d \operatorname{vol} M$ and K is the Gauss curvature.

The integral (4.1) can also be calculated as an integral over S^{N-1}, by counting the number of times every point $z^* \in S^{N-1}$ is covered by the map η (see Figure 3(a)). A vector $z^* \in S^{N-1}$ is covered as often as the unit vector z^* is normal for a point $x \in M$ at $f(x) \in f(M)$ to the immersed surface. This is the case for a point $x \in M$ if and only if the composition $z \circ f$ with the linear function z on E^N whose gradient vector is z^*, has a critical point at x. At the same point $f(x)$, the hyperplane with equation $z = c = z(f(x))$ is tangent to $f(M)$. By Sard's theorem almost all linear functions z give functions $z \circ f$ on M, with nondegenerate critical points only and whose number μ_z is finite. We can neglect the other linear functions and find from (4.1)

$$\tau(f) = c_{N-1} \int_{z^* \in S^{N-1}} \mu_z \, d \operatorname{vol} S^{N-1}. \tag{4.4}$$

The variable z in the integrand runs over all linear functions z with unit gradient vectors z^*. If we consider μ_z as a random variable (function) on S^{N-1} with the homogeneous probability measure, then $\tau(f)$ in (4.4) is an average value over S^{N-1} or an *expectation value* (symbol $\mathcal{E}$):

$$\tau(f) = \mathcal{E}_z \mu_z. \tag{4.5}$$

μ_z is the *number of critical points* of the function $z \circ f$ on the manifold M. For a given immersion or embedding $f: M^n \to E^N$, let the minimum of the number of critical points z of all functions $z \circ f$ be denoted

$$\min_z \mu_z. \tag{4.6}$$

Let the minimum of the number of critical points $\mu(\varphi)$ for all *smooth nondegenerate* functions $\varphi: M \to R$ be called

$$\min_\varphi \mu(\varphi) = \gamma. \tag{4.7}$$

Denote the sum of the Z_2-Betti-numbers of M by

$$\beta = \sum_j \beta_j, \qquad \beta_j = \operatorname{Rank} H_j(M, Z_2).$$

As

$$H_0(M, Z_2) = H_n(M, Z_2) = Z_2,$$

we find

$$\beta \geqslant 2.$$

By Morse theory (Milnor [56])

$$\gamma \geqslant \beta. \qquad (4.8)$$

Combining the definitions and these remarks with (4.5), we obtain

Theorem 2 (essentially already in Chern and Lashof [13].[3] *The total absolute curvature of a smoothly immersed closed manifold $f: M^n \to E^N$ in Euclidean space E^N is*

$$\tau(f) \geqslant \gamma \geqslant \beta \geqslant 2. \qquad (4.9)$$

Corollary. *The total absolute curvature of a closed surface M with Euler characteristic $\chi(M)$ in E^N is* (Kuiper [39])

$$\tau(f) \geqslant \gamma = \beta = 4 - \chi(M) \geqslant 2 \qquad ((4.9); n = 2)$$

where for $N = 3$

$$\tau(f) = \int_M \frac{|K d\sigma|}{2\pi}.$$

Given a smooth closed manifold $M = M^n$, we are interested in getting small values for the absolute total curvature $\tau(f)$ for immersions or embeddings, possibly with side conditions. We recall:

Theorem 3 (Kuiper [36], but for a complete proof see Wilson [70]). *The greatest lower bound of $\tau(f)$ for all smooth immersions f of M^n in all Euclidean spaces is*

$$\inf_f \tau(f) = \gamma.$$

Proof (idea). Let $g: M \to E^N$ be a smooth immersion, $\varphi: M \to R = E^1$ a nondegenerate function with γ critical points. Consider the immersions

$$f_u = g \times u\varphi : M \to E^N \times E^1 = E^{N+1}$$

defined by

$$f_u(x) = (g(x), u\varphi(x)), \qquad u > 0.$$

Then (look at the normal map near critical points of φ)

$$\lim_{u \to \infty} \tau(f_u) = \gamma$$

and the theorem is proved.

---

[3] Part of this theorem is already expressed in Chern's paper [12] of 1955.

Definition. The smooth immersion or embedding $f: M \to E^N$ has *minimal total absolute curvature* in case

$$\tau(f) = \gamma.$$

We can formulate a general problem: Given a smooth closed n-manifold M, is there an immersion or embedding f substantially in E^N with minimal $\tau(f) = \gamma$, and for which N? If so, what can one say about f?

The first result in this direction is the

Theorem of Fenchel [19].[4] *The total absolute curvature of an immersed circle in E^N is*

$$\tau(f) = \oint \frac{|\rho \, ds|}{\pi} \geqslant \gamma = \beta = 2,$$

with equality if and only if f is an embedding onto a plane convex curve.

The theorem of Fenchel was generalized in the

Theorem of Chern and Lashof [13]. *The total absolute curvature of a smooth immersion f of a closed manifold M^n substantially in E^N is*

$$\tau(f) \geqslant 2.$$

$\tau(f) < 3$ *implies that M^n is homomorphic to S^n; the equality $\tau(f) = 2$ implies that $N = n + 1$ and f is an embedding onto the boundary of a convex body in E^{n+1}.*

Both theorems above are generalized in Theorem 11, Section 10.

5. Total Absolute Curvature and Tightness

The relation between tightness and the total absolute curvature of smooth immersed closed manifolds is expressed in (see (4.7) and (4.8))

Theorem 4. *The smooth manifold immersion $f: M^n \to E^N$ is tight if and only if* (a) $\gamma = \beta$, *and* (b) *f has minimal absolute total curvature:*

$$\tau(f) = \gamma = \beta. \tag{5.1}$$

Connected with this theorem is the following interesting open

Problem 1. *Is there any smooth immersion of a closed manifold M for which $\gamma \neq \beta$, in some Euclidean space with minimal absolute total curvature $\tau(f) = \gamma \neq \beta$?*

We know no example.

Proof of Theorem 4. Suppose f is tight, and let the linear function $z: E^N \to R$ produce the nondegenerate function $zf: M \to R$ with critical points assumed at

---

[4] For $N = 3$; generalized for $N \geqslant 3$ by Borsuk [6]. Milnor [55] defined $\tau(f) \leqslant \infty$ for any topologically embedded S^1 in E^3. He and Fary [18] proved that $\tau(f) > 4$ if $f(S^1)$ is knotted. This was sharpened for complicated knots by Fox [23]. See also Langevin and Rosenberg [50], Wintgen [72].

different levels

$$c_1 < c_2 < \cdots < c_\nu.$$

Let $h(z, c)$ be the half space of E^N with equation

$$z \leqslant c.$$

By tightness we have injective homomorphisms

$$H_*\big(f^{-1}h(z, c)\big) \to H_*(M)$$

and therefore also the homomorphisms in the sequence

$$0 \to H_*\big(f^{-1}h(z, c_1)\big) \to \cdots \to H_*\big(f^{-1}h(z, c_{j-1})\big) \to H_*\big(f^{-1}h(z, c_j)\big) \to H_*(M)$$

are *injective*. The ranks of these groups form a nondecreasing sequence. But from differential topology (see Milnor [57]) we know that $f^{-1}h(z, c_j)$ is obtained from $f^{-1}h(z, c_{j-1})$ by "attaching one handle" whose dimension is the index k of the critical point at level c_j, and thickening it. The function near this critical point is, in preferred coordinates $\zeta_1, \ldots, \zeta_n$ for M,

$$zf = c_j - \zeta_1^2 - \cdots - \zeta_k^2 + \zeta_{k+1} + \cdots + \zeta_n^2.$$

Then the increase of rank $H_*(f^{-1}h(z, c))$ is *one* for each critical level $c_1, \ldots, c_\nu$, and the total rank dimension is

$$\nu = \operatorname{rank} H_*(M) = \beta.$$

This being true for almost every z, we obtain $\tau(f) = \beta$, $\gamma = \operatorname{rank} H_*(M) = \beta$.

If on the other hand $\tau(f) = \gamma = \beta$ and zf is nondegenerate, then each of its β critical points must give a positive increase of $H_*(f^{-1}h(z, c))$ in order to add up to $H_*(M) = \beta$. Then, again by differential topology $H_*(f^{-1}h(z, c)) \to H_*(M)$ is injective for almost all z, and hence for all $z \in S^{N-1}$. Tightness follows. $\square$

Remark. Definition of $\tau(f)$ for continuous maps of compact spaces $f : X \to E^N$ into E^N (Compare Kuiper [44]): We first define the "number of critical points" $\mu(\varphi)$ of a function $\varphi : X \to R$. Let $H_*(\varphi_{t_2}, \varphi_{t_1})$ be the relative homology group of the pair

$$(\varphi_{t_2}, \varphi_{t_1}), \quad t_1 < t_2, \quad \varphi_t = \{x \in X : \varphi(x) \leqslant t\}.$$

Let s be an increasing sequence of real numbers $t_1 < t_2 < t_3 < \cdots < t_l$ (any l). Then

$$\mu(\varphi) = \sup_s \sum_{j=1}^{l} \operatorname{rank} H_*(\varphi_{t_j}, \varphi_{t_{j-1}}) \leqslant \infty.$$

Clearly for f a smooth immersion of a closed manifold and for zf nondegenerate, the definition specializes to the earlier one.

$$\mu(zf) = \mu_z.$$

Next we define as before the total absolute curvature, now of a map $f : X \to E^N$ of a compact space, as the *average* or *expectation value*

$$\tau(f) = \mathcal{E}_z \mu_z = \mathcal{E}_z \mu(zf),$$

in case this makes sense, and ∞ otherwise. Observe that $\tau(f)$ is finite for a simplexwise linear map of a finite simplicial complex X.

A part of Theorem 2 then survives for maps $f: X \to E^N$ in

Theorem 5. *The total absolute curvature of a continuous map* $f: X \to E^N$ *of a compact space X into E^N is*

$$\tau(f) \geqslant \beta, \qquad \beta \leqslant \infty. \tag{5.2}$$

For $\beta < \infty$ the minimum can be obtained (take a map of X into one point of E^N), and f is tight if and only if it has minimal curvature $\tau(f) = \beta < \infty$. Recall that $\beta = \infty$ for the tight embedding of the taut Swiss cheese of Section 3.

6. Total Absolute Curvature and Smooth Tight Immersions (TPP) of Closed Surfaces in E^N

We concentrate known results (Kuiper [39, 40]) in

Theorem 6

(a) *No tight immersion (not even a continuous immersion) exists in E^3 for the projective plane P, nor for the Klein bottle K.*
(b) *No tight immersion in E^3 is known for $P \# T$, the projective plane with one handle, with Euler characteristic $\chi = -1$.*
(c) *For all other surfaces M there are tight smooth immersions in E^3*

$$\tau(f) = \frac{|k \, d\sigma|}{2\pi} = 4 - X(M).$$

Only the construction for $M = P \# T \# T$ is complicated (Kuiper [40]). See Figure 1(d) for $K \# T$.

(d) *$N = 4$. There exist tight smooth substantial embeddings in E^4 for all surfaces, except of course for the 2-sphere, and except that none is known for the Klein bottle K.*
(e) *For $N \geqslant 5$ the remarkable result is that the only tight smooth substantially immersed surface is the projective plane P embedded onto the algebraic Veronese surface in E^5, described below, and unique but for projective transformations [40].*
(f) *There exists a tight smooth immersion of the torus T in E^3 which is not an embedding.*

There remain, after many years, two important open problems:

Problem 2. *Is there a tight immersion of $M = P \# T$, $\chi(M) = -1$, $\beta(M) = 5$, in E^3?*

Even a continuous example would be interesting.

Problem 3. *Is there a smooth tight immersion of the Klein bottle in E^4?*

A tight polyhedral embedding exists (Section 9).

If we enlarge the class of smooth maps of surfaces into E^3 from immersions to stable maps in the sense of Thom, then stable singularities at isolated points are admitted. We have

Theorem 6′ (Kuiper [46]). *All closed surfaces have stable smooth tight maps f in E^3.*

Although the Gauss curvature K explodes at stable singularities, the integrals are convergent in the formula

$$\tau(f) = \int \frac{|K\,d\sigma|}{2\pi} = 4 - \chi(M).$$

An example for the projective plane is given by

$$f : (x, y, z) \rightarrow (u, v, w) = (x^2 - y^2, 2xy, yz). \tag{6.1}$$

Here (x, y, z) are homogenous coordinates for P, normalized by $x^2 + y^2 + z^2 = 1$, and u, v, w are coordinates in E^3.

The image is all, except part of the line $v = w = 0$, of the *Steiner surface* with equation

$$v^2(u^2 + v^2 + 2w^2 - 1) + 2w^2(2u + 2u^2 + u^2 + 2w^2) = 0.$$

It has stable singularities at $(u, v, w) = (0, 0, 0)$ and $(1, 0, 0)$. The topology of the image is there a cone on a figure eight. Tight top cycles are in the supporting planes $u = 1 \pm w\sqrt{2}$. (See Figure 4.) The intersection with any hyperplane

$$\alpha u + \beta v + \gamma w = (u + v + w)\delta$$

is clearly a conic section in P, and divides it in at most two parts, whence TPP and tightness. Two planes in E^3, π_1 and π_2, determine a pencil of planes through $\pi_1 \cap \pi_2$ and a pencil of conic sections in P. There are at most three singular conics (line pairs) in a pencil. They correspond to at most three tangent planes in the pencil of planes. It follows that *the geometrical class* (see Section 1 and Part II) *of our stable map $f : P \rightarrow E^3 \subset P^3$ is three.*

We next present the *Veronese surface V* by quadratic forms in homogeneous coordinates $(x, y, z) = (a_1, a_2, a_3)$ for P. It is tight for the same reasons as above.

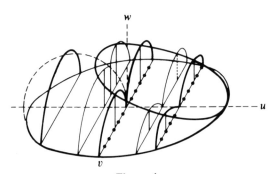

Figure 4.

In view of the generalizations in the next paragraph, we choose the description of V as a subspace of the space of 3×3 real symmetric matrices $A = {}^t A$, namely those which are orthogonal projections on lines through the origin $0 \in R^3$. The set of such lines is P. We get

$$V = \{ A : A = {}^t A = A^2, \text{trace}\, A = 1 \},$$

where

$$A = \begin{bmatrix} a_{11} & a_{12} & a_{13} \\ a_{21} & a_{22} & a_{23} \\ a_{31} & a_{32} & a_{33} \end{bmatrix} = \begin{bmatrix} a_1 \\ a_2 \\ a_3 \end{bmatrix} (a_1, a_2, a_3)$$

with matrix multiplication, and

$$\text{trace}\, A = \sum a_{ii} = \sum a_i^2 = 1, \tag{6.2}$$

$$\text{trace}\, A^2 = \sum (a_{ij})^2 = \sum (a_i a_j)^2 = \sum \left(a_i^2 \right)^2 = 1. \tag{6.3}$$

Observe that V is contained not only in the E^5 with equations $A = {}^t A$, trace $A = 1$, but also in the round 4-sphere $S^8 \cap E^5$, where S^8 has the equation $\sum (a_{ij})^2 = 1$. Recall that by the definitions, the manifold $\nu(V)$ of unit normal vectors at points of $V \subset E^5$ covers the unitsphere $S^4 \subset E^5$ by the map η at most points precisely $3 = 4 - \chi(P)$ times. It can be seen that the exceptional points of S^4 form again a Veronese surface, whose complement is covered three times by η. As before, we find that in any pencil of hyperplanes of E^5, there are at most three that intersect P in singular conic sections (line pairs). They correspond to tangent hyperplanes. Hence *the geometrical class of $V \subset E^5$ is three.*

For some of our immersed or embedded tight closed surfaces, we can decide that the class is $\beta = 4 - \chi$. But we have not decided the open

Problem 4. *Let $f : M^n \to E^N$ be a tight smooth immersion of the closed manifold M^n in E^N. Is the class equal to $\tau(f) = \beta$? In particular, can this be proved for surfaces, $n = 2$?*

This problem concerns the question whether in particular any pencil of hyperplanes whose common $(N - 2)$-plane has a nonvoid intersection with the convex hull $\mathcal{H}f(X)$ contains not more than $\beta(X)$ tangent hyperplanes. For the other pencils this is true by tightness. For which pencils is this number smaller than β?

MISCELLANEOUS RESULTS ON SMOOTH SURFACES.

Theorem of Pohl and Chen [10]. *Every smooth (C^2 suffices) tight embedding of a torus substantially in E^4 is, except for a projective transformation, the product $f \times g : S_a^1 \times S_b^1 \to E^4$ of two tight (convex) curves $S_a^1 \subset E^2$ and $S_b^1 \subset E^2$.*

Theorem on knotted orientable surfaces in E^3 (Langevin and Rosenberg [50] for genus $g = 1$, Meeks [54] and Morton [59] for any g). *If a smooth embedded closed orientable surface $f : M^2 \to E^3$ has total absolute curvature*

$$\tau(f) = \tau(M) < 2g + 6 = 8 - \chi(M),$$

*then f is regularly isotopic through embeddings to a standard ("unknotted")
embedding.*

The proof is quite complicated.

Problem 5. *Can equality be realized for knotted M^2?*
   I do not believe so.

EXAMPLE (L. Rodriguez). For an immersion of the torus $f: T \to E^3$ that is not
isotopic to an embedding, the total absolute curvature is $\tau(f) > 4$. Values
arbitrarily near to 4 can be obtained by rotating the plane figure eight about a
vertical axis.

7. Examples of Tight, Even Taut, Smooth n-Manifolds in E^N

Around 1970 many examples of tight embeddings of well-known spaces were
found—for example, for the projective spaces over $F = R, C, K$ the quaternions,
and for the projective plane over O the Cayley numbers, also called octaves (Tai
[66]). Also the embedding $SO(n) \subset R^{n^2}$ given by the orthogonal matrices of
determinant 1 is tight (Wilson [71]). Kobayashi [29] found that *all compact
homogeneous Kähler manifolds have tight embeddings.* He obtained many more
examples [30], and in particular, with Takeuchi [31], the large class of R-space
embeddings. Ferus [22] found that all symmetric submanifolds of E^N are exactly
the R-space embeddings of certain symmetric spaces, and tight. Here a submanifold $M \subset E^N$ is called symmetric if it is invariant under that element of
$g_x \in O(n)$ that fixes any given point $x \in M$ as origin, gives a reflection about the
origin of its tangent space, and induces the identity on a normal space at x.
   We will describe without proof (see the references for details) two classes of
tight embeddings of homogeneous spaces: the Grassmann manifolds and some
classical groups. In each of these cases the image which has our interest as a
projective submanifold as far as tightness is concerned, *lies in fact in a round
hypersphere* $S^{N-1} \subset E^N$. This implies that the half-space intersections in E^N are
the same as the intersections with round N-balls in E^N, that is, with spherical
$(N-1)$-balls $b \cap S^{N-1}$ in S^{N-1}. In the conformal geometry of the $(N-1)$-
sphere $= E^{N-1} \cup \infty$, which can be considered as a one-point compactification of
E^{N-1} with its conformal (Möbius-group, inversive) structure, we use the

Definition. An embedding or map $f: X \to S^k = E^k \cup \infty$ is called *taut* if for any
round k-ball b (bounded by a $(k-1)$-sphere) in S^k, the induced homomorphism
in Čech homology with Z_2-coefficients

$$H_*\left(f^{-1}(b)\right) \to H_*(X) \qquad (7.1)$$

is injective. From this definition being in terms of round balls it follows that
tautness of a set $X \subset S^k$ or a map $f: X \to S^k$ is a *conformal property*: If
$\sigma: S^k \to S^k$ is a conformal isomorphism (of the Möbius group, which can
incidentally be obtained from those projective transformations of $P^N \supset E^N$ that
leave $S^{N-1} = S^k$ invariant) and f is taut, then so is $\sigma \circ f$. Clearly also, if
$f: X \to E^k \subset E^k \cup \infty = S^k$ is taut, then f is tight.

EXAMPLE 1. (Grassmann manifolds over F). Here F means R, C, or K, and F is of real dimension $d = 1$, 2, or 4 respectively. The Grassmann manifold $F\mathcal{G}_{p,q}$ consists of all p-dimensional F-subspaces of F^m, $m = p + q$. We identify a p-plane for short with the $m \times m$ matrix A which is "symmetric" ($A = {}^t\overline{A}$) and defines the orthogonal projection $A^2 = A$ onto that p-plane. Hence trace $A = p$ and

$$F\mathcal{G}_{p,q} = \{A : {}^tA = A = A^2, \text{trace } A = p\} \subset F^{m^2}. \tag{7.2}$$

For the octaves, $F = O$, and $p = 1$, $q = 2$, $m = 3$; the formula (7.2) defines a tight octave plane in E^{26}, but it does not represent subspaces of O^3. (See Section 12 and Freudenthal [24, 25]).

The conditions

$$A = {}^t\overline{A} \quad (\text{in coordinates} \quad a_{ij} = \bar{a}_{ji})$$

and

$$\text{trace } A = p \quad \left(\text{in coordinates} \sum_j a_{jj} = p\right)$$

are linear in the dm^2 real coordinates of F^{m^2}, so that the image lies in fact in a linear subvariety

$$E^N, \quad N = m + d\frac{m(m-1)}{2} - 1$$

(A is determined by m real numbers $a_{jj} = {}^ta_{jj}$ along the diagonal with sum p, and $\frac{1}{2}m(m-1)$ elements of F above the diagonal). The image lies also in the hypersphere S^{N-1} with the extra equation

$$p = \text{trace } A = \text{trace } A^2 = \text{trace } A\,{}^t\overline{A} = \sum a_{ij}\bar{a}_{ij} = \sum|a_{ij}|^2 = p. \tag{7.3}$$

Being tight (see references), it is therefore taut as well.

It can be shown that the linear function

$$\sum_{j=1}^m 3^j a_{jj}$$

is nondegenerate on the Grassmann manifold. It takes different critical values at its critical points, which are those matrices A that consist of p numbers 1 on the diagonal, and 0 in all other places. (These are clearly orthogonal projectors.) There are

$$\binom{p+q}{p} = \beta$$

such matrices. *The total absolute curvature of the tightly embedded Grassmann manifold $F\mathcal{G}_{p,q}$ of real dimension $n = dpq$ is then*

$$\tau(f) = \tau(F\mathcal{G}_{p,q}) = \binom{p+q}{p}. \tag{7.4}$$

For the projective spaces this is

$$\tau(f) = \tau(F\mathcal{G}_{1,n}) = \tau(FP(n)) = n + 1. \tag{7.5}$$

The geometrical class is probably equal to this value for all these cases; see Section 12.

EXAMPLE 2 (Unitary groups FU_m). Here we consider the groups FU_m of orthogonal matrices ($F = R$), unitary matrices ($F = C$), and unitary quaternion matrices ($F = K$; $RU_m = O(m)$, $KU_m = Sp(m)$) with their natural embeddings in $F^{m^2} = R^{dm^2}$ given by

$$FU_m = \{A : A\,{}^t\overline{A} = 1\} \subset S^{N-1} \subset E^N, \qquad N = dm^2 \qquad (7.6)$$

The dimension of FU_m is (see Chevalley [14])

$$\dim\{A : A + {}^t\overline{A} = 0\} = d\frac{m(m-1)}{2} + (d-1)m.$$

It can be shown that the linear function

$$\sum_{j=1}^{m} 3^j a_{jj}$$

is nondegenerate on $FU_m \subset E^N$. It takes its different critical values at points (matrices) A that consist of numbers 1 and -1 along the diagonal and 0 elsewhere (clearly elements of FU_m). There are 2^m such matrices. *The total absolute curvature of the manifold $FU_m \subset E^N$* of dimension $n = dm(m-1)/2 + (d-1)m$ is therefore

$$\tau(f) = \tau(FU_m) = 2^m. \qquad (7.7)$$

For the orthogonal group this is the total curvature of O_m, which space consists of two components according to the value of the determinant of A, $+1$ or -1. Restricting to one component one has for the total absolute curvature of the special orthogonal group

$$\tau(f) = \tau\left(SO(m) \subset R^{m^2}\right) = 2^{m-1}. \qquad (7.8)$$

Being tight, the manifold FU_m is taut, as it lies in the hypersphere with equation

$$m = \text{trace } 1 = \text{trace } A\,{}^t\overline{A} = \sum a_{ij}\bar{a}_{ij} = \sum |a_{ij}|^2 = m.$$

Special cases are given in Table 1.

Table 1

n	M^n	$\subset E^N$
2	$R\mathcal{G}_{1,2} = RP(2)$	$\subset S^4 \subset E^5$ (Veronese)
n	$R\mathcal{G}_{1,n} = RP(n)$	$\subset S^{\frac{1}{2}(n^2+3n-2)} \subset E^{\frac{1}{2}n(n+3)}$
3	$SO(3) = RP(3)$	$\subset S^8 \subset E^9$
4	$C\mathcal{G}_{1,2} = CP(2)$	$\subset S^7 \subset E^8$
6	$SO(4) = (S^3 \times S^3)/Z_2$	$\subset S^{15} \subset E^{16}$
4	$CU_2 = U(2)$	$\subset S^7 \subset E^8$
8	$K\mathcal{G}_{1,2} = KP(2)$	$\subset S^{13} \subset S^{14}$

Although tight (taut) embeddings are now known for many metric compact homogeneous manifolds, the following remains an interesting open problem.

Problem 6. *Find more tight (taut) embeddings of compact homogeneous (possibly Riemannian homogeneous, possibly symmetric) manifolds and/or find homogeneous manifolds that have no tight imbedding into Euclidean space.*

Candidates are SU_m, the quadratic hypersurfaces in RP_m, and CP_m, the flag manifolds.

8. Substantial Codimension Restrictions and the Nonexistence of Tightness for Certain Smooth Manifolds

Let $f: M^n \to E^N$ be a tight substantial smooth immersion of a closed manifold M^n in E^N. Let z_1 be linear and $z_1 f$ nondegenerate with maximal value at a point $m \in M^n$. For convenience we assume $f(m) = 0 \in R^N = E^N$. The linear functions on E^N that vanish on the tangent space $T(M)$ form an $(N - n)$-dimensional vector space W. For each $z \in W$ the 2-jet $\mathcal{J}^2(zf)$ of zf at m is a quadratic function in local coordinates $u_1, \ldots, u_n$ for M^n. So there is a linear map

$$\eta : W \to Z$$

into the linear space Z of quadratic functions in n variables, of dimension $\frac{1}{2}n(n + 1)$. It is easy to prove (Kuiper [43, Chapter 3]) that already *TPP implies that η is injective, and hence*

$$N - n \leqslant \tfrac{1}{2}n(n + 1). \tag{8.1}$$

Next suppose $\beta_j = H_j(M, Z_2) = 0$ and f is tight again. Then, by injectivity for any nondegenerate function zf,

$$H_j(\{x \in M : zf \leqslant c\}, Z_2) = 0$$

for all $c \in R$. Therefore this number cannot jump, and zf has no critical point of index j (see Section 4).

Let $c(\beta_0, \ldots, \beta_n)$ be the maximal dimension of a linear family of quadratic functions φ in n variables $u_1, \ldots, u_n$ such that it contains a positive definite function, but such that no function of the family has index k when $\beta_k = 0$, for any k. We have

Theorem 7 (Kuiper [43]). *The dimension of a tight substantial smooth immersion $f: M^n \to E^N$ is*

$$N - n \leqslant c(\beta_0, \ldots, \beta_n) \leqslant \tfrac{1}{2}n(n + 1). \tag{8.2}$$

This follows immediately from the injectivity of the map $\eta : W \to Z$, because $\eta(W)$ has dimension $c(\beta_0, \ldots, \beta_n)$ at most.

The equality $N - n = \frac{1}{2}n(n + 1)$ is realized for the standard embedding as a Veronese manifold of $RP_n = R\mathcal{G}_{1,n}$ given in Example 1 of Section 7. A remarkable generalization of our theorem on the projective plane is

Theorem 8 (Little and Pohl [52]). *Let $N - n = \frac{1}{2}n(n + 1)$. If the smooth tight immersion $f: M^n \to E^N$ is substantial and has the two-piece property (TPP; tightness is not needed), then $M^n = RP_n$ and f is the standard embedding of $M^n = RP_n$ of Example 1, Section 7.*

For the sphere S^n we have $c(\beta_0, \ldots, \beta_n) = c(1,0,0, \ldots, 0,1) = 1$ and the codimension of a tight smooth immersion is 1. Moreover the image is the boundary of a convex body. Consequently we have the

Theorem. *There is no tight smooth immersion of an exotic sphere Σ^n in a Euclidean space.*

For a $(p-1)$-connected closed $2p$-manifold one can compute (see Kuiper [43])

$$c(\beta_0, \ldots, \beta_{2p}) = c(1,0, \ldots, 0,1,0, \ldots, 0,1)$$

$$= \begin{cases} 4 & \text{for } n = 2p = 4, \\ 6 & \text{for } n = 2p = 8, \\ 10 & \text{for } n = 2p = 16, \\ 2 & \text{for } n = 2p \notin \{4,8,16\}. \end{cases}$$

Examples for the first three cases are the projective planes $FP_2 = F\mathcal{G}_{1,2}$ for $F = C, K,$ and O. See Part II.

This puts a strong restriction on the existence of tight immersions for $(p-1)$-connected $2p$-manifolds. For example, *the total space M^{18} of the nontrivial S^9-bundle over S^9 has no tight immersion in Euclidean space* [43]. In this connection we state the open

Problem 7 (A bold conjecture). *Let M^{2p} be a smooth $(p-1)$-connected manifold of dimension $2p \geqslant 4$. If M^{2p} has a smooth tight immersion in a Euclidean space, then M is diffeomorphic with CP_2, KP_2, OP_2 or $S^p \times S^p$.*

Although almost all closed surfaces are known to have tight smooth immersions in Euclidean spaces, and although we have seen many examples in higher dimensions, tightness still should be rare. We state this in

Problem 8. *Show that in some sense "most" closed n-manifolds have no tight smooth embeddings in Euclidean spaces.*

We mention some related theorems on total absolute curvature for certain smooth manifolds with side conditions.

Theorem of Moore [58]. *Let $f: M^n \subset E^{n+2}$ be a closed smooth orientable submanifold of codimension two, whose sectional curvature is positive on every tangent 2-plane at every point $m \in M^n$. Then M^n is a homotopy sphere.*

The more general case where only nonnegative sectional curvature is assumed was approached by Baldin and Mercuri. They try to prove that $f(M)$, if not a homotopy sphere, must be a metric product of tight spheres $S^p \subset E^{p+1}$ and $S^q \subset E^{q+1}$, $p + q = n$. They do get this conclusion under the assumption $p \neq 1 \neq q$.

Theorem of Ferus [20]. *A smooth immersion of codimension 2 of an exotic sphere* $f : \Sigma^n \to E^{n+2}$ *has total absolute curvature*

$$\tau(f) \geqslant 4.$$

The following is a good geometrical problem.

Problem 9. *Can* $\tau(f) = 4$ *be attained for an exotic sphere in codimension* 2?

I do not believe so.

9. Tight Polyhedral and Topological Immersions of Surfaces

Banchoff [1] started the study of tight polyhedral surfaces and introduced the notion of the two-piece property. He found that *the substantial codimension for closed surfaces has no limit for increasing genus.* We first describe his examples. Without saying so, we always mean *substantial embeddings.*

The following are *Banchoff's polyhedral surfaces*:

(a) The *tight Möbius band* in E^4 is obtained by a triangulation with the minimal number of five vertices. Let σ_p be the p-dimensional simplex in E^p and $\mathrm{Sk}_q\sigma_p$ its q-skeleton, the union of its subsimplexes of dimension $\leqslant q$. We put the five vertices in general position in E^4 (that is, in $\mathrm{Sk}_0\sigma_4$) and fill in five triangles. In this way we obtain, as can be immediately seen, a tight embedding f_1 of the Möbius band M, and

$$\mathrm{Sk}_1\sigma_4 \subset f_1(M) \subset \mathrm{Sk}_2\sigma_4 \subset \sigma_4 \subset E^4.$$

Suitable projections in E^4 are also embeddings and tight. See Figure 5 (a, b).

(b) *The tight projective plane* P in E^4 is obtained from a triangulation of P by 6 vertices. Then an embedding $f(P)$ is obtained by placing the 5 vertices in $\mathrm{Sk}_0\sigma_5$, and we find

$$\mathrm{Sk}_1\sigma_5 \subset f(P) \subset \mathrm{Sk}_2\sigma_5 \subset \sigma_5 \subset E^5.$$

Clearly any linear function has one maximum and one minimum on $\mathrm{Sk}_1\sigma_5$ and hence on $f(P)$, and f is therefore tight. See Figure 5 (a, b).

(c) *The TPP Möbius band* $f_2(M)$ in E^5 is obtained from the tight projective plane by deleting the interior of one triangular face. Clearly TPP is preserved by this operation, but *not tightness*. If in Figure 5(b) we delete the interior of triangle $(3, 5, 6)$, then a half space h which just avoids triangle $(1, 2, 4)$ will meet $f_2(M)$ in $h \cap f_2(M)$, which is seen to be homotopy-equivalent to the wedge of two circles, $S^1 \vee S^1$. Then injectivity is not possible in $H_1(h) \cap f_2(M) \to H_1(M) = Z_2$, and $f_2(M)$ is not tight.

(d) A *tight Klein bottle* K in E^5 can be obtained by observing that $K = P \# P$, the connected sum. This connected sum is embedded by connecting the boundaries $\partial f_1(M)$ and $\partial f_1'(M)$ of two parallel tight Möbius bands in parallel E^4's in E^5, by a cylinder of parallelograms or rectangles. A tight embedding in E^4 can be obtained in the same way from the tight Möbius band of Figure 5(b) in E^3. See Figure 5 (c) for the relevant polyhedral subdivision of K. Banchoff has another tight Klein bottle in E^5 based on a triangulation with 8 vertices, the minimal number for K.

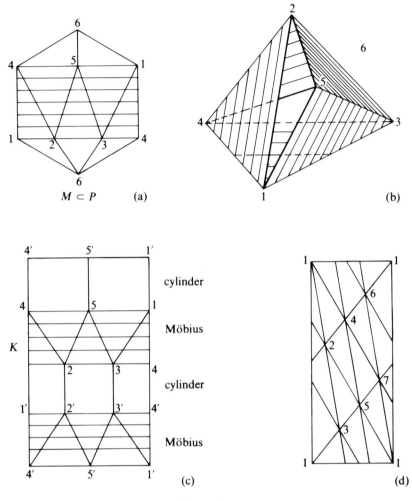

Figure 5.

(e) *The tight torus T in E^5* (Figure 5(d)). With the triangulation with seven vertices, the minimal number, the unique tight torus $f(T)$ is obtained in the same way, and

$$\mathrm{Sk}_1 \sigma_6 \subset f(T) \subset \mathrm{Sk}_2 \sigma_6 \subset \sigma_6 \subset E^6.$$

(f) If $(p-2)(p-3)$ is divisible by 12 and $g = (p-2)(p-3)/12$, then there is (see Ringel [62]) a triangulation obtained from an embedding

$$\mathrm{Sk}_1(\sigma_p) \subset T^{\# g}$$

of the connected sum of g tori (a sphere with g handles). This yields a *tight embedding* $f: T^{\# g} \subset E^p$ with $\mathrm{Sk}_1(\sigma_p) \subset f(T^{\# g}) \subset \mathrm{Sk}_2(\sigma_p) \subset E^p$. For example, there is a tight embedding of the sphere with 776 handles ($\chi = -1550$) substantially in the simplex σ_{99} with 100 vertices in E^{99}. Compare a very nice forthcoming paper of Kühnel [35].

(g) *Pohl* found that deleting the interior of one triangle of the tight torus in E^6 of example (e) gives a bounded surface which is TPP but not tight. It is a 2-disc D with one handle, $D \# T$.

Theorem of Banchoff. *All the polyhedral examples* (a), (b), (c), (d), (e) *of polyhedral tight* (*TPP in case* (c)) *surfaces are unique* (*the only ones modulo projective transformations*).

In [43] we observed that tightness of manifolds is a notion that does not require smooth or polyhedral structure. We proved the conclusion of the theorem of Chern and Lashof on spheres for topological immersions, and we showed in [45] that *any tight substantial topological immersion of the Möbius band in E^4 is necessarily the polyhedral Banchoff example* (a).

Then Kuiper and Pohl proved in [47] that *a tight topological embedding of the projective plane P substantially in E^5 is an embedding onto either the algebraic Veronese surface* (*Section 6*) *or the Banchoff surface in* (b). This was quite hard to prove, and the proof uses heavily the fundamental Theorem 1, as do all proofs on the subject. Pohl has proved since that *a tight topological torus substantially in E^6 is Banchoff's example* (e), *and a topological TPP embedding of $D \# T$ in E^6 is necessarily example* (g). As a matter of fact, Banchoff's polyhedral arguments for this carry easily over to the topological case.

We summarize the results of this section in

Theorem 9. *Banchoff's surfaces* (a), (b), (c), (e), *Pohl's surface* (g), *and the Veronese surface of Section 6 are* **all** *TPP or tight topological substantial embeddings of the given surfaces in the given Euclidean spaces.*

TPP and tight embeddings and immersions of manifolds with boundary have been studied by L. Rodriguez [63] in the smooth case, by W. Kühnel [34, 35] also for polyhedral embeddings, and by W. Lastufka [51] for topogical embeddings of surfaces with boundary. We reproduce Kühnel's tight embedding $f: D \# T \subset E^3$ with $\tau(f) = 3$ in Figure 6.

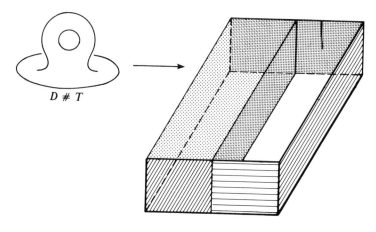

Figure 6.

In Section 11 we deal with tight continuous maps of surfaces in E^N. We have reserved the present section for polyhedral tight surfaces, and we have seen that with large genus the substantial codimension can be large. Then of course the product of such a tight surface and a tight polyhedral n-sphere in E^{q+1} gives a tight polyhedral embedding of a higher-dimensional manifold with the same high substantial codimension. We conjecture that such a splitting *of a surface* is necessary and that otherwise no high codimension can be reached. We formulate in this connection the simplest problem:

Problem 10. *Is a tight topological embedding of RP_3 in substantial codimension six necessarily smooth? Recall that there is a smooth tight embedding in E^9.* *

10. Tight Maps. Further Important Consequences of the Definition

The following theorem expresses the relation between tightness and convexity. We use a convenient

Definition. A compact space with the Z_2-homology of a point is called a *homology point*.

Theorem 10.[5] *Let the compact space X be a homology point, and let $f: X \to E^N$ be tight and substantial. Then $f(X)$ is convex and $\dim X \geqslant \dim f(X) = N$. Moreover $f(X')$ is convex for every top* set X'.*

Proof. The proof is by induction on N. We assume the statement true for all $N' < N$. By the fundamental Theorem 1, the assumptions and hence the conclusions apply to all top sets $X_z \subset X$, as their images lie in lower-dimensional linear subvarieties $E^{N'}$. So $f(X_z) = \mathcal{H}f(X_z)$ is convex for all z. As

$$\partial\mathcal{H}f(X) = \bigcup_z \mathcal{H}f(X_z),$$

this set equals

$$\partial\mathcal{H}f(X) = \bigcup_z f(X_z) \subset f(X).$$

Let (see Figure 7)

$$W = f^{-1}(\partial\mathcal{H}fX). \tag{10.1}$$

We will prove the isomorphisms

$$H_*(W) \xrightarrow{\cong} H_*(\partial\mathcal{H}fX) \xrightarrow{\cong} H_*(S^{N-1}).$$

We introduce the relation

$$R_1 \subset W \times S^{N-1},$$

which assigns to every unit vector $z^* \in S^{N-1}$ (the gradient of the linear function $z: E^N \to R$) the top set $X_z \subset W \subset X$. The compact set R_1 has the natural

---

*Kühnel claims he has a *PL*-example (added September 1980).

[5] A somewhat related problem was treated by Aumann and Kosinski [33].

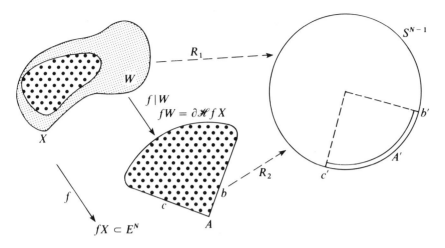

Figure 7.

projections p_W and p_S in the diagram

$$W \xleftarrow[p_W]{} R_1 \xrightarrow[p_S]{} S^{N-1}.$$

The inverse image $p_S^{-1}z*$ for any $z* \in S^{N-1}$ is homeomorphic with $X_z \subset W$, and $H_*(X_z) = H_*(pt)$ by the assumptions and Theorem 1. So all inverse images of points in S^{N-1} are homology points. The inverse image $p_W^{-1}(w)$ for $w \in W$ is homeomorphic with the convex set

$$\{z* : w \in X_z\} \subset S^{N-1},$$

whose homology is again that of a point (see the point A and the convex set A' in Figure 7, for example).

By the theorem of Vietoris and Begle (see Spanier [65, p. 344])[6] applied to p_W and p_S, we find induced isomorphisms

$$H_*(W) \xleftarrow{\cong} H_*(R_1) \xrightarrow{\cong} H_*(S^{N-1}).$$

Let $R_2 \subset \partial \mathcal{H} fX \times S^{N-1}$ be the relation which assigns to every unit vector $z* \in S^{N-1}$ the image top set $f(X_z) \subset \partial \mathcal{H} fX$. Again we find natural projections and isomorphisms

$$H_*(\partial \mathcal{H} f X) \xleftarrow{\cong} H_*(R_2) \xrightarrow{\cong} H_*(S^{N-1}).$$

Finally by composition we obtain the isomorphism induced by $f: W \to \partial \mathcal{H} fX$:

$$H_*(W) \xrightarrow{\cong} H_*(\partial \mathcal{H} fX) \simeq H_*(S^{N-1}).$$

As $\partial \mathcal{H} fX \subset fX$, we have a commutative diagram

$$
\begin{array}{ccc}
H_{N-1}(W) & \longrightarrow & H_{N-1}(\partial \mathcal{H} fX) = Z_2 \\
\downarrow & & \downarrow \\
H_{N-1}(X) = 0 & \longrightarrow & H_{N-1}(f^X)
\end{array}
$$

---

[6]Which is here seen to hold also for relations instead of maps.

and it yields the result that $\partial \mathfrak{K} fX \subset fX$ bounds in fX. Then every interior point of $\mathfrak{K} fX$ must be in fX, and fX is convex. From the homology relations follows that $\dim X \geqslant N$, and the theorem is proved. $\square$

It would be interesting to decide on

Problem 11. *Is $f^{-1}(y)$ a homology point for every point $y \in fX$ under the conditions of Theorem 10?*

For later use we prove a lemma related to this problem:

Lemma 10a. *Assume the same for f as in Theorem 10. Then the set $f^{-1}(H) \subset X$ is a homology point or empty for every hyperplane $H \subset E^N$.*

Also for any y in the interior of the convex set $fX = \mathfrak{K} fX$, the complement of $f^{-1}(y)$, $X \setminus f^{-1}(y)$, is connected.

Proof. Let h and h^- be the half spaces with intersection $H = h \cap h^-$. Then $f^{-1}(h) \cup f^{-1}(h^-) = f^{-1}(E^N) = X$, $f^{-1}(h)$, and $f^{-1}(h^-)$ are all homology points. Then so is $f^{-1}(H) = f^{-1}(h) \cap f^{-1}(h^-)$ by the Mayer–Vietoris theorem. For the second part of the lemma we chose any $x \in X$ with $f(x) \neq y$. Then there is a half space $h \subset E^N$ with $f(x) \in h$, $y \notin h$. The set $f^{-1}(h)$ meets $W = f^{-1}(\partial \mathfrak{K} fX)$ and is connected, by tightness. Then x and W can be connected inside $f^{-1}(h) \subset (X \setminus f^{-1}(y))$. The lemma is proved. $\square$

The following theorem generalizes an important part of the theorems of Fenchel and of Chern and Lashof (Section 2) concerning embeddings.

Theorem 11a. *Let $f: X = S^n \to E^N$, $N \geqslant n + 1$, be a continuous tight substantial map. Then $N = n + 1$, and $f(S^n) = \partial \mathfrak{K} fS^n$ is the boundary of a convex $(n + 1)$-dimensional body.*

Proof. Every top set X_z of f is a proper part of S^n and is therefore a homology point by tightness. Theorem 10 applies, and fX_z is convex. We then obtain relations R_1 and R_2 with analogous properties to the previous ones in a diagram where $W = f^{-1}(\partial \mathfrak{K} fX) \subset S^n$:

$$X = S^n \supset W \; - - \overset{R_1}{-} - \to S^{N-1}$$

$$fX \supset \partial \mathfrak{K} f^X$$

As before, we conclude that up to isomorphisms

$$H_*(W) \overset{\cong}{\to} H_*(\partial \mathfrak{K} fX) \simeq H_*(S^{N-1}).$$

But $H_{N-1}(W) = 0$ for $N - 1 > n$, and it can only be Z_2 for $N - 1 = n$, and then only in case W is all of S^n. Therefore $N - 1 = n$ and $W = X = S^n$, and Theorem 11a is proved. $\square$

We next solve Problem 11 for $m \leqslant 2$ in

Theorem 11b. *The continuous tight map f in Theorem 11a is a cellular map $f: S^n \to fS^n = \partial \mathcal{K} f S^n \subset E^{n+1}$ for $n \leqslant 2$.*

Recall that a *compact set* $Z \subset M$ in an n-manifold M is called *cellular* if it is the intersection of a nested sequence of open sets U_j, $j = 1, 2, \ldots$, each homeomorphic to R^N:

$$U_j \supset \overline{U}_{j+1} \supset U_{j+1} \cdots \supset \bigcup_{k=1}^{\infty} U_k = Z.$$

The identification of Z with one point then gives a homeomorphic manifold $M/Z \approx M$.

A map $f: M \to N$ between manifolds is called *cellular* if $f^{-1}(y)$ is a cellular set for each $y \in fM \subset N$. Siebenmann [64] proved that surjective cellular maps between manifolds are just the limits of homeomorphisms. In manifolds of dimension $n \leqslant 2$, every compact set that is a homology point is known to be cellular. See the survey articles of Lacher [49] and Edwards [15] for these and related matters.

Proof. Assume the conditions of Theorem 11a, and $n \leqslant 2$. Every E^0-top*-set is a point y, and $f^{-1}(y)$ is a homology point; hence it is cellular by tightness. If X' is an E^1-top*-set and $y \in fX'$, then Lemma 10a can be applied to the restriction $f | X'$, with the point y equal to $X' \cup H$ for some "hyperplane H". So $f^{-1}(y)$ is again a homology point and cellular in S^2.

This takes care of $n = 1$. Cellular sets are here points or arcs, and a tight map $f = h \circ g: S^1 \to fS^1 = \partial \mathcal{K} fS^1 \subset E^2$ is then the composition of a cellular map: $g: S^1 \to S$ and a tight embedding $h: S^1 \to fS^1 \subset E^2$. The above arguments take also care of the case $n = 2$, except for the case of a point y in the interior of an E^2-top-set image $y \in fX_z = \mathcal{K} fX_z \in E^2$. We can however apply the second part of Lemma 10a to the tight map $f | X_z$, and conclude that $f^{-1}(y)$ has a connected complement in S^2. Therefore $H_1(f^{-1}(y)) = 0$. There remains only to prove that $f^{-1}(y)$ is connected: $H_0(f^{-1}(y)) = 0$. We do this in the following hard way.

Suppose $f^{-1}(y)$ is not connected. Every point $y_1 \in \partial \mathcal{K} fX_z = \partial fX_z$ lies in some E^j-top*-set for $j \leqslant 1$ (consider top sets of fX_x). It has therefore a cellular inverse $f^{-1}(y_1)$. The quotient space obtained by the above arguments, by dividing S^2 by the equivalence that is the identification of these sets $f^{-1}(y_1)$, is homeomorphic to S^2 by cell-like-map theory. We obtain a new map with the same image, which we again call f and which restricts to a homeomorphism on the new topological circle $W = f^{-1}(\partial fX_z)$. Then f maps the topological disc X_z with boundary W onto the convex disc fX_z with boundary $fW = \partial fX_z$ in E^2. See Figure 8(a).

Take $y \in E^2$ as the origin for orthogonal coordinates u and v. Let α and β be the lines with equations $v = 0$ and $u = 0$ respectively. The sets $f^{-1}(\alpha)$ and $f^{-1}(\beta)$ are homology points by Lemma 10a. Each meets W in exactly two (end) points, in view of our modification of f. We now pinch $f^{-1}(\beta)$ into one point. Then we obtain, from X_z with boundary the topological circle W, a pair of discs D_1 and

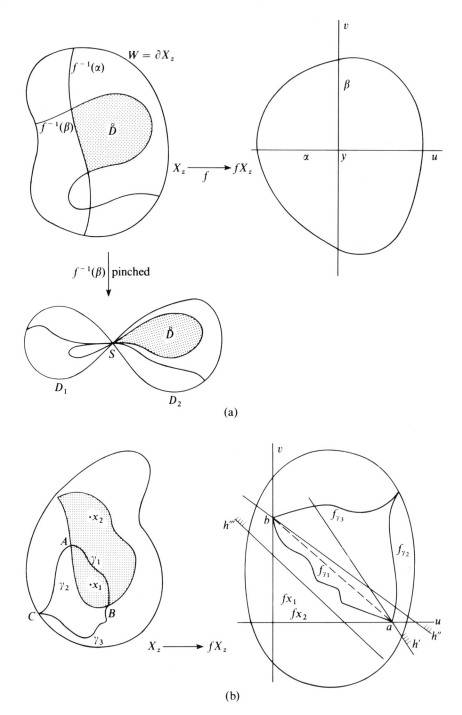

Figure 8.

D_2 connected in one point $D_1 \cap D_2 = s$, with boundaries obtained from W by identifying the pair of points $f^{-1}(\beta) \cap W$ into one point. In the point s all of $f^{-1}(y)$ is absorbed, of course, but no point of $f^{-1}(\alpha)\backslash f^{-1}(y)$. The point s disconnects the image of $f^{-1}(\alpha)$ in at least three parts because $f^{-1}(y)$ has at least two components. Then $f^{-1}(\alpha)$ divides either D_1 or D_2 in at least three open parts, one of which has no boundary parts in W.

The boundary ∂D of such an open part—an open disc, denoted by $\overset{\circ}{D}$, in X_z—is completely contained in $f^{-1}(\alpha) \cup f^{-1}(\beta)$. We now concentrate our attention on this component $\overset{\circ}{D}$, and observe that $f\partial D \subset \alpha \cup \beta$. (See Figure 8(b).) As by tightness no linear function in u and v has a maximum inside D, there must exist $a \in \alpha \cap f(\partial D)$ and $b \in \beta \cap f(\partial D)$ with $u(a) \neq 0, v(b) \neq 0$. Say $u(a) = v(b) = 1$.

Let γ_1 be an arc in $\overset{\circ}{D}$ which connects a point A such that $f(A) = a$ on the boundary ∂D to a point B such that $f(B) = b$, also on the boundary ∂D. And let $2\delta > 0$ be the minimal value of $u + v$ on the image $f(\gamma_1)$. Introduce a half space h''' by the inequality $u + v \geqslant \delta, \alpha$ half space h' by $(u - 1) + (1 - \epsilon)v \geqslant 0$, and a half space h'' by

$$u + (1 + \epsilon)(v - 1) \geqslant 0,$$

where $\epsilon > 0$ is so small that neither h' nor h'' has a point in common with the difference $fX_z\backslash h'''$. See Figure 8(b). Choose also two points x_1 and x_2 in $\overset{\circ}{D}$ on either side of γ_1, but very near to $f^{-1}(y)$, so that $f(x_1)$ and $f(x_2)$ are outside h'''. Choose finally a point $C \in W \cap f^{-1}(h'')$. As $f^{-1}(h')$ is cellular, A can be connected to C inside $f^{-1}(h')$ by (the shape of) an arc γ_2. An analogous arc γ_3 connects B with C inside $f^{-1}(h'')$.

By tightness the (shape of a) circle $\gamma_1 \cup \gamma_2 \cup \gamma_3$ can be contracted inside the cellular set $f^{-1}(h'')$. But, looking inside $\overset{\circ}{D}$, it has to pass by either x_1 or x_2 during this contraction, and neither is in $f^{-1}(h''')$, a contradiction. So $f^{-1}(y)$ is connected and thus cellular. This being the case for every $y \in f(S^2)$, Theorem 11b is proved. $\square$

For general n the tight map f in Theorem 11a need not be cellular. This we see in an

EXAMPLE. Let $M \subset S^4$ be a contractible 4-manifold with a non-simple-connected homology three-sphere ∂M as smooth boundary. Such manifolds were found by Mazur [53] and Poenaru [60]. Make a collar $C = \partial M \times I, I = [0, 1]$, on the outside of M in S^4, and let P be the closure of the remaining part. (See Figure 9.) P and M are diffeomorphic. Take in ∂M a 3-ball D^3, and in C the corresponding cylinder $D^3 \times I$.

Recall first that the closure of the complement of D^3 in ∂M is a three-manifold with boundary, which is a homology point but not simply connected. Call it $F = (C\backslash D^3)$.

Next we define a map $f: S^4 \to \partial K$ onto the boundary ∂K of a convex body K in E^5 with the following properties: $f(F \times t)$ is a *point* y_t, for $t \in I$, and the points

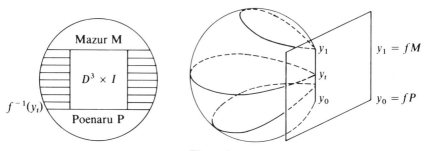

Figure 9.

y_t form a straight line segment in ∂K with end points $y_0 = f(M)$ and $y_1 = f(P)$. On the remaining part of S^4, the interior of $D^3 \times I$, we extend f to be a homeomorphism onto the remaining part of ∂K, which does not have to fulfill further conditions.

Then $f: S^4 \to E^5$ is tight, but $f^{-1}(y_t)$ is not simply connected for $0 < t < 1$. Hence f is not cellular.

11. Some Tight Maps of Surfaces

Theorem 12. *Any tight substantial map $f: M \to E^N$, $N \geqslant 4$, of the Möbius band $X = M$ is a cellular map onto Banchoff's polyhedral Möbius band with five vertices in E^4. Any tight substantial map $f: p \to E^N, N \geqslant 5$, of the real projective plane $X = P$ is a cellular map onto the Veronese surface or onto Banchoff's polyhedral projective plane with six vertices in E^5.*

Proof. Let $f: X \to E^N$ be tight and substantial as given in the theorem. Denote by j the smallest integer for which there exists some E^j-top*-set X' that is not a homology point. If there is no such j, take $j = N$ and let $X' = X$. All top sets of $f \,|\, X'$ (f restricted to X') are homology points, and they have convex images by Theorem 10. Then $W = f^{-1}(\partial\mathcal{H}fX')$ is, as before, a homology $(j - 1)$-sphere in X'. This certainly excludes $j = N \geqslant 4$, so that $X' \neq X$. As now $X' \neq X$, it excludes $j - 1 \geqslant 2$. We obtain therefore

$$j \leqslant 2.$$

Next let our X' with $H_1(X') \neq 0$ be an E^j-top*-set with image $f(X') \subset E^j$. Let η be a projection in E^N, parallel to E^j, into some E^{N-j}. Then $\eta \circ f: X \to E^{N-j}$ equals $g \circ q$, where $q: X \to Y = X\backslash X'$ is the identification of the top* set X' to one point, and where $g: Y \to E^{N-j}$ is then uniquely defined by the above equality. Also g is tight, because f was tight. If $X = M$, then Y is a homology point of dimension 2, and $g(Y)$ is convex of dimension $\leqslant 2$, substantial in E^{N-j} by Theorem 10. Then $N - j \leqslant 2$. As $j \leqslant 2$, $N - j \leqslant 2$, $N \geqslant 4$, then $N = 4, j = 2$.

If $X = P$, then Y has the homology of S^2 and is of dimension 2. The arguments for the proof of Theorem 11 yield that $g(y)$ spans a k-plane of dimension $k = 3$ at most. From $j \leqslant 2$, $N - j \leqslant 3$, $N \geqslant 5$, it follows that $N = 5$, $j = 2$. We conclude for $X = M$ and for $X = P$:

Lemma 12a. $j = 2$.

We now continue the proof of the theorem. In both cases $X = M$ and $X = P$ we denote by $X_0 \subset X$ the set of points which are in E^i-top*-sets for $i = 0$ and $i = 1$, or in E^2-top*-sets that are homology points. Then as in the proof of Theorem 11b, we deduce that f restricted to X_0 is cellular. For all points in X_0 we can divide by the cellular equivalence, so that $f | X_0$ can be assumed to be an embedding.

Observe that in X_0 we find all top sets of E^2-top *-sets and also the whole circle $\partial \mathcal{K} f X'$, that carries an essential 1-cycle of X', in case X' is not a homology point. We then refer to the (hard) proof due to Kuiper and Pohl [45, 47], concerning topological embedding of X, where only such points and cycles (in X_0) are used (and there are enough!) to conclude to the standard models. Following those proofs, we see that X_0 must exhaust X, $X_0 = X$, and the conclusion of Theorem 12 follows. □

Part II. Manifolds of Geometrical Class Three

12. Examples

(See (7.2))

Theorem 13a. *All standard models* $M_0 = M_0^{2d} \subset E^{3d+2}$ *of projective planes over* R, C, K, *and* O *respectively for* $d = 1, 2, 4, 8$, *have geometrical class three.*

All standard models $F\mathcal{G}_{1, m-1} \subset E^N$ *for projective* $(m - 1)$-*spaces* FP_{m-1} *over* R, C, *and* K *have geometrical class* m. *All are therefore tight as well.*

Proof for projective spaces over R, C, and K. The orthogonal projector A (See (7.2)) on a one-dimensional F-subspace of F^m can be expressed in terms of any unit vector q in that F-subspace, as a matrix product over F:

$$A = \begin{bmatrix} q_1 \\ \vdots \\ q_m \end{bmatrix} (\bar{q}_1 \cdots \bar{q}_m) = q^t \bar{q}, \qquad {}^t\bar{q}q = \|q\|^2 = 1. \tag{12.1}$$

Clearly $A^2 = A = {}^t\bar{A}$. These operators A are the points of the projective space

$$M_0 = F\mathcal{G}_{1, m-1} = FP_{m-1} = \{ A = q^t\bar{q} : \|q\| = 1 \} \subset E^N \subset F^{m^2} = R^{dm^2},$$

which is embedded substantially in E^N, $N = m + 1 + \frac{1}{2}dm(m - 1)$. It is the base space of a fibre bundle with total space the $(dm - 1)$-sphere

$$S^{dm-1} = \{ q \in F^m : \|q\| = 1 \} \subset R^{dm}. \tag{12.2}$$

The fibre of q is the "great" $(d - 1)$-sphere

$$S^{d-1}(q) = \{ (q_1 b, \ldots, q_m b) : b \in F, |b| = 1 \} \tag{12.3}$$

The geometrical class of the submanifold M_0 is the greatest number of tangent hyperplanes in R^{dm^2} or in E^N that occur in pencils of hyperplanes. Such a pencil is given by equations

$$\varphi_1 - \lambda\varphi_2 = 0, \qquad \lambda \in R \cup \infty, \tag{12.4}$$

where φ_1 and φ_2 are functions of degree at most one in the coordinates of R^{dm^2}. The hyperplane is tangent for a value λ_0 if and only if the restriction of the function φ_1/φ_2 to M_0 has a critical value λ_0. We can exclude the critical value $\lambda_0 = \infty$ by a suitable projective change of the parameter λ. After restriction to M_0 of (12.1), the functions φ_1 and φ_2 are expressed linearly in the real coordinates of the elements $a_{ij} = q_i\bar{q}_j$ of A in (12.1). The functions φ_1 and φ_2 therefore lift to quadratic functions in the md real coordinates $q_{11}, \ldots, q_{1d}, \ldots, q_{md}$ of the vector $q = (q_1, \ldots, q_m)$. We denote these last functions by $[\varphi_1]$ and $[\varphi_2]$.

Next we observe that the functions φ_1/φ_2 and $[\varphi_1]/[\varphi_2]$ have a critical value λ if and only if the quadratic hypersurface in real projective $(md - 1)$-space given by the equation

$$[\varphi_1] - \lambda[\varphi_2] = 0 \tag{12.5}$$

in homogeneous coordinates $q_{11}, \ldots, q_{mm}$ is singular. This is the case if and only if the coefficients of the quadratic form on the left in (12.5) (which can be assumed to be homogeneous because $1 = \sum q_{rs}^2$) has determinant

$$\det([\varphi_1] - \lambda[\varphi_2]) = 0. \tag{12.6}$$

Equation (12.6) has md roots, counting multiplicities. But $[\varphi_1]/[\varphi_2]$ is constant over each great $(d - 1)$-sphere (12.3), because $b\bar{b} = |b|^2 = 1$ implies

$$(q_i b)(\overline{q_j b}) = q_i b\bar{b}\bar{q}_j = q_i\bar{q}_j . \tag{12.7}$$

For almost every pencil (12.4), each root of (12.6) has multiplicity d. Therefore there are at most $dm/d = m$ different roots, corresponding to at most m real tangent hyperplanes. For almost every pencil of *parallel* hyperplanes this number cannot be less then $\beta(M_0) = m$. Therefore the geometrical class is m. Moreover $\tau(M_0) = \beta(M_0) = m$ and M_0 is tight. $\square$

The *octave plane* was discovered by G. Hirsch [27] as a 16-dimensional manifold in which there is a set of 8-spheres fulfilling the axioms for "lines" in a "projective plane". Freudenthal [24] gave the algebraic model (7.2). We now use the methods of Tai [66] (whose demonstration was not quite correct) and prove first

Proposition 13b. *The octave-plane model* $M_0 = M_0^{16} \subset E^{26}$ *in (7.2) is tight.*

Recall that the octaves or Cayley numbers form a nonassociative algebra O of dimension 8 over R, containing R. Any octave $q \notin R$ generates a subalgebra ismorphic to C. Any two octaves q_1 and q_2 generate an associative subalgebra isomorphic to K, if not to C or R.

Proof of the proposition. We define the model M_0^{16} in a way which agrees with Freudenthal's definition (7.2):

$$OP_2 = M_0 = \{ A = q'\bar{q} : {}'q\bar{q} = \|q\| = 1, (q_1 q_2) q_3 = q_1(q_2 q_3)\} \subset E^{26}, \quad (12.8)$$

where

$$q = \begin{bmatrix} q_1 \\ q_2 \\ q_3 \end{bmatrix}, \quad {}'\bar{q} = (\bar{q}_1, \bar{q}_2, \bar{q}_3),$$

q_2, q_2, and $q_3 \in O$ octaves. By the *associativity assumption* $(q_1 q_2)q_3 = q_1(q_2 q_3)$, q_1, q_2, and q_3 generate an associative subalgebra isomorphic to R, C, or K in O. Clearly then $A = {}'\bar{A}$ and $A = A^2$, but A cannot be considered as an operator on O. There is no Hopf fibre bundle either, but supernumerary coordinates

$$(q_1, q_2, q_3) \quad \text{and} \quad (q_1 b, q_2 b, q_3 b) \quad (12.9)$$

represent the same point $A \in M_0$ in case the octave b, $b\bar{b} = 1$, generates with q_1, q_2, and q_3 an associative subalgebra of O. Those unit octaves b form a 3-sphere in case q_1, q_2, and q_3 generate a quaternion algebra, and a 7-sphere in case q_1, q_2 and q_3 generate a complex (C) or real (R) algebra.

M_0 lies in the unit sphere with equation

$$\text{Trace } X^2 = \sum_{i=1}^{3} \zeta_i^2 + 2 \sum_{j=1}^{3} |x_j|^2 = i \quad (12.10)$$

of the "Euclidean" space E^{26} of matrices

$$X = {}'\bar{X} = \begin{bmatrix} \zeta_1 & x_3 & \bar{x}_2 \\ \bar{x}_3 & \zeta_2 & x_1 \\ x_2 & \bar{x}_1 & \zeta_3 \end{bmatrix}, \quad \zeta_i \in R, \quad X_i \in O, \quad \text{trace } X = \sum_i \zeta_i = 1. \quad (12.11)$$

An example of a "line" in OP_2 in this model M_0 is the *round 8-sphere*, obtained by taking $q_1 = 0$. It has equations

$$\zeta_1 = 0, \zeta_2 + \zeta_3 = x_2 = x_3 = 0 \quad \text{and} \quad \zeta_2^2 + \zeta_3^2 + 2x_1{}^2 = 1.$$

Any other "line" can be seen to be congruent to this one by a "unitary transformation" of E^{26} (see below).

We find three coordinate charts together covering OP_2 by those choices of b in (12.9) that make q_1, q_2, or q_3 real positive. For example for $a_{11} = q_1 \bar{q}_1 \neq 0$ we can *assume* $q_1 = |q_1| > 0$. As $q_1 q_1 + q_2 q_2 + q_3 q_3 = 1$, a coordinate chart is then given by the open 16-ball

$$\{(q_2, q_3) : q_2 \bar{q}_2 + q_3 \bar{q}_3 < 1\} \subset O^2 = R^{16} \quad (12.12)$$

By cyclic permutation of q_1, q_2, q_3 we obta!n the other two coordinate charts. For each chart there are certain "unitary" transformations U which are isometries and leave invariant M_0, as well as that chart of E^{26}:

$$X \to U X' \bar{U}, \quad U' \bar{U} = 1.$$

For the first chart given above they are of the following kinds: (O-dilation about

$(1, 0, 0)$,

$$U = \begin{bmatrix} 1 & 0 & 0 \\ 0 & a & 0 \\ 0 & 0 & \bar{a} \end{bmatrix}, \qquad |a| = 1, \quad a \in O, \tag{12.13}$$

and real rotation about $(1, 0, 0)$

$$U = \begin{bmatrix} 1 & 0 & 0 \\ 0 & \cos\alpha & -\sin\alpha \\ 0 & \sin\alpha & \cos\alpha \end{bmatrix}, \qquad \alpha \in R. \tag{12.14}$$

Their effects on the given chart coordinates are respectively

$$U : (q_2, q_3) \to (aq_2, \bar{a}q_3) \tag{12.13'}$$

and

$$U : (q_2, q_3) \to (q_2 \cos\alpha - q_3 \sin\alpha, q_2 \sin\alpha + q_3 \cos\alpha). \tag{12.14'}$$

Such transformations for all three charts generate the 52-dimensional Lie group F_2 of "unitary" transformations of E^{26} that induce isometries on (M_0^{16}, E^{26}).

We want to show that almost every linear function φ on E^{26} has three critical points on $M^{16} \subset E^{26}$. Freudenthal [24] uses elements of the above group F_2 to bring any φ into diagonal form:

$$\varphi = \lambda_1 \zeta_1 + \lambda_2 \zeta_2 + \lambda_3 \zeta_3 = \lambda_1 q_1 \bar{q}_1 + \lambda_2 q_2 \bar{q}_2 + \lambda_3 q_3 \bar{q}_3, \tag{12.15}$$

Assuming the real numbers λ_1, λ_2, and λ_3 different, we find in the first coordinate chart for φ (with $q_1 = |q_1| > 0$) the expression

$$\varphi = \lambda_1 (1 - q_2 \bar{q}_2 - q_3 \bar{q}_3) + \lambda_2 q_2 \bar{q}_2 + \lambda_3 q_3 \bar{q}_3$$

with exactly *one critical point*, namely at $q_2 = q_3 = 0$, and with critical value λ_1. It is analogous for the other charts. So φ has exactly the three critical points $(1, 0, 0)$, $(0, 1, 0)$, and $(0, 0, 1)$, and Proposition 13b is proved. $\square$

There remains to prove the harder

Proposition 13 c. The octave-plane model $M_0 \subset E^{26}$ has geometrical class three.

Proof. As before, we consider a pencil of hyperplanes

$$\varphi_1 - \lambda \varphi_2 = 0, \qquad \lambda \in R \cup \infty. \tag{12.4}$$

Suppose $\lambda = 0$ and $\lambda = \infty$ give tangent hyperplanes $\varphi_1 = 0$ and $\varphi_2 = 0$ at points A_1 and A_2 in M_0 respectively. So we assume at least two tangent hyperplanes. The R-linear functions φ_1 and φ_2 in the real components of $\{q_i \bar{q}_j\}$ are homogeneous quadratic over R in the real components of q_1, q_2, and q_3. They can be assumed to take the values $\varphi_1(A_1) = \varphi_2(A_2) = 0$. After suitable unitary transformations φ_1 can be again assumed to be in the diagonal form (12.15) in coordinates for which $(q_1, q_2, q_3) = (1, 0, 0)$ is A_1. The points $(0, 1, 0)$ and $(0, 0, 1)$ in M_0 are called B and C (see Figure 10). With $\varphi_1(A_1) = 0$ we obtain the expression

$$\varphi_1 = \lambda_2 q_2 \bar{q}_2 + \lambda_3 q_3 \bar{q}_3, \qquad \lambda_2, \lambda_3 \in R. \tag{12.16}$$

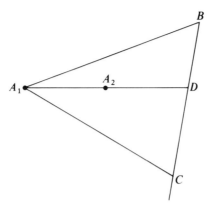

Figure 10

Assuming the projective plane structure (see Freudenthal [24]), let D be the intersection of the "lines" A_1A_2 and BC in OP_2. If $D \neq B$, we place B at the point D as follows. An O-dilatation (12.3) sends $D = (0, q_2, q_3)$ into $D' = (0, aq_2, \bar{a}q_3)$ and leaves (12.16) invariant. We want D' to be a "real point", assuming that D is not yet real:

$$(aq_2)(\bar{a}q_3)^{-1} \in R.$$

By octave calculus we have

$$(aq_2)(\bar{a}q_3)^{-1} = (aq_2)(q_3^{-1}a) = a(q_2q_3^{-1})a.$$

As the octave $z = q_2q_3^{-1}$ generates an algebra isomorphic to C, we can solve for the number a from $|a| = 1$ and $aza \in R$ in that algebra:

$$a = (z/|z|)^{-1/2}$$

So we can assume now D real. Its coordinates are, for example,

$$D = (0, \cos\alpha, \sin\alpha), \qquad \alpha \in R.$$

Applying the real rotation (12.14), we find a new coordinate system (q_1, q_2, q_3) for which the "line" A_1A_2 is given by $q_3 = 0$, whereas φ_1 has the new expression

$$\varphi_1 = \lambda_2(q_2\cos\alpha - q_3\sin\alpha)(\bar{q}_2\cos\alpha - \bar{q}_3\sin\alpha)$$
$$+ \lambda_3(q_2\sin\alpha + q_3\cos\alpha)(\bar{q}_2\sin\alpha + \bar{q}_3\cos\alpha).$$

Substituting $Q_1 = q_1/|q_3|$, $Q_2 = q_2/|q_3|$, we obtain

$$\varphi_1 = q_3\bar{q}_3\left(\gamma_1 Q_2\bar{Q}_2 + \gamma_2(Q_2 + \bar{Q}_2) + \gamma_3\right)$$

for suitable constants γ_1, γ_2, and γ_3 in R. We complete the affine space of the 16 real coordinates underlying the sets of pairs of octaves (Q_1, Q_2) to a real projective space RP_{16} with nonhomogeneous coordinates $(u_1, \ldots, u_8, v_1, \ldots, v_8)$ corresponding to (Q_1, Q_2) and with homogeneous coordinates $(u_1, \ldots, u_8, v_1, \ldots, v_8, s)$, say. Then we observe that the factor $\gamma_1 Q_2\bar{Q}_2$

$+ \gamma_2(Q_2 + \overline{Q}_2)s + \gamma_3 s^2$ of φ_1 involves only nine linear functions of the homogenous coordinates. This does not change if we apply any "unitary transformation" that preserves the third coordinate chart, that is,

$$U = \begin{pmatrix} b & & \\ & \overline{b} & \\ & & 1 \end{pmatrix} \quad \text{or} \quad U = \begin{pmatrix} \cos\alpha & \sin\alpha & \\ \sin\alpha & \cos\alpha & \\ & & 1 \end{pmatrix}.$$

Observe that also $q_3\overline{q}_3$ is invariant under these last transformations. As the same reasoning applies to φ_2 as to φ_1, we may assume coordinates q_1, q_2, q_3, for which the projective line $A_1 A_2$ has equation $q_3 = 0$. Moreover we can assume that φ_1 as well as φ_2 is, except for a factor $q_3\overline{q}_3$, a homogenous quadratic function in the seventeen coordinates $u_1, \ldots, u_8, v_1, \ldots, v_8, s$, involving only nine linear functions of these coordinates. The equation for tangent hyperplanes (unknown λ)

$$\det\left(\frac{\varphi_1}{q_3\overline{q}_3} - \frac{\lambda\varphi_2}{q_3\overline{q}_3} \right) = 0$$

has seventeen roots. But $\lambda = 0$ and $\lambda = \infty$ each has multiplicity eight. To see this for $\lambda = 0$, assume the nine linear functions of coordinates to be the coordinates $v_1, \ldots, v_8, s$ and look at the 17×17 matrix of coefficients of the quadratic function $\varphi_1/q_3\overline{q}_3$. There remains one root, corresponding to one more tangent hyperplane at some point outside the line $A_1 A_2$ ($q_3 = 0$). There cannot be another tangent hyperplane at a point on the "line" $A_1 A_2$, because this "line" is a round 8-sphere in E^{26}.

Thus the geometrical class is three, and Proposition 13c is proved. □

In Sections 13 and 14 we will show that the models for projective planes of Section 12 are nearly unique as tight smooth submanifolds of codimension $d + 2$.

13. Tight Manifolds Like Projective Planes

First recall that if $f: M^n \subset E^N$ is a closed smooth submanifold of geometrical class two, substantially embedded in E^N, then its total absolute curvature is $\tau(f) = 2$, and it is the boundary of a convex body in E^{n+1} by Theorem 11. Class three is the next case. If M is not a sphere and f is a smooth embedding of geometrical class three, then the total absolute curvature is $\tau(f) = 3$, and almost every linear function on E^N restricts to a nondegenerate function on M with three critical points. Closed manifolds M^n that admit nondegenerate continuous functions with exactly three critical points are called *manifolds like projective planes*. Eells and Kuiper [16] gave many examples of M^{2d}, necessarily all of dimensions $2d = 2, 4, 8$ or 16. They are obtained from R^{2d} under compactification by a d-sphere. The projective planes $M^{2d} = FP_2$ for $F = R, C, K, O$ are special cases. They have the models of Section 12 as tight embeddings $f: M_0^{2d} \subset E^{3d+2}$, $\tau(f) = 3$. We prove in this Section:

Theorem 14. *If $f: M^{2d} \to E^N$ is a tight substantial continuous embedding of a manifold like a projective plane, then*

$$N \leqslant 3d + 2.$$

Proof. Let $X = fX$ be an E^j-top*-set of the given tight embedding f, which is not homologous to a point and for the smallest possible value j. The (tight) top sets of X are homology points, and convex by Theorem 10. Their union is a $(j - 1)$-sphere $S^{j-1} = \partial \mathcal{H}X \subset X$. If S^{j-1} bounds in M^{2d} in homology, then it also does so in X by tightness. Then every point inside S^{j-1} belongs to X : X is convex, a contradiction. As S^{j-1} does not bound in M^{2n} and as it cannot represent a $2d$-cycle, it must represent a d-cycle in M^{2d}, and

$$j = d + 1.$$

Keeping in mind the same X in an E^{d+1}, let η be a projection of E^N parallel to that E^{d+1} into some E^{N-d-1}. Let $q : M^{2d} \to Y = M^{2d}/X$ be the identification of X into one point, and let $g : Y \to E^{N-d-1}$ be the unique map defined by

$$\eta \circ f = g \circ q : M^{2d} \to E^{N-d-1}.$$

Then g is a tight map of the homology $2d$-sphere Y, substantially into E^{N-d-1}. By the arguments in the proof of Theorem 11 we get

$$N - d - 1 \leqslant 2d + 1;$$

hence

$$N \leqslant 3d + 2. \quad \square$$

14. Tight Smooth Submanifolds M^{2d} Like Projective Planes. Uniqueness

Theorem 15. *Let* $f : M^{2d} \subset E^{3d+2}$ *be a tight smooth substantial embedding of a closed manifold with Morse number* $\beta(M) = 3$ *("like a projective plane"). Then* M^{2d} *is algebraic. It is the union of its* E^{d+1}-top-sets, *smooth d-spheres* S^d *that are quadratic d-manifolds. They constitute the "lines" of a projective plane structure in* M^{2d}. *For* $d = 1,2$ *the manifold* M^{2d} *equals the model* M_0^{2d}, *of geometrical class three, up to a real projective transformation in* E^{3d+2}.

The proof will develop in a sequence of six lemmas. We assume the conditions of the theorem for the embedded $fM = M = M^{2d} \subset E^{3d+2}$, and we assume the standard model $M_0 = M_0^{2d} \subset E^{3d+2}$ with all its global isometrices given. A point $p \in M$ is called a *nondegenerate extreme point* of M in case some linear function z_1 on E^{3d+2} has a nondegenerate maximal value on M at p. We now formulate

Lemma 1. *The two-jet* $\mathcal{G}_p^2 (M)$ *of* M *at a nondegenerate extreme point* p *of* M *is unique up to affine transformations. In particular,* M *and* M_0 *can be assumed osculating at* p *once a suitable affine transformation of* E^{3d+2} *is applied to* M.

Proof. We use coordinates

$$(u_1, \ldots, u_{3d+2}) \quad \text{for } E^{3d+2}, \tag{14.1}$$

with p at the origin and $u_{2d+1} = \cdots = u_{3d+2} = 0$ defining the tangent space $T_p(M)$; $d = 1, 2, 4,$ or 8. The functions $u_{2d+1}, \ldots, u_{3d+2}$ are the basis of the linear space W of those linear functions that vanish on $T_p(M)$. The map $\eta : W \to Z$ assigns to any $z \in W$ the 2-jet $\mathcal{G}_p^2 (zf)$, a homogenous quadratic function in $'y = (u_1, \ldots, u_{2d})$, which can be expressed as

$$\eta(z) = {'y}Ay,$$

where $A = {'A}$ is a symmetric $2d \times 2d$ matrix and $'$ stands for matrix transpose. Already by the two-piece property, η is injective. By tightness all quadratic forms in $\eta(W)$ have index 0, d, or $2d$ (Section 8, Theorem 7). As p is a nondegenerate extreme point, the indices 0 and $2d$ do occur. It was proved in Kuiper [43] (see Hurwitz [28]) that such linear families of real quadratic functions are unique if they have the greatest possible dimension, which is $d + 2$.

Also for a suitable linear choice of the coordinates $u_1, \ldots, u_{2d}$, the following normal form for the family $\eta(W)$ can be built:

$${'y}Ay,$$

$$A = \begin{pmatrix} w_1 1 & B \\ {'B} & w_2 1 \end{pmatrix}$$

$$= w_1 A_1 + w_2 A_2 + s_1 A_3 + \cdots + s_d A_{d+2}$$

$$(14.2)$$

where as before $'y = (u_1, \ldots, u_{2d})$, and B is the upper left $d \times d$ matrix in

$$\begin{pmatrix}
s_1 & -s_2 & -s_3 & -s_4 & -s_5 & -s_6 & -s_7 & -s_8 \\
s_2 & s_1 & -s_4 & s_3 & -s_6 & s_5 & -s_8 & s_7 \\
s_3 & s_4 & s_1 & -s_2 & -s_7 & s_8 & s_5 & -s_6 \\
s_4 & -s_3 & -s_2 & s_1 & s_8 & s_7 & -s_6 & -s_5 \\
s_5 & s_6 & s_7 & -s_8 & s_1 & -s_2 & -s_3 & s_4 \\
s_6 & -s_5 & -s_8 & -s_7 & s_2 & s_1 & s_4 & s_3 \\
s_7 & s_8 & -s_5 & s_6 & s_3 & -s_4 & s_1 & -s_2 \\
s_8 & -s_7 & s_6 & s_5 & -s_4 & -s_3 & s_2 & s_1
\end{pmatrix}$$

As $\eta : W \to \eta(W)$ is bijective, we can choose the coordinates $u_{2d+1}, \ldots, u_{3d+2}$ in (12.17) so that

$$\mathcal{J}_p^2(u_{2d+j}, M) = {'y}A_j y, \qquad j = 1, \ldots, d + 2. \qquad (14.3)$$

By the existence of this unique normal form for $\mathcal{J}_p^2(M)$, Lemma 1 is now proved. $\square$

Observe that the coordinates $u_1, \ldots, u_{3d+2}$ are not uniquely determined by (14.2) and (14.3) in view of the isometries of M_0. The 2-jet $\mathcal{J}_p^2(M)$ is also invariant under linear transformations of the kind

$$u'_i = u_i + \sum_j \alpha_{ij} u_{2d+j}, \qquad i \in \{1, \ldots, 2d\},$$

$$u'_{2d+j} = u_{2d+j}, \qquad j \in \{1, \ldots, d + 2\}. \qquad (14.4)$$

In calculations we will sometimes use the following notation

$${'y} = (u_1, \ldots, u_d, v_1, \ldots, v_d) \quad \text{instead of} \quad (u_1, \ldots, u_{2d}), \qquad (14.5)$$

and

$$\|u\|^2 = \sum_{j=1}^{d} u_j^2, \qquad \|v\|^2 = \sum_{j=1}^{d} v_j^2.$$

As an explicit illustration of (14.2) and (14.3) we obtain then, for

$$z = w_1 u_{2d+1} + w_2 u_{2d+2} + s_1 u_{2d+3} + \cdots + s_d u_{3d+2} \in W, \qquad (14.6)$$

with real constants $w_1, w_2, s_1, \ldots, s_d$, and $d = 4$ (it is analogous for $d = 1, 2$, and 8)

$$\begin{aligned}
\mathcal{G}_p^2(z, M) = \eta(z) &= w_1(u_1^2 + u_2^2 + u_3^2 + u_4^2) + w_2(v_1^2 + v_2^2 + v_3^2 + v_4^2) \\
&+ 2s_1(u_1 v_1 + u_2 v_2 + u_3 v_3 + u_4 v_4) \\
&+ 2s_2(-u_1 v_2 + u_2 v_1 - u_3 v_4 + u_4 v_3) \\
&+ 2s_3(-u_1 v_3 + u_2 v_4 + u_3 v_1 - u_4 v_2) \\
&+ 2s_4(-u_4 v_1 - u_2 v_3 + u_3 v_2 + u_4 v_1).
\end{aligned} \qquad (14.7)$$

The function z in (12.22) has a degenerate singularity on M at p in case (see (14.27))

$$\det A = \left(w_1 w_2 - \sum_{j=1}^{d} s_j^2\right)^d = 0. \qquad (14.8)$$

To prove this formula for $\det A$, observe that it holds for $s_2 = s_3 = \cdots = s_d = 0$, and observe as well that it is invariant under orthogonal tranformations on $(s_1, \ldots, s_d)$ by the special form of B.

The quadratic cone $C^* \subset W$ defined by (14.8) divides its complement in W into three parts:

$$w_1 w_2 - \sum_{j=1}^{d} s_j^2 < 0, \quad zf \text{ has index } d \text{ at } p;$$

$$> 0, \quad w_1 + w_2 < 0, \qquad zf \text{ is maximal at } p;$$

$$> 0, \quad w_1 + w_2 > 0. \qquad zf \text{ is minimal at } p.$$

The rays (straight lines) of C^* are the points of a d-sphere $S^d(C^*)$. The function zf is *degenerate maximal* on M at p if

$$w_1 w_2 - \sum_{j=1}^{d} s_j^2 = 0, \qquad w_1 + w_2 < 0.$$

The supporting half spaces given by $z \geq 0$ for such z bound (envelop) a *dual quadratic cone* $C \subset E^{3d+2}$ with equations

$$C : \left\{ \begin{aligned} u_{2d+1} u_{2d+2} - \sum_{j=3}^{d+2} u_{2d+j}^2 &= 0 \\ u_{2d+1} + u_{2d+2} &\geq 0 \end{aligned} \right\}. \qquad (14.9)$$

It consists of half $(2d + 1)$-planes through p, all bounded by the same $2d$-plane $T_p(M)$ in E^{3d+2}. They are parametrized by the points of $S_d(C^*)$. (See Figure 11.)

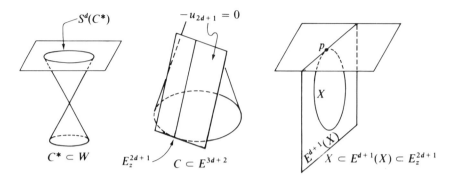

Figure 11.

We can now formulate:

Lemma 2. *Let zf be degenerate singular at p with maximal value zero. Then the top set $X = M_z$ carries an essential d-cycle of M, and it is a E^{d+1}-top-set.*

Proof. There is a group of isometries of the model (M_0, E^{3d+2}) with a subgroup that fixes the point p. It is so large that it acts transitively on the rays of C^*. We described this group in Section 12 for the (hardest) case of the plane. The group of isometries of (p, M_0, E^{d+2}) yields automorphisms of $\mathscr{S}_p^2(M) = \mathscr{S}_p^2(M_0)$ and automorphisms of the family (14.2). Therefore we may *assume* the special case

$$z = -u_{2d+1}. \tag{14.10}$$

The top set

$$X = M_z = M \cap (-u_{2d+1} \geqslant 0)$$

lies then in the cone C given by (14.9) and hence in the $(2d + 1)$-plane

$$E_z^{2d+1} : 0 = u_{2d+1} = 0 = u_{2d+j}, j = 3, \ldots, d + 2. \tag{14.11}$$

Observe that X is the limit for $\epsilon \downarrow 0$ of half-space intersections

$$X_\epsilon = M \cap (-u_{2d+1} + \epsilon u_{2d+2} \geqslant -\epsilon).$$

But the function on M

$$-u_{2d+1} + \epsilon u_{2d+2} = \sum_{i=1}^{d} u_i^2 + \epsilon \sum_{j=1}^{d} v_j^2 + \text{h.o. terms}$$

has at p a nondegenerate critical point of index d with value 0. Then by Morse theory and the definition of tightness we conclude that X_ϵ carries an essential d-cycle of

$$H_d(X_\epsilon, Z_2) \hookrightarrow H_d(M, Z^2).$$

By continuity of Čech homology the limit X also carries this d-cycle. We will decide later that the connected set X is a manifold. We begin and prove first that

X has a tangent d-plane at p. We use the notation of (14.5) and the $\mathcal{O}$-*symbol of Landau* defined as follows. If $y \geqslant 0$, then $x = \mathcal{O}(y)$ means that $c > 0$ exists for which $|x| \leqslant cy$. We will use the symbol also in equations (but we do it with care).

Recall that on M at p

$$\mathcal{J}_p^2(u_{2d+1}) = u_1^2 + \cdots + u_d^2 = \|u\|^2.$$

Hence

$$u_{2d+1} = u_1^2 + \cdots + u_d^2 + \mathcal{O}\left(\left(u_1^2 + \cdots + u_d^2 + v_1^2 + \cdots + v_p^2\right)^{3/2}\right)$$

$$= \|u\|^2 + \mathcal{O}\left(\left(\|u\|^2 + \|v\|^2\right)^{3/2}\right).$$

On the connected top set X near p we find by substituting $u_{2d+1} = 0$

$$\|u\|^2 + \mathcal{O}\left(\left(\|u\|^2 + \|v\|^2\right)^{3/2}\right) = 0.$$

By $\mathcal{O}$-calculus it follows that

$$\|u\|^2 = \mathcal{O}(\|v\|^3), \qquad u = \mathcal{O}(\|v\|^{3/2}), \qquad u_i = \mathcal{O}(\|u\|) = \mathcal{O}\left((\|v\|)^{3/2}\right). \quad (14.12)$$

We conclude that X at p has a tangent d-plane $T_p(X)$ in $T_p(M)$ with equations

$$u_1 = \cdots = u_d = 0 = u_{2d+1} = \cdots = u_{3d+2} = 0.$$

Next we will prove that X near p is contained in some smooth d-manifold neighborhood $U \subset M$ with equations

$$u_i = \zeta_i(v_1, \ldots, v_d), \qquad i = 1, \ldots, d, \quad (14.13)$$

in the local coordinates $u_1, \ldots, u_d, v_1, \ldots, v_d$ for M. Assume U is an open d-ball. We deal with the case $d = 4$ only. The other cases are analogous. Assuming $\|v\| \neq 0$, we get from (14.7) the equations

$$\tfrac{1}{2}\|v\|^{-1} u_{2d+3} = \|v\|^{-1}\left[u_1 v_1 + u_2 v_2 + u_3 v_3 + u_4 v_4 + \text{h.o. terms}\right],$$

$$\tfrac{1}{2}\|v\|^{-1} u_{2d+4} = \|v\|^{-1}\left[-u_1 v_2 + u_2 v_1 - u_3 v_4 + u_4 v_3 + \text{h.o. terms}\right],$$

$$\tfrac{1}{2}\|v\|^{-1} u_{2d+5} = \|v\|^{-1}\left[-u_1 v_3 + u_2 v_4 + u_3 v_1 - u_4 v_2 + \text{h.o. terms}\right],$$

$$\tfrac{1}{2}\|v\|^{-1} u_{2d+6} = \|v\|^{-1}\left[-u_1 v_4 - u_2 v_3 + u_3 v_2 + u_4 v_1 + \text{h.o. terms}\right].$$

The coefficients of u_1, u_2, u_3, u_4 on the right-hand sides form an orthogonal matrix with determinant 1. So the right-hand sides define a local diffeomomorphism from $(R^4, 0)$ to $(R^4, 0)$ depending smoothly on $v_1, \ldots, v_4$. We can solve for $u_1, \ldots, u_4$, substitute $u_{2d+3} = u_{2d+4} = u_{2d+5} = u_{2d+6} = 0$, and obtain, as desired, (14.13).

The points in U that belong to X are those for which moreover $u_{2d+1} = 0$. In fact we will prove that all of U belongs to X. For that we consider the analogous top set M_{z+} for $z^+ = -u_{2d+2}$ and obtain $X^+ = M_{z+}$. It has a tangent d-plane at r with equations

$$v_1 = \cdots = v_d = u_{2d+i} = 0, \qquad i = 1, \ldots, d + 2,$$

in a $(2d + 1)$-plane E_{z+}^{2d+1}. The top sets X and X^+ meet only in the unique intersection point p of their tangent d-planes, inside $E_z^{2+1} \cap E_{z+}^{2d+1} = T_p(M)$ $\subset E^{3d+2}$.

If U has a point not belonging to X, then we can homotope $X \cap U$, away from the boundary of U, to a position where $X \cap X^+$ becomes empty. This contradicts the fact that the self-intersection of the essential d-cycle in $H_d(M, Z^2)$ is nonzero. So $U \subset X$, and $X = M_z \subset M$ near p is a *smooth* d-manifold with tangent space $T_p(X)$.

Let W' be the linear space of linear functions z' on E_z^{2d+1} that vanish on $T_p(X)$. To any $z' \in W'$ we assign the 2-jet

$$\eta'(z') = \mathcal{J}_p^2(z'X),$$

which is a quadratic form in the local coordinates $v_1, \ldots, v_d$ for X. By tightness $H_*(X, Z_2) = H_*(S^d, Z_2)$, and all quadratic forms $\eta'(z')$ must have index 0 or d. Each must then be a multiple of

$$\mathcal{J}_p^2(u_{2d+2}, X) = v_1^2 + \cdots + v_d^2.$$

Then there are constants λ_i such that

$$\mathcal{J}_p^2(u_i, X) = 2\lambda_i \mathcal{J}_p^2(u_{2d+2}, X), \qquad i = 1, \ldots, d.$$

Again by tightness, the topset $X = M_z$ must lie in the $(d+1)$-plane with equations

$$
\begin{aligned}
u_i - 2\lambda_i u_{2d+2} &= 0, \\
u_{2d+1} = u_{2d+j} &= 0, \qquad j = 3, \ldots, d+2.
\end{aligned}
\tag{14.14}
$$

And Lemma 2 is proved. □

We introduce new coordinates by

$$
\begin{aligned}
u'_{2d+1} &= u_{2d+1} + u_{2d+2}, \\
u'_{2d+2} &= u_{2d+1} - u_{2d+2}, \\
u'_{2d+j} &= u_{2d+j}, \qquad j = 3, \ldots, d+2, \\
u'_i &= u_i + \sum_{j=2}^{d+2} \beta_{ij} u_{2d+j}, \qquad i = 1, \ldots, d,
\end{aligned}
\tag{14.15}
$$

and we formulate a corollary which we use in the proof of Lemma 5:

Lemma 3. *We can assume that* $2(d+1)$ E^{d+1}*-top-sets through p, for M and for M_0, have the same $(d+1)$-planes*

$$E^{d+1}(M_z) = E^{d+1}((M_0)_z), \tag{14.16}$$

They are top sets for the following linear functions in the new coordinates:

$$z = u'_{2d+1} \pm u'_{2d+j}, \qquad j = 2, \ldots, d+2.$$

Their $(d+1)$-planes are orthogonal to $T_p(M) = T(M_0)$.

Proof. For $z = -u'_{2d+1} - u'_{2d+2} = -2u_{2d+1}$ this follows from (14.7). Sufficient conditions in the equations for the affine transformation (14.15) are

$$\beta_{i2} = 2\lambda_i, \qquad i = 1, \ldots, d.$$

Then $E^{d+1}(M_z)$ has the new equations

$$u'_i = 0, \qquad i = 1, \ldots, d,$$

and $E^{d+1}(M_z)$ is orthogonal to $T_p(M)$ in the standard Euclidean metric with respect to the new coordinates.

We can do the same for M_0 as for M and create the desired coincidence (14.16). The analogous result for $z = -u'_{2d+1} + u'_{2d+2} = -2u_{2d+2}$ is obtained in the coordinates

$$v_j = u_{d+j}, \qquad j = 1, \ldots, d.$$

This determines the numbers

$$\beta_{i2}, \qquad i = d + 1, \ldots, 2d.$$

Now u'_{2d+2} can take the place of $u'_{2d+j}, j = 3, \ldots, d + 2$, by a suitable orthogonal transformation that permutes $u'_{2d+2}, \ldots, u'_{3d+2}$. For

$$z = -u'_{2d+1} \pm u'_{2d+j}$$

we obtain therefore, by the same arguments, suitable values for β_{ij}, $i = 1, \ldots, 2d$. The choices so made for $j = 2, \ldots, d + 2$, do not interfere with each other. This takes care of Lemma 3. $\square$

Lemma 4. *The top set* $X = M_z$ *(as in Lemma 2) is a smooth d-sphere and a quadratic hypersurface in the* $(d + 1)$-*plane it spans,* $E^{d+1}(X)$.

Proof. All topsets of X are tight homology points and hence convex. Therefore $\partial \mathcal{H} X$ is part of X, and it carries the essential d-cycle in X.

For any second nondegenerate extreme point of M, $q \in X$, we find another quadratic hypercone C_q, consisting of half $(2d + 1)$-planes, each carrying a d-cycle $\partial \mathcal{H} Y \subset E^{d+1}(Y)$. This d-cycle $\partial \mathcal{H} Y$ in $E^{d+1}(Y)$ meets $\partial \mathcal{H} X$ in at least one point r by homology, and in at most one point $r = \partial \mathcal{H} X \cap \partial \mathcal{H} Y \subset C_q \cap E^{d+1}(X)$ by geometry. The union of all points r is the quadratic d-manifold

$$C_q \cap E^{d+1}(X),$$

which then must coincide with $\partial \mathcal{H} X$. Therefore $\partial \mathcal{H} X$ is a quadratic d-manifold, and so is $\partial \mathcal{H} Y$. The tangent space $T_r(M)$ at $r \in \partial \mathcal{H} X$ (any point $r \neq p$) now splits

$$T_r(M) = T_r(\partial \mathcal{H} X) \oplus T_r(\partial \mathcal{H} Y).$$

Hence M meets $E^{d+1}(X)$ along $\partial \mathcal{H} X$ transversally, so that $\partial \mathcal{H} X$ is an isolated component of the connected set X. Consequently $\partial \mathcal{H} X = X$. $\square$

Last stage of the proof of the first part of Theorem 15. The union of all E^{d+1}-top-sets through p (like $X = M_z$), parametrized by $S^d(C^*)$, is a closed submanifold of dimension $2d$ of M^{2d}, in view of the transversal smooth d-spheres Y in M at any point r of any of these topsets. Then this union equals M^{2d}, and M^{2d} is contained in the boundary of its convex hull:

$$M^{2d} \subset \partial \mathcal{K} M^{2d}.$$

M^{2d} is algebraic as the intersection of a sufficient number of quadratic hypercones C_q. Here q can be any point of M playing the role of p. All E^{d+1}-top-sets are then seen to be d-spheres and quadratic hypersurfaces. We call them S^d's. Any two meet in at least one point by homology, and in one point only by geometry. For any two different points in M there is one S^d containing them, so the axioms for a projective plane are satisfied. $\square$

In order to prove the last part of Theorem 15, we need two more lemmas.

Lemma 5 (not for $d = 4$ and $d = 8$). *Let $C_p(M)$ and $C_p(M_0)$ be the cones from the point p over the sets M and M_0 in E^{3d+2}, including the tangent $2d$-plane $T_p(M)$. Assume the affine transformations of Lemma 1 and Lemma 3. Let $d = 1$ or $d = 2$. Then*

$$C_p(M) = C_p(M_0).$$

Note. We are convinced that the conclusion also holds for $d = 4$ and 8, but our arguments are not yet complete. Talking with Banchoff has been helpful.

Proof. E^{3d+2} is completed to a real projective space P^{3d+2} in which it is the complement of a hyperplane. Take a P^{3d+1} in P^{3d+2} that does not contain p. Define the intersections

$$IC_p(M) = C_p(M) \cap P^{3d+1},$$

$$IC_p(M_0) = C_p(M_0) \cap P^{3d+1},$$

$$P^{2d-1}(\infty) = T_p(M) \cap P^{3d+1}.$$

The symbol ∞ is used for this P^{2d-1} as well as for its projective subspaces.

In Lemma 3 we found $2(d+1)$ E^{d+1}-top-sets lying in the same $(d+1)$-planes for M as for M_0. They meet P^{3d+1} in $2(d+1)$ P^d's called

$$P_j^d, \quad j = 1, \ldots, 2(d+1).$$

Their intersections with $P^{2d-1}(\infty)$ are P^{d-1}'s called $P_j^{d-1}(\infty), j = 1, \ldots, 2(d+1)$.

From our knowledge about M and M_0 we can deduce various intersection properties concerning $IC_p(M)$ and $IC_p(M_0)$. Projection with center p of the $(d+1)$-planes of E^{d+1}-top-sets of M not through p gives P^{d+1}'s in P^{3d+1}. They form a family Ω (respectively Ω_0 for M_0). Any one meets P_j^d, for $j = 1, \ldots, 2(d+1)$, and any two different ones meet each other in exactly one point of $IC_p(M)$ (respectively $IC_p(M_0)$). All points of $IC_p(M)$ (respectively $IC_p(M_0)$) except those in $P^{2d-1}(\infty)$ are so obtained.

Let Ω_+ be the family of *all* P^{d+1}'s that meet (nonempty intersection) P_j^d for

$j = 1, \ldots, 2(d + 1)$. Clearly $\Omega \subset \Omega_+$ and $\Omega_0 \subset \Omega_+$. It is therefore sufficient to prove the equality

$$\Omega_+ = \Omega_0,$$

which concerns only the model space $IC_p(M_0)$. We will prove this for $d = 1$ and 2.

Take any $x \in P_1^d$, $y \in P_2^d$; $x, y \notin P^{2d-1}(\infty)$. By the geometry of M_0 as a "projective plane" there is exactly one P^{d+1} in $\Omega_0 \subset \Omega_+$ through x and y. It is sufficient to prove that there is no other one in Ω_+. Chose a $P^{3d-1} \subset P^{2d+1}$ which contains $P^{2d-1}(\infty)$, and hence $P_j^{d-1}(\infty)$ for $j = 1, \ldots, 2(d + 1)$, but which does not contain x or y. We project our problem from the real line through x and y into P^{3d-1}. The projection of P^{d+1}'s through x and y yields P^{d-1}'s in P^{3d-1}.

So now we must find all P^{d-1}'s in P^{3d-1} that meet $2d$ P^d's namely the projections of P_j^d into P^{3d-1} for $j = 3, \ldots, 2(d + 1)$.

For $d = 1$ the problem is to find all intersection points (P^0's) of two lines (P^1's) in a plane (P^2). The two lines are different because they have distinct points $P_3^0(\infty)$ and $P_4^0(\infty)$ in $P^1(\infty)$. So the number of intersection points is one, as desired.

For $d = 2$ we must study all lines that meet four distinct planes, say α, β, γ, and δ, in P^5. There is exactly one line that meets α and β, through any point $z \in P^5$ not in $\alpha \cup \beta$. Therefore by projective geometry the lines that meet α, β, and γ (respectively α, β, and δ) determine a projective bijection

$$\tau_\gamma : \alpha \to \beta \qquad (\text{respectively} \quad \tau_\delta : \alpha \to \beta).$$

Solutions (lines that meet α, β, γ, and δ) correspond one–one to fixed points of the projective transformation

$$\tau_\delta^{-1} \tau_\gamma : \alpha \to \alpha.$$

Such a projective transformation has in general (and in our case) one or three (real) fixed points.

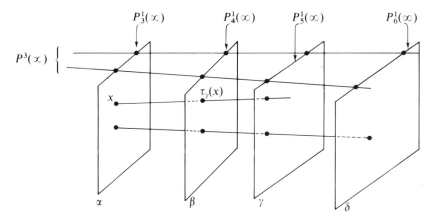

Figure 12.

But let us see what happens inside $P^3(\infty)$. The planes α, β, γ, δ meet $P^3(\infty)$ in the lines $P_j^1(\infty)$ for $j = 3, 4, 5, 6$ which are in general position in $P^3(\infty)$. This implies that there are *two* lines in $P^3(\infty)$ that meet these four lines. They give two of the fixed points in $P^3(\infty) \cap \alpha = P_3^1(\infty)$ (by the same arguments which we used above, applied to the four lines). So there remains *one* valid solution, a line that meets α, β, γ, δ outside $P^3(\infty)$, as desired, and again $\Omega_+ = \Omega_0 = \Omega$. $\square$

For $d = 4$ and 8, this method gets too complicated.

Lemma 6. *If $\oint_p^2(M) = \oint_p^2(M_0)$ and $C_p(M) = C_p(M_0)$, then there exists a projective transformation τ of P^{3d+2} that preserves every line through p, and is given by equations*

$$u_i' = u_i \bigg/ \left[1 + \sum_{j=1}^{3d+2} \alpha_j u_j\right], \qquad i = 1, \ldots, 3d+2, \qquad (14.17)$$

such that $\tau(M) = M_0$.

Proof. By the assumptions, outside the projective $2d$-plane of $T_p(M)$ there are projections (bijective diffeomorphisms) p, p_1, and p_0 in the following diagram:

$$M \xrightarrow{\ \ p\ \ } M_0$$

with p_1 and p_0 and $IC_p(M) = IC_p(M_0)$.

By Lemma 4 any 2-plane section $S^1 \subset M$, not through p, is a conic section. Then $p_1(S_1)$ is also a plane conic section. And $p_0^{-1}p_1(S^1) \subset M_0$ is a smooth closed curve in P^3, which lies on the quadratic cone from p over S^1. It is an intersection of quadratic surfaces that are on M_0. And it is bounded away from p. Then $p_0^{-1}p_1(S^1) = p(S^1)$ must also be a plane conic section, by a small calculation.

By induction on i, we then easily find that any P^{i+1}-section of an $S^d \subset M$, a smooth i-sphere and quadratic i-manifold S^i, has the same kind of i-manifold as its image under $p : M \to M_0$, for $i = 1, 2, \ldots, d$. Hence p sends S^d's on M to S^d's on M_0.

We can now define τ by its action on a choice of $3d + 2$ points, representing $3d + 2$ linearly independent vectors in $(E^{3d+2}, 0)$. Take three S^d's in M, with neither S_a^d nor S_b^d through p, and S_c^d through p. Call the three intersection points (assumed different) $x_1 = S_a^d \cap S_b^d$, $x_2 = S_a^d \cap S_c^d$, $x_3 = S_b^d \cap S_c^d$. Take d more points on S_a^d, d more points on S_b^d, and $d - 1$ more points on S_c^d.

Let τ be the unique projective transformation (12.31) which sends these $3d + 2$ points into $3d + 2$ points of M_0. Then as S^d's are mapped under p onto S^d's of M_0, we conclude

$$\tau(S_a^d) \subset M_0,$$

$$\tau(S_b^d) \subset M_0.$$

Taking into account that $\oint_p^2(M) = \oint_p^2(M_0)$, we also find, first considering conic sections through p in S_c^d, that

$$\tau\left(S_c^d\right) \subset M_0.$$

Next let an S^d in M meet S_a^d, S_b^d, and S_c^d in three different points $y_1, y_2, y_3 \neq p$. The plane section through y_1, y_2, and y_3 is a conic section S^1, and we can conclude $\tau(S^1) \subset M_0$.

The projective transformations of the "projective plane" M_0 that keep fixed three "lines" S_a^d, S_b^d, and S_c^d act transitively on the points of their complement. Therefore the conic sections S^1 on M_0 obtained above cover all of M_0, and all points of $\tau(M)$ are points of M_0: $\tau(M) = M_0$. □

For $d = 1$ and 2, the assumptions in Lemma 6 are true by Lemma 5, and Theorem 15 is proved.

Problem 12. Theorem 15 does not remain true for $d = 1$ if we assume only topological embeddings (see Theorem 9). Does it remain true for $d = 2, 4, 8$?

REFERENCES

[1] T. F. Banchoff, Tightly embedded 2-dimensional polyhedral manifolds. *Amer. J. Math.* **87**, 462–472 (1965).

[2] ———, Critical points and curvature for embedded polyhedra. *J. Differential Geometry* (1967).

[3] ———, The spherical two-piece property and tight surfaces in spheres. *J. Differential Geometry* **4**, 193–205 (1970).

[4] ———, The two-piece property and tight n-manifolds-with-boundary in E^n. Trans. Amer. Math. Soc **161**, 233–267 (1971).

[5] ———Tight polyhedral Klein bottles, projective planes and Moebius bands. *Math. Ann.* **207**, 233–243 (1974).

[6] K. Borsuk, Sur la courbure totale des courbes. *Ann, de la Soc. Math. Pol.* **20**, 251–265 (1947).

[7] S. Carter and A. West, Tight and taut immersions. *Proc. London Math. Soc.* **25** 701–720 (1972).

[8] T. Cecil, Taut immersions of noncompact surfaces into a euclidean 3-space. *J. Differential Geometry* **11**, 451–459 (1976).

[9] T. E. Cecil and P. J. Ryan, Focal sets, taut embeddings and the cyclides of Dupin. *Math. Ann.* 236(1978), 177–190.

[10] C. S. Chen and W. F. Pohl, On the classification of tight surfaces in a euclidean 4-space. Preprint.

[11] C. S. Chen, On the tight isometric immersions of codimension two. *Amer. J. Math.* **94**, 974–990 (1972).

[12] S. S. Chern, La geometrie des sous-variétés d'un espace euclidien à plusieurs dimensions. *Enseignement Math.* **40**, 5.12 (1955).

[13] S. S. Chern and R. K. Lashof, On the total curvature of immersed manifolds I, II. *Amer. J. Math.* **79**, 306–313 (1957); *Mich. Math. J.* **5**, 5.12 (1958).

[14] C. L. Chevalley, *Theory of Lie groups.* Princeton U. P., (1946).

[15] R. D. Edwards, The topology of manifolds and cell-like maps. Preprint I.H.E.S., 1979, 28 pp.

[16] J. Eells and N. H. Kuiper, Manifolds which are like projective planes, *Publ. Math. I.H.E.S.* **14** 128–222 (1962).

[17] S. Eilenberg and N. Steenrod, *Foundations of Algebraic Topology.* Princeton U. P., 1952.

[18] I. Fary, Sur la courbure totale d'une courbe gauche faisant un noeud. *Bull. Soc. Math. France* **77**, 128–138 (1949).

[19] W. Fenchel, Über die Krummung und Windung geschlossener Raumkurven. *Math. Ann.* **10 1**, 238–252 (1929).

[20] D. Ferus, Über die absolute Totalkrümmung höher-dimensionaler Knoten. *Math. Ann.* **171**, 81–86 (1971).

[21] ———, *Totale Absolutkrümmung in Differentialgeometrie und Topologie.* Lecture Notes in Mathematics No. 66, Springer, Berlin, Heidelberg, New York, 1968.

[22] ———, Symmetric submanifolds of euclidean space. preprint, 1979.

[23] R. H. Fox, On the total curvature of some tame knots. *Ann. Math.* **52** 258–261 (1950).

[24] H. Freudenthal, Zur ebenen Oktavengeometrie. *Proc. Akad. Amsterdam A* **56**; *Indag. Math.* **15**, 195–200 (1953).

[25] ———, Oktaven, Ausnahmegruppen und Oktavengeometrie. Preprint, Math. Inst. Univ. Utrecht, 1951.

[26] P. Haupt and H. Künneth, *Geometrische Ordnungen*, Springer, Berlin, 1967.

[27] Guy Hirsch, "La Géométrie projective et la topologie des espaces fibrés. In *Topologie Algébrique. Colloque Int. CNRS* 12, Paris, 1949.

[28] Hurwitz, Über die Komposition der quadratischen Formen. In *Math. Werke II*, p. 641; *Math. Ann.* **88** 1–25 (1923).

[29] S. Kobayashi, Imbeddings of homogenous spaces with minimum total curvature, *Tohoku Math. J.* **19** 63–70 (1967).

[30] S. Kobayashi, Isometric imbeddings of compact spaces. *Tohoku Math. J.* **20** 21–25 (1968).

[31] ———, and M. Takeuchi, Minimal imbeddings of R-spaces, *J. Differential Geometry*, **2**, 230–215 (1968).

[32] ———, and K. Nomizu, *Foundations of Differential Geometry II.* 1969, note 21, p. 361.

[33] A. Kosinski, A theorem on Families of Acyclic Sets and its Applications. *Pacific J. Math* **12**, 317–325 (1962).

[34] W. Kühnel, Total curvature of manifolds with boundary in E^n. *J. London Math Soc.* (2) **15**, 173–182 (1977).

[35] ———, Tight and 0-tight polyhedral embeddings of surfaces. *Invent. Math.* **58**, 161–177 (1980).

[36] N. H. Kuiper, Immersions with minimal total absolute curvature. In *Coll. de Géométrie Diff.*, Centre Belge de Recherches Math., Bruxelles, 1958, pp. 75–88.

[37] ———, Sur les immersions á courbure totale minimale, In *Seminaire de Topologie et Géométrie Différentielle Dirigé par Ch. Ehresmann*, Faculte des Sciences, Paris, 1959, pp. 1–5.

[38] ———, La courbure d'indice k et les applications convexes. In *Séminaire de Topologie et Géométrie Différentielle Dirigé par Ch. Ehresmann*, Faculté des Sciences, Paris, 1960, pp. 1–15.

[39] ———, On surfaces in euclidean three space. *Bull Soc. Math. Belg.* **12**, 5–12 (1960).

[40] ———, Convex immersions of closed surfaces in E^3, *Comm. Math. Helv.* **35**, 85–92 (1961).

[41] ———, On convex maps. *Nieuw Archief voor Wisk* **10**, 147–164 (1962).

[42] ———, Der Satz von Gauss Bonnet für Abbildungen im E^N. *Jahr. Ber. DMV.* **69**, 77–88 (1967).

[43] ———, Minimal total absolute curvature for immersions. *Invent. Math.* **10**, 209–238 (1970).

[44] ———, Morse relations for curvature and tightness. In *Proc. Liverpool Singularities Symp. II* (C. T. C Wall, Ed.) Springer Lecture Notes in Mathematics 209, 1971, pp. 77–89.

[45] N. H. Kuiper, Tight topological embeddings of the moebius band. *J. Diff. Geometry* **6**, 271–283 (1972).

[46] ———, Stable surfaces in euclidean three space. *Math. Scand.* **36**, 83–96 (1975).

[47] ———, and W. F. Pohl, Tight topological embeddings of the real projective plane in E^5. *Invent. Math.* **42**, 177–199 (1977).

[48] ———, Curvature measures for surfaces in E^N. Lobachevski Colloquium. Kazan, USSR (1976).

[49] R. C. Lacher, Cell-like maprings and their generalizations. *Bull. AMS* **83**, 495–552 (1977).

[50] R. Langevin and H. Rosenberg, On curvature integrals and knots. *Topology* **15**, 405–416 (1976).

[51] W. Lastufka, on TPP immersions. Thesis, Univ. of Minnesota, 1979.

[52] J. A. Little and W. F. Pohl, On tight immersions of maximal codimension. *Invent. Math.* **13**, 179–204 (1971).

[53] B. Mazur, A note on some contractible 4-maniflods. *Ann. of Math.* (2) **73**, 221–228 (1961).

[54] W. Meeks, *Lectures on Plateau's problem* (July 1978), Escola de Geometria Differencial DO CEARA Brasil, 1979, 53 pp.

[55] J. W. Milnor, On the total curvature of knots. *Ann. of Math.* **52** (1950).

[56] ———, On the total curvature of closed space curves. *Math. Scand.* **1**, 289–296 (1953).

[57] ———, *Morse Theory*, Ann. Math. Stud. 51, Princeton, 1963.

[58] J. D. Moore, Codimension two submanifolds of positive curvature. *Proc. AMS* **70**, 72–78 (1978).

[59] H. R. Morton, A criterion for an embedded surface in R^3 to be unkotted. preprint, Liverpool, 1976; in *Proc. Conf. on Low Dimensional Topology, Sussex, 1977*, to appear.

[60] V. Poenaru, Les Décompositions de l'hypercube en produit topologique. *Bull. Soc. Math. France* **88**, 113–129 (1960).

[61] M. Retberg, Komplexe Mannigfaltigkeiten in einen euklidischen Raum. Diplomarbeit, Univ. Bielefeld, 1978.

[62] G. Ringel, Map colour theorem. In *Grundlehre der Math. Wiss.*, Band 209, Springer 1974.

[63] I. L. Rodriguez, The two piece property and convexity for surfaces with boundary. *J. Diff. Geometry* **11**, 235–250 (1976).

[64] L. C. Siebenmann, Approximating cellular maps by homeomorphisms. *Topology* **II**, 271–294 (1973).

[65] E. Spanier, *Algebraic Topology*. McGraw-Hill, 1966.

[66] S. S. Tai, On the minimum imbeddings of compact symmetric spaces of rank one. *J. Diff. Geom.* **2**, 55–66 (1968).

[67] Eberhard Teufel, Total krümmung und totale Absolutkrümmung in der spharischen Differentialgeometrie und Differentialtopologie. Thesis, Stuttgart, 1979.

[68] A. Weinstein, Positively curved n-manifolds in R^{n+2}. *J. Diff. Geom.* **4**, 1–4 (1970).

[69] T. J. Willmore, Tight immersions and total absolute curvature. *Bull. London Math. Soc.* **3**, 129–151 (1971).

[70] J. P. Wilson, The total absolute curvature of immersed manifolds. *J. London Math. Soc.* **40**, 362–366 (1966).

[71] J. P. Wilson, Some minimal imbeddings of homogeneous spaces. *J. London Math. Soc.* 335–340 (1969).

[72] P. Wintgen, Zur Integralkrümmung verknoteter Sphären, Thesis, Humboldt-Universität, Berlin, 1976.

Geometry of Quadrics and Spectral Theory[1]

J. Moser*

1. Introduction

a. Background

In this paper we are concerned with integrable Hamiltonian systems. This concept goes back to classical analytical dynamics of the last century. Briefly these are nonlinear systems of ordinary differential equations described by a Hamiltonian function and possessing sufficiently many integrals (or conserved quantities) so that they are more or less explicitly solvable by quadrature. Therefore these systems played a crucial role in the last century before more qualitative methods for differential equations were developed at the turn of the century. Subsequently interest in these systems decreased, partly due to the realization that the existence of global integrals can be established only for exceptional Hamiltonian systems.

In the last 15 years the subject of integrable Hamiltonian system has regained considerable interest with the discovery of some partial differential equations which can be viewed as such systems with infinite degrees of freedom. In this case the integrals form an infinite sequence of conserved functionals. The most celebrated example is the Korteweg–deVries equation: $u_t + uu_x + u_{xxx} = 0$. Extensive investigations of this equation have led to surprising links with scattering theory, spectral theory, complex analysis of hyperelliptic curves and their θ-functions, and differential geometry.

The purpose of this paper is to establish a connection of some classical integrable Hamiltonian systems with the elementary geometry of quadrics. The motivation starts with the following observation: The classical approach to finding the relevant integrals was based on solving the Hamilton–Jacobi equation by separation of variables (Stäckel [19], Jacobi [6]). This required the appropriate choice of variables and computational skill. A case in point is Jacobi's integration of the geodesics on an ellipsoid or C. Neumann's study [14] of a mass point moving on a sphere under the influence of a linear force.

[1] This work was partially supported by the NSF Grant MCS76-01986.
*Courant Institute of Mathematical Sciences, New York, NY, USA; presently at ETH-Zentrum, CH 8092 Zürich, Switzerland.

On the other hand, in the recent studies of partial differential equation the integrals were found as eigenvalues of some linear operators which depend on the solution of the partial differential equation but have the feature that their spectrum is conserved for each solution of the partial differential equation considered. Thus under the time evolution of this equation the linear operator changes in such a way that its spectrum remains fixed, i.e., it undergoes an isospectral deformation. The eigenvalues, viewed as functionals, represent the integrals. This approach of using isospectral deformation of a linear operator has been developed by P. D. Lax in connection with the Korteweg–deVries equation and has been applied by other investigators to many other examples.

The question arises naturally whether all intergrable Hamiltonian systems can be described by isospectral deformation. The question is shifted from finding the integrals of the systems, provided they exist, to finding the linear operator whose spectrum is preserved. We will not attempt to answer this question in any generality, but consider some classical examples, such as Jacobi's geodesic flow on the ellipsoid, and construct an isospectral deformation for them. The relevant matrix turns out to be symmetric, and we will give a geometrical interpretation for the eigenvalues and eigenvectors. This does not lead to new results for this old problem, but to an interesting geometrical interpretation of the eigenvalues and eigenvectors of these operators. In the course of this investigation we will see that our approach also is applicable to the Korteweg–deVries equation, thus establishing a link between this partial differential equation and the theory of confocal quadrics.

b. Geodesics on an Ellipsoid

We begin directly to illustrate our approach with the geodesic flow on an ellipsoid, which had been first integrated by Jacobi. In December 1838 he wrote to his friend and colleague Bessel: "Yesterday I solved the equations for the geodesic lines on an ellipsoid with three different axes by quadrature. These are the simplest formulae of the world, Abelian integrals, which turn into elliptic integrals if two of the axes become equal."[2] This quotation shows how much the Abelian integrals were in vogue at the time; below we will see how this theory ties in with isospectral manifolds.

If A is a positive definite symmetric n-by-n matrix with distinct eigenvalues and $x \in R^n$ an n-vector, then we write the equation for the $(n-1)$-dimensional ellipsoid as

$$\langle A^{-1}x, x \rangle = 1 \tag{1.1}$$

and the differential equations of the geodesics as

$$\frac{d^2x}{dt^2} = -\nu A^{-1}x, \qquad \nu = \frac{\langle A^{-1}y, y \rangle}{|A^{-1}x|^2}, \qquad y = \frac{dx}{dt}, \tag{1.2}$$

---

[2] Translated from Leo Koenigsberger, *Carl Gustav Jacob Jacobi*, Teubner Verlag, Leipzig 1904, p. 251.

where we restrict ourselves to solutions which lie on the ellipsoid. Here $\langle \, , \, \rangle$ denotes the inner product in R^n.

For this problem it will turn out that the relevant isospectral matrices L, which we give here without motivation, are of the form

$$L(x, y) = P_y(A - x \otimes x)P_y, \tag{1.3}$$

where the tensor product $x \otimes y$ denotes the matrix $(x_i y_j)$, and P_y the orthogonal projection into the orthogonal complement of the vector y. Thus the symmetric matrix $L(x, y)$ depends on two vectors $x, y \in R^n$, where, however, the length of $y \neq 0$ is irrelevant.

If we identify x with the position on the ellipsoid and set $y = dx/dt$, then the eigenvalues $\lambda_1, \lambda_2, \ldots, \lambda_n$ of L are preserved under the geodesic flow (1.2). Actually one eigenvalue, say λ_n, is equal to zero and belongs to the eigenvector y of L. But the other $n - 1$ eigenvalues are nontrivial algebraic integrals of (1.2). It is better to form the symmetric functions of the λ_j and look at the characteristic polynomial $l(z) = \det(zI - L)$ of L, which is a polynomial of x, y. In fact, the ratio

$$\frac{|y|^2}{z} \frac{\det(zI - L)}{\det(zI - A)} = \Phi_z(x, y) \tag{1.4}$$

is a rational function of z with poles at the eigenvalues $\alpha_1, \alpha_2, \ldots, \alpha_n$ of A and zeros at $\lambda_1, \ldots, \lambda_{n-1}$, the nontrivial eigenvalues of $L(x, y)$. As a function of x, y the function $\Phi_z(x, y)$ is a quartic polynomial. The partial-fraction expansion of $\Phi_z(x, y)$ is

$$\Phi_z = \sum_{j=1}^{n} \frac{G_j(x, y)}{z - \alpha_j},$$

where the $G_j(x, y)$ are also quartic polynomials of x, y which are integrals for the flow (1.2). Actually only $n - 1$ of them are independent on the ellipsoid, since there the relation

$$0 = \Phi_0 = - \sum_{j=1}^{n} \alpha_j^{-1} G_j(x, y)$$

holds.

We wish to indicate the connection with confocal quadrics to the ellipsoid (1.1), which are given by the equation

$$\langle (z - A)^{-1} x, x \rangle + 1 = 0.$$

We will set

$$Q_z(x, y) = \langle (z - A)^{-1} x, y \rangle, \qquad Q_z(x) = Q_z(x, x), \tag{1.5}$$

and denote the quadric

$$Q_z(z) + 1 = 0$$

by $\mathfrak{A}_z$.

To interpret geometrically the eigenvalue equation $\Phi_z(x, y) = 0$ of (1.4) we first establish the identity

$$\Phi_z(x, y) = Q_z(y)(1 + Q_z(x)) - Q_z^2(x, y),$$

so that for fixed z, x this represents a quadratic form. The equation

$$\Phi_z(x, y) = 0$$

represents the quadratic cone of tangents to $\mathfrak{A}_z$ going through the point x, after the point x is translated to the origin. Secondly one has

$$\Phi_z(x + sy, y) = \Phi_z(x, y),$$

so that Φ_z is constant along any line $x = x_0 + sy$, $y \neq 0$. From these facts one sees that for a given line $x = x_0 + sy$, the roots $z = \lambda_1, \lambda_2, \ldots, \lambda_{n-1}$ of the equation

$$\Phi_z(x_0, y) = 0$$

are such that the above line is tangent to the confocal quadrics $\mathfrak{A}_{\lambda_j}$ ($j = 1, 2, \ldots, n-1$). Generically a line in R^n touches just $n - 1$ confocal quadrics—and the set of lines tangent to $\mathfrak{A}_{\lambda_1}, \ldots, \mathfrak{A}_{\lambda_{n-1}}$ forms a normal congruence; see Bianchi [2].

Thus *the "isospectral" manifold of matrices $L(x, y)$ with a fixed distinct spectrum $\lambda_1, \lambda_2, \ldots, \lambda_{n-1}$ is identified with the normal congruence of common tangents to $n - 1$ confocal quadrics $\mathfrak{A}_{\lambda_j}$ ($j = 1, 2, \ldots, n-1$), which can be considered as a geometrical interpretation of the spectrum of $L(x, y)$.*

Also the eigenvectors ϕ_j of L have a simple geometrical interpretation: The eigenvalue $\lambda_n = 0$ corresponds to $\phi_n = y$ as was mentioned above, while *the other eigenvectors ϕ_j are the normals to $\mathfrak{A}_{\lambda_j}$ at the point of contact of the line $x = x_0 + sy$.* Since $L = L(x, y)$ is symmetric, the n vectors are pairwise orthogonal—which is the content of an old theorem of Chasles (Bianchi [2], Salmon and Fiedler [17]). Under the geodesic flow the orthonormal frame ϕ_j will undergo a motion described by an antisymmetric matrix B, so that

$$\dot{\phi}_j = B\phi_j, \qquad \dot{L} = [B, L].$$

This is the Lax representation of the geodesic flow, where B turns out to be the matrix

$$B = -\left(\alpha_i^{-1}\alpha_j^{-1}(x_i y_j - x_j y_i)\right).$$

The fact that the eigenvalues λ_j are preserved under the geodesic flow means obviously that *the tangents of one geodesic of an ellipsoid will touch the same $n - 2$ quadrics confocal to the ellipsoid*—also a well-known result of geometry.

These results will be derived in Section 3. There are further properties of the isospectral manifold $\mathfrak{M}(\lambda_1, \ldots, \lambda_{n-1})$ of matrices $L(x, y)$ with fixed spectrum, which will be established in Section 4. If we identify the lines $x = x_0 + sy$ on $\mathfrak{M}(\lambda)$ to points and also take the quotient under the reflections $x_j \to \pm x_j$, then we arrive at an $(n - 1)$-dimensional manifold $\mathfrak{M}'(\lambda)$ which is isomorphic to the Jacobi variety of the hyperelliptic curve

$$w^2 = P_{2n-1}(z) = z^{-1}\det(zI - L)\det(zI - A),$$

which is of genus $n - 1$. Thus the Jacobi variety has complex dimension $n - 1$ and is a torus with $2n - 2$ periods. The geodesic flow is linear in the variables of the Jacobi map, and thus the geodesic flow is closely related to Abel's theorem for hyperelliptic integrals. This fact was used by Staude [20] to give a geometrical interpretation of the addition theorem for hyperelliptic integrals.[3]

c. Perturbations of Rank 2

In the above approach the choice of the matrices (1.3) was unmotivated and it is difficult to make the right guess. At present there seems to be no systematic way for finding such isospectral matrices. In this case I owe the essential hint to M. Adler, who suggested looking at matrices of the form $A + x \otimes y - y \otimes x$. The matrices (1.3) can be obtained as limit case of similar matrices, which we will now discuss.

If A is again a fixed symmetric matrix and x, y, ξ, η four n-vectors, we call

$$A + x \otimes \xi + y \otimes \eta$$

a rank-2 perturbation of A. We will study the special case where

$$\xi = ax + by, \qquad \eta = cx + dy,$$

so that

$$L(x, y) = A + ax \otimes x + bx \otimes y + cy \otimes x + dy \otimes y \qquad (1.6)$$

is a matrix which depends on two n-vectors x, y while a, b, c, d are fixed with $\Delta = ad - bc \neq 0$.

We will take this $2n$-parameter family of matrices as starting point, study the algebraic manifold $\mathfrak{M}(\lambda_1, \lambda_2, \ldots, \lambda_n)$ of those x, $y \in R^n$ for which $L(x, y)$ has the fixed spectrum $\lambda_1, \lambda_2, \ldots, \lambda_n$, and investigate the isospectral deformations of these matrices. The basic observation is the following: If we consider the symplectic manifold (R^{2n}, ω) with the symplectic two-form

$$\omega = \sum_{j=1}^{n} dy_j \wedge dx_j,$$

then the eigenvalues of $L(x, y)$ given by (1.6) are in involution,

$$\{\lambda_j, \lambda_k\} = 0,$$

where

$$\{F, G\} = \sum \left(F_{x_j} G_{y_j} - F_{y_j} G_{x_j} \right)$$

denotes the standard Poisson bracket. Again it is better to use the symmetric functions of the eigenvalues of the λ_j or the functions

$$\Phi_z(x, y) = 1 - \frac{\det(zI - L)}{\det(zI - A)},$$

---

[3] I owe this reference to H. Knörrer, Bonn, Germany.

which are quartic polynomials in x, y. With the partial-fraction expansion

$$\Phi_z(x, y) = \sum_{j=1}^{n} \frac{G_j(x, y)}{z - \alpha_j},$$

we have n quartic polynomials G_j which are in involution.

The Hamiltonian vector fields X_H for any Hamiltonian $H = \phi(G_1, G_2, \ldots, G_n)$—or any Hamiltonian depending on the spectrum of L only—is tangential to $\mathfrak{M}(\lambda)$, and

$$X_H = \sum_{j=1}^{n} \frac{\partial H}{\partial G_j} \cdot X_{G_j},$$

where because of

$$\left[X_{G_j}, X_{G_k} \right] = -X_{\{G_j, G_k\}} = 0$$

all these vector fields commute. In particular, the isospectral manifolds are Lagrange manifolds. All these Hamiltonian systems are integrable, by which we mean a vector field having n integrals G_j in involution, for which dG_j are linearly independent in an open dense set. In Section 2 we will show how these Hamiltonian vector fields can be written in the form

$$\frac{d}{dt} L = [B, L].$$

In Section 3 it will be shown that the geodesic flow on an ellipsoid can be derived as the limit case of matrices (1.6) with $a = 0$, $b = -c = \nu$, $d = \nu^2$ for $\nu \to \infty$.

d. Hyperelliptic Curve

In Section 4 we show how $\mathfrak{M}(\lambda)$ is related to the Jacobi variety of a hyperelliptic curve. The manifold $\mathfrak{M}(\lambda)$ is n-dimensional, but after factoring out a 1-dimensional group one is led to an $(n-1)$-dimensional manifold $\mathfrak{M}'$ which is isomorphic to the Jacobi variety of a hyperelliptic curve of genus $n-1$. This generalizes the statement for the geodesic flow on the ellipsoid in which case $\mathfrak{M}'$ is obtained by identifying the straight lines $x \to x + sy$ to points and factoring out the reflections $x_j \to \pm x_j$. The proof is based on introducing a second matrix $M = M(y)$ whose spectrum μ_j together with that of $L = L(x, y)$ determines x, y. In other words, x, y are described by two spectra, each of which is given by a set of functions in involution. The spectrum of M is viewed as a divisor in the Jacobi map. The computation of the symplectic form ω in these variables takes the form

$$\omega = \sum S_{\lambda_j \mu_k} \, d\lambda_j \wedge d\mu_k,$$

where $S = S(\lambda, \mu)$ is the function to which one is also led by the separation of variables of the Hamilton–Jacobi equations. In this way one sees the connection of the two approaches. For details we refer to Section 4.*

---

*In this connection we refer to the forthcoming "Lectures on θ-functions and Their Applications" by David Mumford, held at Bombay and Harvard 1978–1979.

e. Applications

In Section 5 we describe various classical integrable examples for which the integrals can be obtained in terms of the eigenvalues of matrices of the form (1.6). Here we have to distinguish three cases according as the rank of the symmetric part

$$
\begin{bmatrix}
a & \dfrac{b+c}{2} \\
\dfrac{b+c}{2} & d
\end{bmatrix}
$$

of the matrix $\begin{pmatrix} a & b \\ c & d \end{pmatrix}$ is 2, 1, or 0. We give three examples, the last one being illustrated by a subclass of geodesics of the orthogonal group with a left-invariant metric, as it was studied first by Arnold. (For literature see Dikii [4], Manakov [8], Mischenko [11].)

Finally we show how these systems relate to the finite band potential in the Hill's equation, in the periodic and the quasiperiodic case. This is based on a connection between the translation flow for the above finite-band potentials and a mechanical problem of a particle moving on the sphere $|x| = 1$ under the influence of a linear force. This connection was found by E. Trubowitz; it was described previously (Moser [12]). Finally, in the Appendix we describe a class of matrices L involving x_j^{-1} whose eigenvalues are also in involution. They are also integrals for a classical mechanical system discussed in the dissertation of Rosochatius [16].

f. Connection with M. Reid's Result [15]

We want to mention a related result which we learned from a letter of Horst Knörrer. In his unpublished dissertation of 1972 Miles Reid established that the set of $(m-1)$-dimensional linear subspaces of a nonsingular intersection of two quadrics in $P_{2m+1}(C)$ is—as algebraic manifold—isomorphic to the Jacobi variety of a hyperelliptic curve. It is tempting to guess a connection to the above result about the common tangents of $n-1$ confocal quadrics $\mathfrak{A}_{\lambda_1}, \ldots, \mathfrak{A}_{\lambda_{n-1}}$ in C^n. Such a connection really exists, and Knörrer communicated to me a beautiful construction of a 1-to-2^{n-1} mapping of the variety of common tangents to M. Reid's* Jacobi variety for some appropriate quadrics.

g. Final Remarks

The above approach is obviously very unsystematic and relies on lengthy calculation. Why are the eigenvalues of matrices of the form (1.6) in involution? The deeper reasons have still to be revealed. M. Adler, who gave the initial hint for the form of the isospectral matrices required, found a general framework based on the coadjoint representation of certain Kac–Moody algebras in extension of his previous work [1], which allow him to encompass the above examples as special cases of a general theory. Originally it was the plan to publish a joint paper from this point of view; however, since the theory is formidable and

---

*Note added in proof: H. Knörrer on the ellipsoid, *Inv. Math.* **59**, 119–143 (1980).

lengthy, it was necessary to separate off the general approach. Adler's Lie-algebraic approach will appear elsewhere. Here we merely want to show the baffling connection between the spectral theory of the matrices (1.6) and the geometry of quadrics.

I want to express my thanks to Horst Knörrer for informing me about his geometrical construction in connection with M. Reid's work. After presenting this paper in Berkeley,[4] I visited in Warwick and lectured on this topic. I want to thank D. Epstein for his hospitality and Adrian Douaday for interesting discussions. He supplied an elegant alternative proof for the involuntary character of the eigenvalues of (1.6). Because of length restrictions we could not present his argument here. I am grateful to P. Deift for suggestions and for reading the manuscript. Finally, I am particularly indebted to M. Adler, who contributed essential ideas to this work.

2. Perturbation of Rank 2

a. Isospectral Manifolds

We take as a starting point the spectral problem for a perturbation of rank 2 of a symmetric bilinear operator. Let V denote a real (or complex) finite-dimensional vector space, $\langle \ , \ \rangle$ a real inner product, and let A be a matrix symmetric with respect to the inner product, i.e. $\langle Av, w \rangle = \langle v, Aw \rangle$. Moreover, we will assume the eigenvalues of A to be distinct.

A perturbation of rank r is given by

$$Lv = Av + \sum_{\rho=1}^{r} x_\rho \langle \xi_\rho, v \rangle,$$

where $x_1, \ldots, x_r$ and $\xi_1, \ldots, \xi_r$ are two sets of linearly independent vectors in V. We write the above formula in terms of the tensor product $\otimes$ as

$$L = A + \sum_{\rho=1}^{r} x_\rho \otimes \xi_\rho. \tag{2.1}$$

It is well known that the spectrum of L is determined by the formula

$$\frac{\det(z - L)}{\det(z - A)} = \det(I - W_z) \tag{2.2}$$

where W_z is the r-by-r matrix given by

$$W_z = \langle R_z x_\rho, \xi_\sigma \rangle, \quad \rho, \sigma = 1, \ldots, r; \qquad R_z = (zI - A)^{-1}. \tag{2.3}$$

This formula has been extended to infinite-dimensional vector spaces (see Kato [7]); the right-hand side is called the Weinstein–Aronszajn determinant.

We specialize the above to rank two and

$$x_1 = x, \quad x_2 = y; \qquad \xi_1 = ax + by, \quad \xi_2 = cx + dy,$$

---

[4]The lecture presented at the Chern Symposium at Berkeley was based on this Introduction. Since these results have not appeared in print, we decided to include the proofs.

so that

$$L = L(x, y)$$
$$= A + ax \otimes x + bx \otimes y + cy \otimes x + dy \otimes y, \qquad (2.4)$$

where a, b, c, d are constants with determinant $\Delta = ad - bc \neq 0$ and x, y linearly independent vectors of V. This defines a $2n$-dimensional family of matrices in whose spectra we are interested. In particular the n-dimensional foliation given by isospectral matrices will be of interest to us.

The main result of this section is the observation that the eigenvalues of these matrices are "in involution" with respect to the symplectic structure $\sum_1^n dy_j \wedge dx_j$, i.e. the natural symplectic structure of $T^*V = V^* \times V \sim V \times V$. For any two functions $F = F(x, y)$, $G = G(x, y)$ in $C^1(V \times V)$ we define the corresponding Poisson brackets

$$\{ F, G \} = \langle F_x, G_y \rangle - \langle F_y, G_x \rangle,$$

where F_x, F_y are defined by the relation

$$dF = \langle F_x, dx \rangle + \langle F_y, dy \rangle.$$

One calls a family of functions $\mathcal{F}$ "in involution" if for any two of the, $F, G \in \mathcal{F}$, one has

$$\{ F, G \} = 0.$$

Such a family can be extended by closure under composition: If $\phi, \psi \in C^1(R^2, R)$, then

$$\{\phi(F, G), \psi(F, G)\} = \frac{\partial(\phi, \psi)}{\partial(F, G)} \{ F, G \},$$

i.e., if F, G are in involution, so are $\phi(F, G)$, $\psi(F, G)$. Instead of showing the involutary character of the eigenvalues, we will establish this for the symmetric functions of the eigenvalues which are rational in x, y.

If we apply the formulae (2.2), (2.3) to the case (2.4), we obtain a two-by-two matrix

$$W_z = \begin{pmatrix} Q_z(x) & Q_z(x, y) \\ Q_z(x, y) & Q_z(y) \end{pmatrix} \begin{pmatrix} a & c \\ b & d \end{pmatrix}, \qquad (2.5)$$

where

$$Q_z(x, y) = \langle R_z x, y \rangle, \qquad Q_z(x) = Q_z(x, x),$$

and (2.2) becomes

$$\frac{\det(z - L)}{\det(z - A)} = 1 - \operatorname{tr} W_z + \det W_z = 1 - \Phi_z, \qquad (2.6)$$

where

$$\Phi_z(x, y) = aQ_z(w) + (b + c) Q_z(x, y) + dQ_z(y)$$
$$- (ad - bc)\left(Q_z(x)Q_z(y) - Q_z^2(x, y) \right). \qquad (2.7)$$

Thus the eigenvalues of L are the values of z for which the rational function Φ_z takes the value 1. If these eigenvalues λ_j are distinct, the isospectral manifold of matrices (2.4) with spectrum $\lambda_1, \lambda_2, \ldots, \lambda_n$ is given by

$$\{ x, y \mid \Phi_\lambda(x, y) = 1 \text{ for } j = 1, \ldots, n \},\qquad (2.8)$$

and hence it is an algebraic manifold.

b. Isospectral Deformations

Theorem 1. *For any z, z' in the resolvent set of A one has*

$$\{ \Phi_z, \Phi_{z'} \} = 0.$$

i.e. the functions $\Phi_z(x, y)$ are in involution.

This theorem can be verified by a direct but lengthy calculation. We do not present this, since the result will be a consequence of Theorem 2. Clearly this theorem remains valid if multiple eigenvalues of A are permitted.

We extend the class of functions (Φ_z) by forming for any polynomial $f(z)$

$$H(x, y) = \frac{1}{4\pi i} \int_{|z| = R} f(z) \Phi_z(x, y)\, dz,$$

where the circle $|z| = \mathfrak{R}$ contains the spectrum of A. We express this function explicitly by introducing a basis in which

$$A = \operatorname{diag}(\alpha_1, \alpha_2, \ldots, \alpha_n)$$

and set

$$f(\alpha_j) = \beta_j, \qquad \beta = \operatorname{diag}(\beta_1, \beta_2, \ldots, \beta_n).$$

Then one finds

$$2H = a\langle \beta x, x \rangle + (b + c)\langle \beta x, y \rangle + d\langle \beta y, y \rangle$$
$$- \frac{ad - bc}{2} \sum_{i \neq j} \frac{\beta_i - \beta_j}{\alpha_i - \alpha_j} (x_i y_j - x_j y_i)^2. \qquad (2.9)$$

For example, for $\beta_i = \delta_{ik}$ this becomes

$$G_k(x, y) = ax_k^2 + (b + c)x_k y_k + dy_k^2$$
$$- (ad - bc) {\sum_i}' \frac{(x_i y_k - x_k y_i)^2}{\alpha_k - \alpha_i}, \qquad (2.10)$$

where the prime indicates that $i \neq k$. These functions are all in involution, as a consequence of Theorem 1, and the Φ_z are recovered by

$$\Phi_z(x, y) = \sum_{j=1}^{n} \frac{G_j(x, y)}{z - \alpha_j}.$$

and

$$H(x, y) = \frac{1}{2} \sum_{j=1}^{n} f(\alpha_j) G_j(x, y).$$

If $(a, b + c, d) \neq (0, 0, 0)$, then dG_j are linearly independent on an open dense set, while for $a = b + c = d = 0$ we have the relation

$$\sum_{j=1}^{n} G_j = 0.$$

In this case we have n independent commuting functions in $|x|^2$: $G_2, G_3, \ldots, G_n$, for example.

For any choice of the constants $f(\alpha_j) = \beta_j$ the vector field

$$\dot{x} = \frac{\partial}{\partial y} H, \qquad \dot{y} = -\frac{\partial}{\partial x} H \qquad (2.11)$$

is integrable, since $G_1, G_2, \ldots, G_n$ are integrals in involution. Hence the spectrum of the matrix (2.4) is fixed under any of these flows, so that in the case of distinct eigenvalues there exists a nonsingular matrix $U = U(t)$ such that $U^{-1}LU$ is a constant matrix. The infinitesimal version of this statement is that the differential equation (2.11) can be written in the Lax form

$$\frac{d}{dt} L = [B, L] \qquad (2.12)$$

with some matrix B. This is the content of

Theorem 2. *The vector field* (2.11) *with H given by* (2.9) *defines an isospectral deformation of the matrix* (2.4) *given by* (2.12) *where*

$$B = \tfrac{1}{2}(b - c)\beta + (ad - bc)\left(\frac{\beta_i - \beta_j}{\alpha_i - \alpha_j}(x_i y_j - x_j y_i)\right). \qquad (2.13)$$

The diagonal elements of the last matrix are zero.

Corollary. *If* $H = H(G_1, G_2, \ldots, G_n)$, *then the vector field* X_H *corresponds to the isospectral deformation* $\dot{L} = [B, L]$, *where B is of the form* (2.13) *with* $\beta_j = 2\,\partial H / \partial G_j$. *Indeed*

$$X_H = \sum_{j=1}^{n} \frac{\partial H}{\partial G_j} X_{G_j} = \frac{1}{2} \sum \beta_j X_{G_j},$$

where for fixed $G_j = c_j$ *the* β_j *can be considered as constants. For this vector field Theorem 2 gives the statement of the corollary.*

We show that Theorem 1 is a consequence of Theorem 2. From the form (2.12) of the differential equation (2.11) it is plain that the eigenvalues of L, hence any function of the eigenvalues are constant along orbits. Therefore

$$1 - \Phi_z = \frac{\prod (z - \lambda_j)}{\prod (z - \alpha_j)}$$

is a constant of the motion, for any z, or

$$\frac{d}{dt} \Phi_z = \{\Phi_z, H\} = 0.$$

Taking $\beta_j = 2\delta_{jk}$, we get $H = G_k$; hence

$$\{\Phi_z, G_k\} = 0,$$

and hence

$$\{\Phi_z, \Phi_{z'}\} = \sum (z' - \alpha_k)^{-1} \{\Phi_z, G_k\} = 0.$$

The proof of Theorem 2 consists of a calculation, which we break into several steps. Setting

$$s = \tfrac{1}{2}(b + c), \qquad r = \tfrac{1}{2}(b - c),$$

we break L into symmetric and antisymmetric parts:

$$L = A + S + R,$$
$$S = ax \otimes x + s(x \otimes y + y \otimes x) + dy \otimes y,$$
$$R = r(x \otimes y - y \otimes x).$$

With the determinant $\Delta = ad - bc = ad - s^2 + r^2$, we set

$$B = r\beta + \Delta\Gamma, \qquad \Gamma = \left(\frac{\beta_i - \beta_j}{\alpha_i - \alpha_j}(x_i y_j - x_j y_i) \right),$$

the diagonal terms of Γ being zero. The Hamiltonian H is broken up into its quadratic and its quartic part:

$$H = F - \Delta G,$$
$$F = \tfrac{1}{2}a\langle \beta x, x \rangle + s\langle \beta x, y \rangle + \tfrac{1}{2}d\langle \beta y, y \rangle,$$
$$G = \frac{1}{2} \sum_{i<j} \frac{\beta_i - \beta_j}{\alpha_i - \alpha_j}(x_i y_j - x_j y_i)^2.$$

The vector field generated by H is generally denoted by X_H, so that $X_H = X_F - \Delta X_G$.
The differential equations for X_H are given by

$$\frac{d}{dt}\begin{pmatrix} x \\ y \end{pmatrix} = \begin{pmatrix} s & d \\ -a & -s \end{pmatrix}\begin{pmatrix} \beta x \\ \beta y \end{pmatrix} + \Delta\begin{pmatrix} \Gamma x \\ \Gamma y \end{pmatrix}.$$

From these one computes readily the following relations:

$$X_G R = -[\Gamma, R], \qquad X_G S = -[\Gamma, S];$$

hence

$$X_G L = -[\Gamma, R + S] - [\Gamma, L] + [\Gamma, A].$$

Similarly one finds

$$X_F R = r[\beta, S],$$
$$X_F S = -(ad - s^2)[\beta, x \otimes y - y \otimes x];$$

hence

$$X_H L = X_H(R + S) = X_F(R + S) - \Delta X_G(R + S)$$
$$= r[\beta, S] - (ad - s^2)[\beta, x \otimes y - y \otimes x]$$
$$+ \Delta[\Gamma, L] - \Delta[\Gamma, A].$$

Since

$$[\Gamma, A] = -[\beta, x \otimes y - y \otimes x],$$

we obtain

$$X_H L = r[\beta, S] + \left(-(ad - s^2) + \Delta\right)[\beta, x \otimes y - y \otimes x] + \Delta[\Gamma, L]$$
$$= r[\beta, S] + r^2[\beta, x \otimes y - y \otimes x] + \Delta[\Gamma, L]$$
$$= [r\beta, S + R] + \Delta[\Gamma, L]$$
$$= [r\beta + \Delta\Gamma, L] = [B, L],$$

where we have used that A and β commute. This proves Theorem 2, and hence Theorem 1.

c. The Action of Gl(2, R)

The matrices L can be simplified if one subjects them to the linear transformation $(x, y) \rightarrow (\alpha x + \beta y, \gamma x + \delta y)$ with $\alpha\delta - \beta\gamma = 1$. This last condition assures that the transformation is symplectic. If we set

$$V = \begin{pmatrix} \alpha & \beta \\ \gamma & \delta \end{pmatrix}; \quad C = \begin{pmatrix} a & b \\ c & d \end{pmatrix}, \quad \Sigma = \begin{pmatrix} a & s \\ s & d \end{pmatrix},$$

$$C = \Sigma + r\begin{pmatrix} 0 & 1 \\ -1 & 0 \end{pmatrix},$$

then this transformation replaces C by

$$V^T C V = V^T \Sigma V + r\begin{pmatrix} 0 & 1 \\ -1 & 0 \end{pmatrix}.$$

Thus r is invariant as well as the rank ρ and the determinant of Σ. One finds the following normal forms:

(i) $\rho = 2$: $\Sigma = \begin{pmatrix} a & 0 \\ 0 & 1 \end{pmatrix}$, $a \neq 0$
(ii) $\rho = 1$: $\Sigma = \begin{pmatrix} 0 & 0 \\ 0 & 1 \end{pmatrix}$
(iii) $\rho = 0$: $\Sigma = \begin{pmatrix} 0 & 0 \\ 0 & 0 \end{pmatrix}$.

The corresponding form for the matrices L are

(i) $L = A + (ax \otimes x + y \otimes y) + r(x \otimes y - y \otimes x)$, $a \neq 0$
(ii) $L = A + y \otimes y + r(x \otimes y - y \otimes x)$
(iii) $L = A + r(x \otimes y - y \otimes x)$.

All these cases occur in examples of mechanics and geometry, as we will see in the following sections. Case (i) describes the motion of a particle on an ellipsoid $\langle A^{-1}x, x \rangle = 1$ under the influence of the central force ax. Case (ii) is related to the geodesic flow on an ellipsoid, to the motion of a particle on the sphere $|x| = 1$ under the influence of the external force Ax, and to the Toda lattice in the periodic case (see van Moerbeke [22]), as was shown recently by P. Deift and E. Trubowitz. Case (iii) describes special orbits of the geodesic flow on the orthogonal group $O(n)$ with a left-invariant metric. In the next section we describe the connection with the geodesic flow on the ellipsoid, which is the oldest candidate for an integrable system.

d. Trace Formulae

We present another representation of the basic functions (2.9) which are in involution for all choices of the diagonal matrix β. For this purpose we introduce a parameter ϵ and consider the matrix

$$L_\epsilon = A + \epsilon R + \epsilon^2 S$$

so that the matrix

$$C = \begin{pmatrix} a & b \\ c & d \end{pmatrix}$$

is replaced by

$$C = \epsilon^2 \begin{pmatrix} a & s \\ s & d \end{pmatrix} + \epsilon r \begin{pmatrix} 0 & 1 \\ -1 & 0 \end{pmatrix}.$$

Set

$$\tilde{\Phi}_z = aQ_z(x) + (b+c)Q_z(x,y) + dQ_z(y)$$
$$- r^2 \left(Q_z(x)Q_z(y) - Q_z^2(x,y) \right),$$

and for any polynomial $g(z)$ with $g'(\alpha_j) = \beta_j$ define

$$2\tilde{H} = \frac{1}{2\pi i} \int \tilde{\Phi}_z g'(z)\, dz$$
$$= a\langle \beta x, x \rangle + (b+c)\langle \beta x, y \rangle + d\langle \beta y, y \rangle$$
$$- \frac{r^2}{2} \sum_{i \neq j} \frac{\beta_i - \beta_j}{\alpha_i - \alpha_j} (x_i y_j - x_j y_i)^2.$$

The functions $\tilde{\Phi}_z, \tilde{H}$ are obtained from Φ_z, H (see (2.7), (2.9)) by replacing $\Delta = ad - bc$ with r^2 in the coefficient of the last term. This is an insignificant change, since this coefficient can always be normalized to 1 by rescaling if it is not zero. Therefore these functions $\tilde{H}$ are also in involution for any two choices of the diagonal matrix β.

The following formula is due to M. Adler. For any polynomial $g(z)$ one has

$$\text{tr } g(L_\epsilon) = \text{tr } g(L_0) + 2\epsilon^2 \tilde{H} + O(\epsilon^3),$$

or

$$\frac{1}{4} \left(\frac{d}{d\epsilon} \right)^2 \text{tr } g(L_\epsilon) \bigg|_{\epsilon=0} = \tilde{H}. \tag{2.14}$$

To prove this result we use the formula

$$\text{tr}(z - L_\epsilon)^{-1} - \text{tr}(z - L_0)^{-1} = \omega^{-1} \frac{d\omega}{dz},$$

where

$$\omega(z) = \det(I - W_z).$$

This follows from (2.2) and the identity

$$\log \det X = \operatorname{tr} \log X$$

for any nonsingular matrix X with appropriate definition of the branch (see Kato [7]). Hence

$$\operatorname{tr} g(L_\epsilon) - \operatorname{tr} g(L_0) = \frac{1}{2\pi i} \int g(z) \omega^{-1} \frac{d\omega}{dz} \, dz,$$

and since in our case

$$\omega(z) = \det(I - W_z)$$
$$= 1 - \epsilon^2 (a Q_z(x) + (b + c) Q_z(x, y) + d Q_z(y))$$
$$\quad + \left(\epsilon^2 r^2 + \epsilon^4 (ad - s^2) \right) \left(Q_z(x) Q_z(y) - Q_z^2(x, y) \right)$$
$$= 1 - \epsilon^2 \tilde{\Phi}_z + O(\epsilon^4),$$

we have

$$\operatorname{tr} g(L_\epsilon) - \operatorname{tr} g(L_0) = -\frac{\epsilon^2}{2\pi i} \int g(z) \frac{d}{dz} \tilde{\Phi}_z \, dz + O(\epsilon^4)$$
$$= \epsilon^2 2 \tilde{H} + O(\epsilon^4),$$

proving the formula (2.14).

3. Connection with Confocal Quadrics

a. Integrals for the Geodesic Flow on the Ellipsoid

In the n-dimensional Euclidean space R^n with inner product $\langle x, y \rangle$ we consider a positive definite symmetric matrix A with distinct eigenvalues. Without loss of generality we can assume $A = \operatorname{diag}(\alpha_1, \alpha_2, \ldots, \alpha_n)$, where $0 < \alpha_1 < \alpha_2 < \cdots < \alpha_n$. Then the equation

$$\langle A^{-1} x, x \rangle = 1 \tag{3.1}$$

defines an ellipsoid. The quadrics $\mathfrak{A}_z$ confocal to this ellipsoid are given by the equation

$$\langle (z - A)^{-1} x, x \rangle + 1 = 0.$$

We introduce the bilinear form

$$Q_z(x, y) = \langle (z - A)^{-1} x, y \rangle, \qquad Q_z(x) = Q_z(x, x), \tag{3.2}$$

so that $\mathfrak{A}_z$ is defined by

$$Q_z(x) + 1 = 0, \tag{3.3}$$

and (3.1) is the quadric $\mathfrak{A}_0$. These confocal quadrics have a number of well-known properties, for which we will give a new interpretation. For example, through any point x with $x_1 x_2 \cdots x_n \neq 0$ pass precisely n confocal quadratics, which moreover intersect each other perpendicularly.

For any given point $x_0 \in R^n$ we ask for the cone of lines which are tangent to a quadric $Q(x) + 1 = 0$ (we supress the subscript z, since it is irrelevant for this question). By an elementary calculation one finds the equation of this cone to be

$$\det \begin{pmatrix} 1 + Q(x) & 1 + Q(x, x_0) \\ 1 + Q(x, x_0) & 1 + Q(x_0) \end{pmatrix}$$

$$= Q(x) - 2Q(x, x_0) + Q(x_0) + Q(x) Q(x_0) - Q^2(x, x_0)$$

$$= 0.$$

Alternatively, if we set $y = x - x_0$ this equation becomes

$$\det \begin{pmatrix} Q(y) & Q(x_0, y) \\ Q(x_0, y) & 1 + Q(x_0) \end{pmatrix}$$

$$= Q(y) + Q(x_0) Q(y) - Q^2(x_0, y)$$

$$= 0,$$

which for fixed x_0 describes a cone with vertex at the origin.

This equation agrees with

$$\Phi_z(x_0, y) = 0$$

if (in the notation of the previous section) we take $a = 0$, $b = -c = 1$, $d = -1$ (or $a = 0$, $b = -c = i$, $d = 1$), showing that the latter equation can be interpreted geometrically as defining the set of lines $x = x_0 + sy$ tangent to $\mathfrak{A}_z$. In particular,

$$\Phi_0(x, y) = 0$$

describes the tangents of the ellipsoid $\mathfrak{A}_0$; we have changed the notation x_0 to x.

The Hamiltonian differential equations

$$\dot{x} = -\frac{\partial}{\partial y} \Phi_o(x, y), \qquad \dot{y} = \frac{\partial}{\partial x} \Phi_o(x, y) \tag{3.4}$$

restricted to $\Phi_0 = 0$ describe the motion of such tangent lines, which is easily interpreted: The point of contact with $\mathfrak{A}_0$ moves along a geodesic while the point x moves perpendicularly to this tangent.

Indeed, if the line through x in direction $y \neq 0$ has the point of contact $x + sy = \xi$ with $\mathfrak{A}_0$, we have

$$Q(x + sy, y) = 0, \qquad \text{or} \quad s = -\frac{Q(x, y)}{Q(y)},$$

and one computes

$$\frac{d}{dt} \xi = \frac{d}{dt} (x + sy) = \dot{x} + s\dot{y} + \dot{s}y$$

$$= \frac{2\Phi_0(x, y)}{Q(y)} A^{-1} y + s\dot{y} = \dot{s}y,$$

since $\Phi_0(x, y) = 0$, and

$$\frac{dy}{dt} = -2Q(y)A^{-1}x + 2Q(x, y)A^{-1}y = -2Q(y)A^{-1}\xi.$$

If we introduce the parameter τ by $d\tau/dt = \dot{s}$, then

$$\frac{d\xi}{d\tau} = y, \qquad \frac{d^2\xi}{d\tau^2} = \frac{dy}{d\tau} = -\frac{2Q(y)}{\dot{s}} A^{-1}\xi,$$

which shows that the point of contact $\xi = \xi(\tau)$ moves on a geodesic. From the relations

$$\langle \dot{x}, y \rangle = 2\Phi_0(x, y) = 0, \qquad \langle y, \dot{y} \rangle = 0$$

it follows that $\dot{x}$ is perpendicular to the direction of the line, and that $\langle y, y \rangle$ is a constant.

Thus (3.4) can be viewed as an extension of the geodesic flow on the ellipsoid to a flow in $T^*R^n = R^{2n}$. The geodesic flow is obtained by constraining (3.4) to the symplectic manifold

$$Q_0(x, y) = 0, \qquad |y|^2 = \text{const} > 0,$$

and to the energy manifold $\Phi_0(x, y)$ (see Section 5(c)). The relations $\Phi_0 = 0$, $Q_0(x, y) = 0$ are equivalent to

$$Q_0(x) + 1 = 0, \qquad Q_0(x, y) = 0, \qquad \text{if } Q_0(y) \neq 0$$

which describes the tangent bundle of the ellipsoid, where the constrained flow takes place.

To establish the geodesic flow as an integrable one it suffices to show that the extended flow (3.4) is integrable (see Section 5). This follows from Theorem 1 and the formula

$$\Phi_z(x, y) = \sum_{j=1}^{n} \frac{G_j}{z - \alpha_j};$$

hence $G_1, G_2, \ldots, G_n$ are integrals of (3.4) which are in involution.

b. Isospectral Deformation

Since the zeros of $1 - \Phi_z(x, y)$ are the eigenvalues of the matrix

$$L(x, y) = A + x \otimes y - y \otimes x - y \otimes y,$$

they also can be viewed as integrals of the motion, and $L(x, y)$ remains similar to itself along the orbits of (3.4). Since the tangents of $\mathfrak{A}_z$ are given by $\Phi_z(x, y) = 0$, it is more natural to construct a matrix $L(x, y)$ whose eigenvalues are the zeros of $\Phi_z(x, y)$. Such a matrix is

$$L(x, y) = P_y(A - x \otimes x)P_y \tag{3.5}$$

where

$$P_y = I - \frac{y \otimes y}{\langle y, y \rangle}$$

is for $|y| = 1$ the projection into the orthogonal complement of y.

To see this we observe that the eigenvalue of

$$L(x, \nu y) = A + \nu(x \otimes y - y \otimes x) - \nu^2 y \otimes y$$

are given by $\Phi_z(x, \nu y) = 1$, or

$$\Phi_z(x, y) = \frac{1}{\nu^2}.$$

This suggests letting ν tend to infinity. Of course, $L(x, \nu y)$ has no limit; in fact one of its eigenvalues $\lambda = \nu^2(|y|^2 + O(\nu^{-2}))$ tends to infinity, and the corresponding eigenvector

$$\phi = y + \nu^{-1}P_y x + O(\nu^{-2})$$

tends to y. Now, if we take ν purely imaginary, then $L(x, \nu y)$ is Hermitian, and on the orthogonal complement of ϕ this matrix becomes

$$L(x, \nu y) - \frac{\phi \otimes \bar{\phi}}{|\phi|^2},$$

which has a limit for $\nu \to \infty$ the matrix (3.5). But it is also easy to verify directly that for the matrix (3.5) one has

$$|y|^2 \frac{\det(z - L)}{\det(z - A)} = -z\Phi_z(x, y). \qquad (3.6)$$

Thus the $n - 1$ roots of $\Phi_z(x, y)$ are eigenvalues of L and the nth eigenvalue is $\lambda = 0$, which corresponds to the eigenvector y. It is clear, then, that the matrix (3.5) undergoes an isospectral deformation under the flow (3.4). We make this more explicit by writing the differential equations (3.4) in the Lax form

$$\frac{d}{dt} L = [B, L]$$

with an appropriate matrix B. We generalize the setup right away and replace the Hamiltonian Φ_0 by

$$H = \frac{1}{2} \sum \beta_j y_j^2 + \frac{1}{2} \sum_{i<j} \frac{\beta_i - \beta_j}{\alpha_i - \alpha_j} (x_i y_j - x_j y_i)^2$$

$$= -\frac{1}{2} \sum_{j=1}^{n} \beta_j G_j \qquad (3.7)$$

with arbitrary constants $\beta_1, \beta_2, \ldots, \beta_n$. For $\beta_j = 2\alpha_j^{-1}$ we obtain $H = \Phi_0$. As in the previous section, we obtain

Theorem 3. *The Hamiltonian system*

$$\dot{x} = H_y, \qquad \dot{y} = -K_x,$$

with H given by (3.7), can be written in the matrix form

$$\frac{d}{dt} L = [B, L], \qquad (3.8)$$

where L is given by (3.5) and

$$B = -\left(\frac{\beta_i - \beta_j}{\alpha_i - \alpha_j} (x_i y_j - x_j x_i) \right), \qquad (3.9)$$

with zero diagonal elements.

Corollary. *If* $H = H(G_1, G_2, \ldots, G_n)$ *then the vector field* X_H *corresponds to the isospectral deformation* (3.8), (3.9) *where*

$$\beta_j = 2 \frac{\partial H}{\partial G_j}.$$

This statement requires a calculation similar to that in the proof of Theorem 2. The differential equations have the form

$$\dot{x} = \beta y + Bx, \qquad \dot{y} = + By,$$

which implies

$$\frac{d}{dt}(x \otimes x) = -\left[B, A - (x \otimes x)\right] + \beta x \otimes y + y \otimes \beta x,$$

$$\frac{d}{dt}(y \otimes y) = \left[B, y \otimes y\right]$$

with $\beta = \mathrm{diag}(\beta_1, \beta_2, \ldots, \beta_n)$. Since $\langle y, y \rangle$ is an integral, the last relation implies

$$\frac{d}{dt} P_y = \left[B, P_y\right]$$

and the first gives for $M_x = A - x \otimes x$ the equation

$$\frac{d}{dt} M_x = \left[B, M_x\right] - \beta x \otimes y - y \otimes \beta x.$$

Combining the last two relations, one finds for $L = P_y(A - x \otimes x)P_y = P_y M_x P_y$, after a short calculation,

$$\frac{d}{dt} L = \left[B, P_y\right]M_x P_y + P_y\left[B, M_x\right]P_y + P_y M_x\left[B, P_y\right]$$

$$= \left[B, P_y M_x P_y\right] = \left[B, L\right],$$

which proves the statement.

c. Interpretation of the Eigenvalues and the Frame of L

According to the last theorem the symmetric matric L given by (3.5) undergoes an isospectral deformation, that is, its eigenvalues are constant and the eigendirections are moved by an orthogonal transformation. We give a geometrical interpretation of the eigenvalues and eigenvectors of L.

For this purpose we introduce the eigenvalues $\mu_1, \mu_2, \ldots, \mu_n$ of the matrix M_x and the eigenvalues $\lambda_1, \lambda_2, \ldots, \lambda_n$ of L, where $\lambda_1 = 0$. Moreover, we set

$$m(z) = \prod_{j=1}^{n} (z - \mu_j), \qquad l(z) = \prod_{j=1}^{n} (z - \lambda_j), \quad \text{and} \quad a(z) = \prod_{j=1}^{n} (z - \alpha_j).$$

Then the formulae of the previous section give

$$1 + Q_z(x) = \frac{\det(z - M)}{\det(z - A)} = \frac{m(z)}{a(z)}, \tag{3.10}$$

$$-z\Phi_z(x, y) = |y|^2 \frac{\det(z - L)}{\det(z - A)} = |y|^2 \frac{l(z)}{a(z)}. \tag{3.11}$$

The first equation shows that *the eigenvalues μ_j of $M = M_x$ are the elliptic coordinates of x.* Indeed, the relation

$$1 + \sum \frac{x_j^2}{z - \alpha_j} = 0 \quad \text{for } z = \mu_1, \mu_2, \ldots, \mu_n$$

defines elliptic coordinates, so that $x \in \bigcap_{j=1}^n \mathfrak{A}_{\mu_j}$. The second equation shows that *the eigenvalues λ_j ($j = 2, 3, \ldots, n$) are those values of z for which the line through x with direction y touches the quadric $\mathfrak{A}_z$.* Indeed, we saw that the equation $\Phi_z(x, y)$ defines the tangents of $\mathfrak{A}_z$. The "general" line touches $n - 1$ confocal quadrics $\mathfrak{A}_{\lambda_j}$ where the λ_j are singled out as the roots of (3.11), as long as $Q_z(y) \neq 0$.

Let the eigenvalues λ_j of $L = L(x, y)$ be distinct. Then the eigenvector for $\lambda_1 = 0$ is $\phi_1 = y$ and *the eigenvectors for $\lambda_j \neq 0$ are the normals to the confocal quadric $\mathfrak{A}_{\lambda_j}$ at the point of contact $\xi_j = x + s_j y$ of the line through x with direction y.*

To prove this statement, we note that the point of contact $\xi_j = x + s_j y$ is determined by

$$Q_\lambda(\xi_j, y) = 0, \quad 1 + Q_\lambda(\xi_j) = 0.$$

Indeed, the first relation expresses the tangency of the line with $\mathfrak{A}_\lambda$, and the second follows from

$$0 = \Phi_\lambda(x, y) = \Phi_\lambda(\xi_j, y) = Q_\lambda(y)(1 + Q_\lambda(\xi_j)) - Q_\lambda^2(\xi_j, y)$$

if $Q_{\lambda_j}(y) \neq 0$.

The normal to $\mathfrak{A}_{\lambda_j}$ at ξ_j is given by

$$\phi = (\lambda_j - A)^{-1} \xi_j.$$

These are the desired eigenfunctions, if $\lambda_j \neq 0$. Indeed, since they are orthogonal to y, the eigenfunction for $\lambda = 0$, we have $P_y \phi = \phi$; hence for $\lambda = \lambda_j$, $\xi = \xi_j$,

$$(\lambda - L)\phi = P_y(\lambda - A + x \otimes x)\phi$$
$$= P_y(\xi + x\langle x, \phi \rangle)$$
$$= P_y(\xi + xQ(x, \xi)),$$

where we abbreviate Q_{λ_j} by Q. Since for $s = s_j$

$$Q(x, \xi) = Q(x + sy, \xi) = Q(\xi) = -1,$$

we conclude

$$(\lambda - L)\phi = P_y(\xi - x) = P_y(sy) = 0,$$

proving the statement.

This implies that the $n - 1$ normals $\phi_j = (\lambda_j - A)^{-1} \xi_j$ ($j = 2, \ldots, n$) and y form an orthogonal frame. This is a well-known theorem of elementary geometry, due to Chasles (Salmon and Fiedler [17]): The normals of two confocal quadrics $\mathfrak{A}_{a_1}$, $\mathfrak{A}_{z_2}$ at the points of contact with a common tangent are perpendicular. This theorem corresponds to the orthogonality of the frame $\Phi_1 = y$, $\Phi_j = (\lambda_j - A)^{-1} \xi_j$. Actually the identity $\{\Phi_{z_1}, \Phi_{z_2}\} = 0$ can be deduced from this geometrical fact (see Moser [12]). Another consequence of this geometrical fact is that the common tangents of $n-1$ confocal quadrics $\mathfrak{A}_{z_1}, \ldots, \mathfrak{A}_{z_{n-1}}$ generically form a normal congruence, i.e. can be viewed as the normals of $(n-1)$-dimensional surfaces (Bianchi [2]). These surfaces are the level surfaces of a function $S = S(x, y)$ which we will determine below (see Section 4).

d. Joachimsthal's Integral

In the classical literature one finds for the geodesic flow in the 3 axial ellipsoid the integral of Joachimsthal. It is given as the product $p \cdot d$ where for a given tangent to the ellipsoid d is the diameter of the ellipsoid in the direction parallel to the tangent and p is the distance from the origin to the tangent plane of the ellipsoid containing the given tangent. We express this integral in terms of the integrals G_1, G_2, G_3 defined by

$$\Phi_z = \sum_{j=1}^{n} \frac{G_j}{z - \alpha_j}. \tag{3.12}$$

One finds

$$\left(\frac{pd}{2}\right)^2 = - \frac{\sum\limits_{j=1}^{3} G_j}{\sum\limits_{j=1}^{3} \alpha_j^{-2} G_j}.$$

Indeed, at a point $\xi \in \mathfrak{A}_0$, i.e., $\langle A^{-1}\xi, \xi \rangle = 1$, the tangent plane is given by $\langle A^{-1}\xi, x \rangle = 1$ and its distance p from $x = 0$ by

$$p = |A^{-1}\xi|^{-1}.$$

For a tangent vector $\xi + \eta$ at ξ we have $\langle A^{-1}\xi, \eta \rangle = 0$ and

$$\left(\frac{d}{2}\right)^2 = \frac{\langle \eta, \eta \rangle}{\langle A^{-1}\eta, \eta \rangle},$$

so that

$$\left(\frac{pd}{2}\right)^2 = \frac{\langle \eta, \eta \rangle}{\langle A^{-1}\eta, \eta \rangle \langle A^{-2}\xi, \xi \rangle}.$$

By comparing the coefficients of z^{-1} in the expansion of (3.12), and taking $(d/dz)\Phi_z$ at $z = 0$, one finds for $x = \xi$, $y = \eta$,

$$\langle \eta, \eta \rangle = \sum_{h=1}^{3} G_j,$$

$$\langle A^{-1}\eta, \eta \rangle \langle A^{-2}\xi, \xi \rangle = - \sum_{j=1}^{3} \alpha_j^{-2} G_j,$$

which gives the stated identity. Clearly this integral has the same interpretation for higher dimensions, but one cannot expect such a geometrical meaning for all integrals.

4. The Hyperelliptic Curve

a. The Isospectral Manifold $\mathfrak{M}(\lambda)$

We investigate the manifold $\mathfrak{M} = \mathfrak{M}(\lambda_1, \lambda_2, \ldots, \lambda_n)$ of $x, y \in R^{2n}$ for which the matrices $L = L(x, y)$ of the form (2.4) have the fixed eigenvalues $(\lambda_1, \lambda_2, \ldots, \lambda_n)$. We will assume that the $2n$ numbers λ_j, α_k are distinct. In this section we will allow all quantities to be complex and will not go into reality considerations. If the vector fields X_{G_j} are linearly independent, they span the tangent space of $\mathfrak{M}$. Furthermore, if we consider $\mathfrak{M}$ as being embedded in the symplectic space (R^{2n}, ω) where $\omega = \sum_{j=1}^{n} dy_j \wedge dx_j$, then $\mathfrak{M}$ is a Lagrange manifold, since

$$\omega(X_{G_j}, X_{G_k}) = -\{G_j, G_k\} = 0.$$

Since $\mathfrak{M}$ can also be characterized by the equations

$$G_k = c_k, \qquad c_k = -\frac{l(\alpha_k)}{a'(\alpha_k)},$$

by Sard's lemma we can assume that the λ_j can be so chosen that the dG_j and hence X_{G_j} are linearly independent on $\mathfrak{M}$, provided $(a, b + c, d) \neq (0, 0, 0)$. This is a restriction, since on the linear space $x_{k*} = y_{k*} = 0$ one has $dG_{k*} = 0$, and thus linear dependence of the dG_j. We impose the restriction of linear independence of the dG_j, so that $\mathfrak{M}$ is a manifold without singularities. It is an algebraic manifold. In this section we will study $\mathfrak{M}$ and relate it to the Jacobi variety of some hyperelliptic curve of genus $n - 1$.

We first note that there are two trivial symplectic group actions on (R^{2n}, ω) leaving $\mathfrak{M}$ invariant—which we will factor out.

The first group is discrete and is generated by the n reflections

$$\tau_k : (x, y) \to (T_k x, T_k y)$$

where $T_k : x_j \to (1 - 2\delta_{jk}) x_j$. Clearly

$$L(T_k x, T_k y) = T_k^{-1} L(x, y) T_k,$$

and therefore the spectrum of $L(x, y)$ is invariant under τ_k, i.e., $\mathfrak{M}$ is invariant under τ_k.

The second group action g^t is one-dimensional and is generated by the Hamiltonian

$$G = \frac{1}{2} \sum_{j=1}^{n} G_j(x, y) = \frac{1}{2}(a\langle x, x \rangle + 2s\langle x, y \rangle + d\langle y, y \rangle).$$

The corresponding vector field X_G is given by the linear differential equations

$$x' = sx + dy, \qquad y' = -ax - sy,$$

with the characteristic exponents $\pm\sqrt{-\vartheta}$, where $\vartheta = ad - s^2$.

Let Γ be the Abelian group generated by the τ_k $(k = 1, 2, \ldots, n)$ and the g^t $(t \in C)$. We are interested in the set of Γ-orbits $\mathfrak{M}/\Gamma$ on $\mathfrak{M}$. If we exclude the points on a quadric cone K, this quotient will be a manifold. For this purpose we construct a cross section which is crossed transversally by the G-orbits. For

$\vartheta = ad - s^2 \neq 0$, i.e. case (i), one can find a quadratic form

$$F = a'\langle x, x\rangle + 2s'\langle x, y\rangle + d'\langle y, y\rangle \tag{4.1}$$

such that

$$X_G F = \{F, G\} = 2\sqrt{-\vartheta}\, F.$$

This is clear because the map $F \to \{F, G\}$ has the eigenvalues $0, 2\alpha, -2\alpha$ where $\alpha = \sqrt{-\vartheta}$. Thus

$$F = F|_{t=0} e^{2\alpha t},$$

and all orbits which do not lie on the cone $F = 0$ cross the section $F = 1$ transversally.

In the case (ii) where $\vartheta = 0$ but a, s, d are not all zero, one can construct a quadratic function (4.1) satisfying

$$X_G F = \{F, G\} = G,$$

so that

$$F = F|_{t=0} + tG.$$

Thus for orbits outside the cone $G = 0$ we have such a cross section in $F = 0$.

Finally, in case (iii), if $a = s = d = 0$ we take $G = \frac{1}{2}\langle x, x\rangle$ and $F = \langle x, y\rangle$ and have a cross section in $F = 0$ for all orbits outside the cone $G = 0$.

Thus if we define the cone K by the equation $F = 0$, $G = 0$, $\langle x, x\rangle = 0$ in case (i), (ii), (iii) respectively, then,

$$(\mathfrak{M} - \mathfrak{M} \cap K)/\Gamma = \mathfrak{M}'$$

is a manifold of dimension $n - 1$. Indeed, any orbit $g^t(x, y)$ intersects the orbit $\tau g^t(x, y)$ (where $\tau \neq \mathrm{id}$ is a product of some $\tau_1, \tau_2, \ldots, \tau_n$) on the linear manifolds $x_{k*} = y_{k*} = 0$ which do not intersect $\mathfrak{M}$. Thus $\mathfrak{M} - \mathfrak{M} \cap K$ is a trivial bundle over $\mathfrak{M}'$ with complex lines as fibers in cases (ii) and (iii). In case (i) the fibers are cylinders, i.e. complex lines identified by the periods $t = i\pi\alpha^{-1}$.

Incidentally, in case (ii) we can choose the eigenvalues $\lambda_1, \lambda_2, \ldots, \lambda_n$ so that $2G = \sum_{k=1}^n G_k = \sum_{k=1}^n c_k \neq 0$, so that $K \cap \mathfrak{M} = \emptyset$. In that case we have $\mathfrak{M}' = \mathfrak{M}/\Gamma$.

To take account of the group generated by the τ_j we use the functions $u_k = x_k^2$, $v_k = x_k y_k$, $w_k = y_k^2$, which are invariant under this group and also characterize equivalence classes. Moreover, all functions $G_j(x, y)$ are quadratic polynomials in the u_k, v_k, w_k, while F and G are linear functions. Since we have the n quadratic relations $u_k w_k = v_k^2$, the manifold $\mathfrak{M}'$ can be viewed as the intersection of $2n$ quadrics in the hyperplane $\mathbb{C}^{3n-1}$ given by $F = \text{const}$.

The object of this section is to prove the following

Theorem 4. *If $\Delta = ad - bc \neq 0$ and the above assumptions hold, then the quotient $\mathfrak{M}' = (\mathfrak{M} - K \cap \mathfrak{M})/\Gamma$ is the Jacobi variety $\mathcal{J}$ of the hyperelliptic curve*

$$w^2 = P(z) = a(z)\big(r^2 a(z) - \Delta l(z)\big), \tag{4.2}$$

with $l(z) = \det(zI - L)$, $a(z) = \det(zI - A)$, of genus $n - 1$. Moreover, if

$$\sum_{k=2}^n \int^{\mu_k} \frac{z^{n-j}\, dz}{\sqrt{P(z)}} = s_j, \qquad j = 2, 3, \ldots, n, \tag{4.3}$$

defines the Jacobi map of the divisor classes $(p_2, \ldots, p_n)$, $p_j = (\mu_j, \sqrt{P(\mu_j)})$, *on the hyperelliptic curve into* C^{n-1}, *then the vector fields* X_{G_j} *are given by*

$$\sum_{k=2}^{n} c_{jk} \frac{\partial}{\partial s_k} \pmod{X_G}$$

with constant coefficients c_{jk}; *i.e., the* G_j-*flow is linear in the* s_k.

Finally, $x_k^2, x_k y_k, y_k^2$ $(k = 1, 2, \ldots, n)$ *restricted to* $\mathfrak{M}'$ *are abelian functions on* $\mathcal{J}$ *and can be expressed as ratios of* θ-*functions on* $\mathcal{J}$.

The main point of this result is that the linear structure on $\mathcal{J}$ given by the X_{G_j} agrees with the linear structure as it is given by Abel's theorem. The solutions of any Hamiltonian $\phi(G_1, G_2, \ldots, G_n) = H$ can be expressed in terms of Jacobi θ-functions plus an exponential $e^{\pm \alpha t}$ or a linear function t coming from the trivial part X_G.

An alternate form of the polynomial $P(z)$ in (4.2) is

$$P(z) = a(z)^2 \Delta \det\left(\text{sym}(C^{-1}) - Q\right),$$

where sym( ) denotes the symmetric part of a matrix and

$$C = \begin{pmatrix} a & b \\ c & d \end{pmatrix}, \qquad Q = \begin{pmatrix} Q_z(x) & Q_z(x, y) \\ Q_z(x, y) & Q_z(y) \end{pmatrix}.$$

We will give the proof of the above statement in the three cases of Section 2: (i) $\vartheta = ad - s^2 \neq 0$, (ii) $\vartheta = 0$ but $d \neq 0$, and (iii) $a = s = d = 0$ but $\Delta = r^2 \neq 0$. We begin with case (i) in detail and then indicate the necessary modifications for the other cases.

b. An Inverse Spectral Problem

We assume $\vartheta = ad - s^2 \neq 0$. Without loss of generality we can take $a = d = 0$ and therefore consider

$$L(x, y) = A + bx \otimes y + cy \otimes x \tag{4.4}$$

with $bc = -\Delta \neq 0$, $b + c \neq 0$, as $\vartheta \neq 0$. In this case our formula become

$$\frac{l(z)}{a(z)} = \frac{\det(zI - L)}{\det(zI - A)} = 1 - \Phi_z(x, y) \tag{4.5}$$

$$\Phi_z(x, y) = (b + c) Q_z(x, y)$$

$$+ bc\left(Q_z(x)Q_z(y) - Q_z^2(x, y)\right). \tag{4.6}$$

If λ_j, α_j are the eigenvalues of L, A respectively, we have

$$l(z) = \prod_{j=1}^{n} (z - \lambda_j), \qquad a(z) = \prod_{j=1}^{n} (z - \alpha_j).$$

To parametrize points on $\mathfrak{M}$ we use the eigenvalues of another matrix, namely

$$M = M(y) = P_y A P_y, \tag{4.7}$$

where $P_y = I - y \otimes y / \langle y, y \rangle$ is—for $\langle y, y \rangle \neq 0$—the projection into the orthogonal complement of y. Thus $M(y)$ has the eigenvalue 0 with the eigenvector y

and $n-1$ other eigenvalues $\mu_2, \mu_3, \ldots, \mu_n$. If we set

$$m(z) = \langle y \rangle^2 \prod_{j=2}^{n} (z - \mu_j), \qquad \langle y \rangle^2 = \langle y, y \rangle = \sum_{k=1}^{n} y_k^2, \qquad (4.8)$$

we have the identity

$$\frac{m(z)}{a(z)} = \frac{\langle y \rangle^2}{z} \frac{\det(zI - M)}{\det(zI - A)} = Q_z(y) = \sum_{j=1}^{n} \frac{y_j^2}{z - \alpha_j}, \qquad (4.9)$$

which defines the $\mu_2, \mu_3, \ldots, \mu_n$ as the zeros of $Q_z(y)$ provided $\langle y \rangle^2 \neq 0$.
    The length $\langle y \rangle$ of y is irrelevant for $M(y)$, and we set

$$\mu_1 = \langle y \rangle^2.$$

We consider the following "inverse spectral problem": Given $\lambda_1, \lambda_2, \ldots, \lambda_n$, $\mu_1, \mu_2, \ldots, \mu_n$, construct the $L = L(x, y)$, $M = M(y)$ so that λ_j are the eigenvalues of L, so that $\mu_2, \ldots, \mu_n$ and 0 are the eigenvalues of M, and so that $\mu_1 = \langle y \rangle^2$. The λ_j, μ_j represent $2n$ variables, which turn out to be sufficient to determine the $2n$ variables x_k, y_k except for branchings and singularities, which will not be investigated here.
    To carry out the construction we require some standard formulae for elliptic coordinates. Since $\mu_1, \ldots, \mu_n$ depend only on the y_k, we will determine the latter first. From (4.9) one reads off that

$$y_j^2 = \frac{m(\alpha_j)}{a'(\alpha_j)}. \qquad (4.10)$$

In the real case when $\alpha_1 < \mu_2 < \alpha_2 < \cdots < \mu_n < \alpha_n$ and $\mu_1 > 0$, the right side of (4.10) is positive and allows for 2^n real vectors y. For $\mu_1 = \langle y \rangle^2 = 1$ the μ_2, $\mu_3, \ldots, \mu_n$ can be viewed as coordinates on the sphere S^{n-1} which form an orthogonal coordinate system.[5] In the following we will derive these properties but ignore the reality conditions.
    By logarithmic differentiation of (4.10) one finds

$$\frac{\partial y}{\partial \mu_k} = \begin{cases} \dfrac{1}{2\mu_1} y & \text{for } k = 1, \\[2mm] \tfrac{1}{2}(\mu_k - A)^{-1} y & \text{for } k \geqslant 2, \end{cases} \qquad (4.11)$$

which implies

$$\left\langle \frac{\partial y}{\partial \mu_k}, \frac{\partial y}{\partial \mu_j} \right\rangle = g_j \delta_{jk},$$

with

$$g_j = \begin{cases} \dfrac{1}{4\mu_1} & \text{for } j = 1, \\[3mm] -\dfrac{m'(\mu_j)}{4a(\mu_j)} & \text{for } j \geqslant 2. \end{cases} \qquad (4.12)$$

[5] E. Rosochatius refers to them as "elliptische Kugelkoordinaten"; see [16], in particular p. 29.

We verify (4.12) only for distinct $\mu_2, \mu_3, \ldots, \mu_n$ and for $\mu_1 \neq 0$, using the resolvent identity

$$(\mu_k - A)^{-1}(\mu_j - A)^{-1} = \frac{-1}{\mu_j - \mu_k}\left((\mu_j - A)^{-1} - (\mu_k - A)^{-1}\right).$$

From (4.11) we find for $j \neq k, j, k \geqslant 2$,

$$\left\langle \frac{\partial y}{\partial \mu_k}, \frac{\partial y}{\partial \mu_j} \right\rangle = \frac{-1}{4(\mu_j - \mu_k)}\left(Q_{\mu_j}(y) - Q_{\mu_k}(y)\right) = 0,$$

and for $j = k \geqslant 2$ we compute

$$\left\langle \frac{\partial y}{\partial \mu_j} \right\rangle^2 = \tfrac{1}{4}\langle(\mu_j - A)^{-2}y, y\rangle = -\frac{1}{4}\frac{d}{dz}Q_z(y)\Big|_{z=\mu_j},$$

which by (4.9) gives the desired result for $j \geqslant 2$. For $j \geqslant 2$ we have have by (4.11)

$$\left\langle \frac{\partial y}{\partial \mu_1}, \frac{\partial y}{\partial \mu_j} \right\rangle = -\frac{1}{4\mu_1}Q_{\mu_j}(y) = 0,$$

$$\left\langle \frac{\partial y}{\partial \mu_1} \right\rangle^2 = \frac{1}{4\mu_1^2}\langle y\rangle^2 = \frac{1}{4\mu_1},$$

which establishes (4.12). Thus we have

$$\sum_{k=1}^{n} dy_k^2 = \sum_{j=1}^{n} g_j\, d\mu_j^2.$$

Having determined y, we turn to the construction of x, which we represent in the orthogonal frame $\partial y/\partial \mu_j$ as

$$x = \sum_{j=1}^{n} X_j \frac{\partial y}{\partial \mu_j}. \tag{4.13}$$

Taking the inner product with $\partial y/\partial \mu_k$, we find for the coefficients X_j, using (4.11), (4.12),

$$g_k X_k = \left\langle x, \frac{\partial y}{\partial \mu_k} \right\rangle = \begin{cases} \dfrac{1}{2\mu_1}\langle x, y\rangle & \text{for } k = 1, \\[2mm] \tfrac{1}{2}Q_{\mu_k}(x, y) & \text{for } k \geqslant 2. \end{cases} \tag{4.14}$$

The terms on the right-hand side can be expressed in terms of λ, μ, thus determining x as a function of λ, μ. To show this we compare the coefficients of z^{-1} as $z \to \infty$ in (4.6):

$$\operatorname{tr}(L - A) = \sum_{j=1}^{n}(\lambda_j - \alpha_j)$$

$$= (b + c)\langle x, y\rangle = 2s\langle x, y\rangle, \tag{4.15}$$

so that

$$\langle x, y\rangle = \frac{1}{2s}\sum_{j=1}^{n}(\lambda_j - \alpha_j).$$

To compute $Q_{\mu_j}(x, y)$ for $j \geqslant 2$, we set $z = \mu_j$ in (4.6), (4.5), using $Q_{\mu_j}(y) = 0$:

$$1 - \frac{l(z)}{a(z)} = (b + c) Q_z(x, y) - bc Q_z^2(x, y)$$

for $z = \mu_2, \mu_3, \ldots, \mu_n$. This is a quadratic equation for $Q_{\mu_j}(x, y)$. If we set $\Delta = -bc$, $2r = b - c$, we obtain for $z = \mu_j$, $j \geqslant 2$,

$$Q_z(x, y) - \frac{b + c}{2bc} = \frac{1}{a(z)\Delta} \sqrt{P(z)},$$

$$P(z) = \left(a(z)r^2 - l(z)\Delta\right)a(z). \qquad (4.16)$$

Thus by (4.13), (4.14), (4.15), (4.16) we can express x, y in terms of λ, μ, solving the inverse spectral problem.

To construct $(\mathfrak{M} - K \cap \mathfrak{M})/\Gamma$ we construct the orbits of $G = \frac{1}{2}\sum_j G_j = s\langle x, y \rangle$, i.e. of

$$x' = sx, \qquad y' = -sy.$$

The function F of (4.1) can be chosen as $F = \langle y \rangle^2$ and

$$F = F|_{t=0} e^{-2st},$$

and the cross section is given by

$$F = \langle y \rangle^2 = 1, \qquad \text{or} \quad \mu_1 = 1.$$

We can take account of the discrete group generated by τ_k by parametrizing $\mathfrak{M}'$ with $u_j = x_j^2$, $v_j = x_j y_j$, $w_j = y_j^2$. We observe that these $3n$ functions are rational symmetric functions of the divisor

$$(p_2, p_3, \ldots, p_n),$$

where $p_j = (\mu_j, \sqrt{P(\mu_j)})$ is a point on the Riemann surface of (4.2). Indeed, by (4.10) $w_j = y_j^2$ is a symmetric polynomial of $\mu_2, \mu_3, \ldots, \mu_n$. By (4.11), (4.12), (4.13), and (4.14) we have

$$v_j = x_j y_j = \sum_{k=1}^{n} X_k \frac{\partial y_j}{\partial \mu_k} y_j$$

$$= \mu_1^{-1}\langle x, y \rangle y_j^2 - \sum_{k=2}^{n} \frac{a(\mu_k)}{m'(\mu_k)} Q_{\mu_k}(x, y)(\mu_k - \alpha_j)^{-1} y_j^2.$$

Using $\langle x, y \rangle = (1/2s)\sum_{j=1}^{n}(\lambda_j - \alpha_j)$, (4.16), and (4.10), it is clear that v_j is also a rational symmetric function of the divisor $p_2, \ldots, p_n$. Finally, the same is true for $u_j = v_j^2 w_j^{-1}$, and therefore for any rational function of u_j, v_j, w_j.

It is well known (see Neumann [13], Siegel [18]) that all rational symmetric functions of $p_2, p_3, \ldots, p_n$ can be represented in terms of θ-functions on the Jacobi variety and thus also u_j, v_j, w_j can all be represented in this way, giving rise to a map of $\mathcal{J} \to \mathfrak{M}'$. For nonspecial divisors $(p_2, p_3, \ldots, p_n)$ and fixed $\lambda_1, \lambda_2, \ldots, \lambda_n$ this map is given by (4.10), (4.13). Since u_j, v_j, w_j are sufficient to distinguish points on $\mathfrak{M}'$, this mapping is an isomorphism, showing that $\mathfrak{M}' \sim \mathcal{J}$.

c. The Symplectic Structure

For the proof of Theorem 4 in case (i), it remains to show that the vector fields X_{G_j} have constant coefficients with respect to the variables s_j of (4.3). For this purpose we express the symplectic form $\omega = \sum_{j=1}^{n} dy_j \wedge dx_j$ in terms of the variables λ_k, μ_k. Since the λ_k as well as the μ_k, being functions of y alone, are in involution, it follows that ω has the form

$$\omega = \sum_{k,j=1}^{n} a_{kj} \, d\lambda_k \wedge d\mu_j,$$

and since $d\omega = 0$ one can—at least locally—find a function $S = S(\lambda, \mu)$ such that

$$a_{kj} = \frac{\partial^2 S}{\partial \lambda_k \, \partial \mu_j}.$$

Therefore it suffices to compute this function S. Setting $\partial_\lambda S = \sum_{k=1}^{n} S_{\lambda_k} \, d\lambda_k$, $\partial_\mu S = \sum_{j=1}^{n} S_{\mu_j} \, d\mu_j$, we write also briefly

$$\omega = \partial_\lambda \partial_\mu S.$$

To compute S we use (4.11) to write the 1-form $\langle x, dy \rangle = \sum_1^n x_k \, dy_k$ as

$$\langle x, dy \rangle = \sum_j \langle x, \frac{\partial y}{\partial \mu_j} \rangle d\mu_j$$

$$= \tfrac{1}{2} \langle x, y \rangle \frac{d\mu_1}{\mu_1} + \frac{1}{2} \sum_{j=2}^{n} Q_{\mu_j}(x, y) \, d\mu_j,$$

or with (4.15), (4.16), and $b + c = 2s$,

$$\langle x, dy \rangle = \frac{1}{4s} \sum (\lambda_k - \alpha_k) \frac{d\mu_1}{\mu_1} + \frac{1}{2} \sum_{j=2}^{n} \left[\frac{s}{bc} + \frac{\sqrt{P(\mu_j)}}{a(\mu_j)\Delta} \right] d\mu_j.$$

Since $\omega = -d\langle x, dy \rangle = \partial_\lambda \partial_\mu S$, we read off that

$$-S = \frac{1}{4s} \sum (\lambda_k - \alpha_k) \log \mu_1 + \frac{1}{2\Delta} \sum_{j=2}^{n} \int^{\mu_j} \frac{\sqrt{P(z)} \, dz}{a(z)}, \qquad (4.17)$$

the intergration being taken along paths on the Riemann surface. Of course, S is determined only up to two additive functions of λ_k and of μ_k alone.

Instead of the $\lambda_1, \lambda_2, \ldots, \lambda_n$ we use as independent variables their symmetric functions $\sigma_1, \sigma_2, \ldots, \sigma_n$ defined by

$$l(z) = z^n + \sigma_1 z^{n-1} + \sigma_2 z^{n-2} + \cdots + \sigma_n,$$

so that

$$\omega = \sum_{k,j} S_{\sigma_k \mu_j} \, d\sigma_k \wedge d\mu_j = \sum_{k=1}^{n} d\sigma_k \wedge d(S_{\sigma_k}). \qquad (4.18)$$

Thus S_{σ_k} are canonically conjugate to the symmetric functions σ_k. For these S_{σ_k} we find

$$-\frac{\partial S}{\partial \sigma_k} = \frac{\delta_{k1}}{4s} \log \mu_1 - \frac{1}{4} \sum_{j=2}^{n} \int^{\mu_j} \frac{z^{n-k}}{\sqrt{P(z)}} dz,$$

which shows that S_{σ_k} for $k = 2, 3, \ldots, n$ agrees, up to the factor $\frac{1}{4}$, with the Abelian differentials of the first kind (4.3).

It remains to rewrite the Hamiltonian vector fields X_H in these variables. Setting

$$H = H(x, y) = \Psi(\sigma, \mu),$$

the differential equation takes the form

$$\begin{pmatrix} 0 & -S_{\sigma\mu}^T \\ S_{\sigma\mu} & 0 \end{pmatrix} \begin{pmatrix} \dot{\mu} \\ \dot{\sigma} \end{pmatrix} = \begin{pmatrix} \Psi_\mu \\ \Psi_\sigma \end{pmatrix}.$$

In particular, if the Hamiltonian Ψ depends only on the σ, i.e. $\Psi_\mu = 0$ and $\det S_{\sigma\mu} \neq 0$, then the system reduces to

$$\dot{\sigma} = 0, \qquad \frac{d}{dt} S_\sigma = \Psi_\sigma$$

with the solutions

$$\sigma = \sigma|_{t=0}, \qquad S_\sigma = S_\sigma|_{t=0} + t\Psi_\sigma.$$

Thus the S_{σ_k} vary linearly on $\mathfrak{M}$. Hence if we set $s_k = 4S_{\sigma_k}$ for $k = 1, 2, \ldots, n$, then the vector field X_H for a Hamiltonian $H(x, y) = \Psi(\sigma)$ becomes

$$X_H = 4 \sum_{k=1}^{n} \Psi_{\sigma_k} \frac{\partial}{\partial s_k}.$$

Since by (4.15)

$$G = s\langle x, y \rangle = -\tfrac{1}{2}\left(\sigma_1 + \sum \alpha_j\right),$$

we have

$$X_G = -2 \frac{\partial}{\partial s_1},$$

so that

$$X_H = 4 \sum_{k=2}^{n} \Psi_{\sigma_k} \frac{\partial}{\partial s_k} + 2\Psi_{\sigma_1} X_G.$$

We apply this remark to $H = G_j$, which are functions of the σ_k alone, to show that X_{G_j} have modulo X_G constant coefficients with respect to $\partial/\partial s_k$. This proves Theorem 4 in case (i).

d. Degenerate Case

In cases (ii) and (iii) we consider the normal form of Section 2,

$$L = A + r(x \otimes y - y \otimes x) + dy \otimes y, \qquad r \neq 0,$$

where $d = 1$ or $d = 0$. The spectrum of L is given by the zeros of $1 - \Phi_z$, where

$$\Phi_z = dQ_z(y) - r^2\left(Q_z(x)Q_z(y) - Q_z^2(x, y)\right). \tag{4.19}$$

To describe $\mathfrak{M}$ and $\mathfrak{M}'$ we consider again the auxiliary matrix

$$M = P_y A P_y$$

with the spectrum $(0, \mu_2, \mu_3, \ldots, \mu_n)$, and set $\mu_1 = \langle x, y \rangle$. The function $m(z)$ is defined by (4.8).

As before, we show that x, y can be expressed in terms of $\mu_1 = \langle x, y \rangle$ and of the spectra of L, M. For this purpose we merely have to modify the expressions for the right-hand sides of (4.13) and (4.14).

In case $d \neq 0$ we have from (4.19), taking the coefficient of z^{-1} in the expansion at $z = \infty$,

$$\text{tr}(L - A) = \sum_{j=1}^{n} (\lambda_j - \alpha_j) = d\langle y \rangle^2,$$

so that $\langle y \rangle^2$ is a function of the λ_k, while $\mu_1 = \langle x, y \rangle$ is an independent variable on $\mathfrak{M}$. The formulae (4.13), (4.14), and (4.16) are to be replaced by

$$x = \frac{\langle x, y \rangle}{\langle y \rangle^2} y - \frac{1}{2} \sum_{j=2}^{n} g_j^{-1} Q_{\mu_j}(x, y) \frac{\partial y}{\partial \mu_j},$$

$$Q_z(x, y) = \frac{1}{ra(z)} \sqrt{(a(z) - l(z))a(z)} = \frac{1}{a(z)\Delta} \sqrt{P(z)}$$

for $z = \mu_2, \mu_3, \ldots, \mu_n$. This latter formula agrees with (4.16), since $\Delta = r^2$.

These equations, together with (4.10), again give x, y in terms of λ, μ, solving the inverse problem for $d = 1$. The mapping of $\mathfrak{M}'$ into the Jacobi variety of $w^2 = P(z)$ is the same as before.

We determine the symplectic structure from

$$\langle x, dy \rangle = \mu_1 \frac{\langle y, dy \rangle}{\langle y \rangle^2} - \frac{1}{2} \sum_{j=2}^{n} Q_{\mu_j}(x, y) d\mu_j$$

$$= d\left(\frac{\mu_1}{2} \log\langle y \rangle^2 \right) - \frac{1}{2} \log\langle y \rangle^2 d\mu_1$$

$$- \frac{1}{2\Delta} \sum_{j=2}^{n} a^{-1}(\mu_j) \sqrt{P(\mu_j)} \, d\mu_j,$$

leading to the symplectic form $\omega = \partial_\lambda \partial_\mu S$ with

$$S = \frac{\mu_1}{2} \log\left(\sum (\lambda_k - \alpha_k) \right) + \frac{1}{2\Delta} \sum_{j=2}^{n} \int^{\mu_j} \frac{\sqrt{P(z)} \, dz}{a(z)}, \qquad (4.20)$$

where $P(z) = r^2(a - l)l$ is a polynomial of degree $2n - 1$, and the genus of the curve $w^2 = P(z)$ is again $n - 1$. One branch point of the corresponding Riemann surface is at ∞, and n of them are the eigenvalues λ_k of L.

Finally, in the case (iii) with $d = 0$ one has

$$\sum_{j=1}^{n} (\lambda_j - \alpha_j) = d\langle y \rangle^2 = 0,$$

so that only $n - 1$ of the λ_j, say $\lambda_2, \lambda_3, \ldots, \lambda_n$, are independent variables. Similarly M has only $n - 1$ independent eigenvalues $\mu_2, \ldots, \mu_n$, while $\mu_1 = 0$. As additional variables we could use $\mu_1 = G = \frac{1}{2}\langle x, x \rangle$ and $F = \langle x, y \rangle$ and proceed as above. This is in accordance with our reduction at the beginning of this section. Notice that the first term in S in (4.17) and (4.20) reflects the exponential and linear behavior of μ_1 in t.

e. Limit Cases

The isospectral manifold of $L = L(x, y)$ is defined by

$$\Phi_\lambda(x, y) = 1, \quad j = 1, 2, \ldots, n.$$

In Section 3 we were interested in the manifold

$$\Phi_\lambda(x, y) = 0, \quad j = 1, 2, \ldots, n - 1,$$

which represented the spectrum of $P_y(A - x \otimes x)P_y$. This manifold can be obtained as a limit case from the former, as was shown in Section 3. We want to indicate how one can determine the hyperelliptic curve for this situation, at least for $a = 0, b = -c = -d = 1$.

In Section 3 we saw that the spectrum of

$$L_\nu = L(x, \nu y) = A + \nu(x \otimes y - y \otimes x) - \nu^2 y \otimes y$$

is given by the equation

$$\Phi_z(x, y) = \frac{1}{\nu^2},$$

where $\Phi_z(x, y)$ is the function belonging to L_1. If $\lambda_j(\nu)$ are the eigenvalues of L_ν, we set

$$l_\nu(z) = \prod_{j=1}^{n} (z - \lambda_j(\nu))$$

and have the hyperelliptic curve

$$w^2 = P_\nu(z) = \left(r^2 a(z) - \Delta l_\nu(z)\right)a(z).$$

We assume again $(a, b, c, d) = (0, 1, -1, -1)$. In the limit $\nu \to \infty$ one root, say $\lambda_1 = \lambda_1(\nu)$, tends to infinity, and $\nu^{-2}l_\nu(z)$ has a limit, namely a polynomial of degree $n - 1$ having the eigenvalues of $P_y(A - x \otimes x)P_y$ as roots. Let $l^{(0)}(z)$ be the polynomial of degree $n - 1$ and highest coefficient 1 having these roots. Then we conclude

$$\Phi_z(x, y) = \frac{1}{\nu^2}\left(1 - \frac{l_\nu(z)}{a(z)}\right) \to \kappa \frac{l^{(0)}(z)}{a(z)}$$

with some factor κ independent of z. This factor is determined by the asymptotic behavior of Φ_z for large z as

$$\kappa = (a\langle x, x \rangle + 2s\langle x, y \rangle + d\langle y, y \rangle) = 2G(x, y).$$

Thus

$$\Phi_z(x, y) = 2G(x, y) \frac{l^{(0)}(z)}{a(z)},$$

so that $zl^{(0)}(z)$ is the characteristic polynomial of $P_y(A - x \otimes x)P_y$.

We will not repeat the above construction in this case, but note that as auxiliary variables one could use the eigenvalues of either

$$M = P_y A P_y \quad \text{or} \quad N = A - x \otimes x.$$

The eigenvalues of the latter equation represent the elliptic coordinates of x, while the eigenvalues of M represent the orthogonal coordinates of y on the sphere $|y| = 1$ which we used previously. In any event the isospectral manifold $\mathfrak{M}$ of $P_y(A - x \otimes x)P_y$ leads in this case to the Jacobi variety of the hyperelliptic curve

$$w^2 = a(z)l^{(0)}(z)$$

$$= z^{-1} \det(z - A) \det(z - L).$$

This case is particularly pleasant, since the branch points of this polynomial are given by the eigenvalues of A and L where the trivial zero eigenvalue of L is omitted.

In this case the manifold[6] $\mathfrak{M}' = (\mathfrak{M} - \mathfrak{M} \cap K)/\Gamma = \mathfrak{M}/\Gamma$ can be interpreted geometrically. It is the manifold of common tangents to $n - 1$ confocal quadrics $\mathfrak{A}_{\lambda_1}, \mathfrak{A}_{\lambda_2}, \dots, \mathfrak{A}_{\lambda_{n-1}}$, where those 2^n tangents are to be identified which go into each other by the reflections $x_j \to \pm x_j$. Thus we obtain the result that the common tangents to $n - 1$ confocal quadrics form a 2^n-fold covering of the Jacobian variety (see Staude [20]).

5. Examples of Integrable Flows

a. Constrained Systems

In the following examples we will have to constrain a Hamiltonian system

$$\dot{x} = H_y, \qquad \dot{y} = -H_x \tag{5.1}$$

in the symplectic space (R^{2n}, ω) $\omega = \sum_{j=1}^{m} dy_j \wedge dx_j$, to a symplectic submanifold. For our purposes it will suffice to describe the submanifold M by $2r$ equations

$$M : F_1(x, y) = \cdots = F_{2r}(x, y) = 0.$$

If

$$\det(\{F_j, F_k\}) \neq 0 \quad (j, k = 1, 2, \dots, 2r), \tag{5.2}$$

then the manifold M is symplectic with $\omega|_M$ the two-form restricted to TM.

The vector field (5.1), which we will denote also by X_H, need not be tangential to M, but we can construct such a vector field by replacing H with H_M, the restriction of H to M. Then the function H_M on the symplectic manifold (M, ω_M) defines a vector field X_{H_M} which is tangential to M. This vector field X_{H_M} will be called the constrained vector field.

There is another way to describe this constrained flow: Since ω_M is nondegenerate in TM, there exists a complementary space $(TM)^\perp$ of TM in R^{2n}, which is orthogonal to TM with respect to the symplectic structure. Now it is easily seen that $X_H - X_{H_M} \in (TM)^\perp$. In other words X_{H_M} is the projection of X_H into TM with respect to the above splitting of R^{2n}.

---

[6] By the remarks at the beginning of this section, we can drop K.

Effectively one can describe the constrained vector field by a Hamiltonian

$$H^* = H - \sum_{j=1}^{2r} \lambda_j F_j(x, y). \qquad (5.3)$$

In order for X_{H^*} to be tangential to M we have to require that

$$0 = X_{H^*} F_k = \{H, F_k\} - \sum \lambda_j \{F_j, F_k\} \qquad \text{on } M,$$

which by (5.2) determines the functions λ_j uniquely on M.

Next we turn to a more special situation where $\mathscr{F}$ denotes a class of functions in involution in (R^{2n}, ω), and consider a vector field X_H constrained to a symplectic manifold

$$M : F_1 = F_2 = \cdots = F_r = 0, G_1 = G_2 = \cdots = G_r = 0, \qquad (5.4)$$

where we assume that

$$F_1, F_2, \ldots, F_r, H \in \mathscr{F}, \qquad (5.5\text{i})$$

$$\det\{F_i, G_j\} \neq 0 \qquad (i, j = 1, 2, \ldots, r). \qquad (5.5\text{ii})$$

Clearly this is a special case of the previous situation since (5.5i, ii) imply (5.2) if we set $F_{j+r} = G_j$ for $j = 1, 2, \ldots, r$. The constrained Hamiltonian (5.3) takes the form

$$H^* = H - \sum_{j=1}^{r} (\lambda_j F_j + \mu_j G_j).$$

Since the F_j and H are in $\mathscr{F}$ and hence in involution, we conclude

$$0 = \sum_{j=1}^{r} \mu_j \{G_j, F_k\}$$

and hence $\mu_j = 0$. Thus the constrained Hamiltonian has the form

$$H^* = H - \sum_{j=1}^{r} \lambda_j F_j,$$

and the corresponding vector field is given by

$$X_{H^*} = X_H - \sum \lambda_j X_{F_j}$$

on M.

This has the consequence that for any function $E \in \mathscr{F}$ we have on M

$$X_{H^*} E = 0,$$

i.e., $E|_M$ is an integral of the constrained vector field. We conclude that in the situation (5.4), (5.5) the constrained flows have the restriction of the functions in $\mathscr{F}$ as integrals. It is evident that these functions in the class $\mathscr{F}_M$, obtained from $\mathscr{F}$ by restriction to M, are in involution.

We will apply this simple device of constraining an integrable system to obtain a new integrable system in the following examples.*

Note added in proof: Other integrable systems have been recognized as constrained systems. See P. Derft, F. Lund and E. Trubowitz, Nonlinear wave equations and constrained harmonic motion, *Comm. Math. Phys.* **74**, 141–188 (1980).

b. A Mass Point on the Sphere $S^{n-1}: |x| = 1$ under the Influence of the Force $-Ax$ (C. Neumann [14])

The differential equation of this system is

$$\ddot{x} = -Ax + \lambda x,$$

where λ is chosen so that $|x| = 1$, $\langle x, \dot{x}\rangle = 0$. We obtain this system by constraining the Hamiltonian

$$H = \tfrac{1}{2}\langle Ax, x\rangle + \tfrac{1}{2}(|x|^2|y|^2 - \langle x, y\rangle^2)$$

to the symplectic submanifold

$$M : F = \tfrac{1}{2}(|x|^2 - 1) = 0, \; G = \langle x, y\rangle = 0.$$

Observe that $\{F, G\} = |x|^2 = 1$ on M. Moreover, if $\mathcal{F}$ is the class of functions generated by

$$\Phi_z(x, y) = Q_z(x) + Q_z(x) Q_z(y) - Q_z^2(x, y),$$

then the expansion at $z = \infty$ takes the form

$$\Phi_z(x, y) = \frac{|x|^2}{z} + \frac{1}{z^2}(\langle Ax, x\rangle + (|x|^2|y|^2 - \langle x, y\rangle^2)) + O\left(\frac{1}{z^3}\right).$$

Hence both F, H belong to $\mathcal{F}$, and the flow restricted to M is given by

$$H^* = H - \lambda F, \qquad \lambda = \{H, G\} = \langle Ax, x\rangle.$$

The equations have on M the form

$$\dot{x} = H_y^* = H_y - \lambda F_y = |x|^2 y = y,$$

$$\dot{y} = -H_x^* = -H_x + \lambda F_x = -Ax - |y|^2 x + \lambda x,$$

or

$$\ddot{x} = -Ax - (|y|^2 - \lambda)x,$$

which is the desired flow if $|y|^2 - \lambda$ is renamed λ.

The functions in $\mathcal{F}$ restricted to M are the desired integrals. In this case we have $a = -1$, $b = -c = 1$, $d = 0$ hence the symmetric part of $\left(\begin{smallmatrix} a & b \\ c & d \end{smallmatrix}\right)$ is $\left(\begin{smallmatrix} a & 0 \\ 0 & 0 \end{smallmatrix}\right)$ of rank 1. If we set

$$\Phi_z = \sum_{j=1}^{n} \frac{G_j(x, y)}{z - \alpha_j},$$

we easily see that

$$2H = \sum_{j=1}^{n} \alpha_j G_j(x, y), \qquad 2F + 1 = \sum_{j=1}^{n} G_j(x, y),$$

which shows explicitly that H, F are functions of the G_j, which are the integrals of this system. These integrals G_j have been found by K. Uhlenbeck (see [21] and Devaney [3]).

c. A Mass Point on the Ellipsoid $Q_0(x) + 1 = 0$ under the Influence of the Force $-ax$ (Jacobi [6])

The motion of a free particle under the influence of the force $-ax$ is described by the Hamiltonian

$$H = \tfrac{1}{2}(|y|^2 + a|x|^2).$$

To describe the restricted motion we introduce the class of functions $\mathcal{F}$ generated by

$$\Phi_z(x, y) = aQ_z(x) + Q_z(y) + Q_z(x)Q_z(y) - Q_z^2(x, y),$$

which are in involution by Section 2. We set

$$F(x, y) = a + \Phi_0(x, y)$$

$$= (1 + Q_0(x))(a + Q_0(y)) - Q_0^2(x, y),$$

$$G(x, y) = Q_0(x, y),$$

and restrict the motion to

$$F(x, y) = 0, \qquad G(x, y) = 0.$$

Since $H, F \in \mathcal{F}$, it follows that the restricted flow is described by

$$H^* = H - \lambda F,$$

and hence has all functions of $\mathcal{F}$ as integrals. It remains to identify this flow with the desired one. For this purpose we note that $G(x, y) = 0$ and $F(x, y) = 0$ implies

$$1 + Q_0(x) = 0 \quad \text{or} \quad a + Q_0(y) = 0.$$

We pick $a + Q_0(y) \neq 0$. This condition is invariant under the flow, as one verifies. Thus we have $1 + Q_0(x) = 0$, $Q_0(x, y) = 0$, which means that x lies on the ellipsoid and y is tangential to it at x. The differential equation becomes

$$\dot{x} = H_y^* = H_y - \lambda F_y = y,$$

$$\dot{y} = -H_x^* = -H_x + \lambda F_x = -ax - 2\lambda(a + Q_0(y))A^{-1}x,$$

or

$$\ddot{x} = -ax - 2\lambda(a + Q_0(y))A^{-1}x,$$

which is the equation of the constrained motion. Thus this system is integrable and $G_j|_M$ are the desired integrals in involution. For $a = 0$ we obtain the geodesic flow on the ellipsoid. Notice that the resulting integrals G_j are obtained from those of the Neumann system (Section 5 (b)) by the symplectic map $(x, y) \to (y, -x)$.

d. Geodesic Flow on the Orthogonal Group (Manakov [8], Mischenko [11])

Arnold studied the geodesic flow on $SO(n)$ under a right-invariant metric, say

$$\int \sqrt{\text{tr}(\dot{U}^T G \dot{U})}\, dt$$

where G is a fixed positive definite symmetric matrix and $U \in SO(n)$. If we define the element $A = \dot{U}U^{-1}$ in the Lie algebra, the Euler equations become

$$\frac{d}{dt}(GA + AG) = [A^2, G].$$

Setting $L = GA + AG + \mu G^2$, $B = A + \mu G$ with any scalar μ, these equations can be written in the Lax form

$$\frac{d}{dt} L = [B, L].$$

These formulas were derived by Manakov [8] to construct integrals of this system from the characteristic polynomial of L.

We should like to consider a $2n$-dimensional subsystem of this flow by setting

$$G = \text{diag}(g_1, g_2, \ldots, g_n),$$

$$L = x \otimes y - y \otimes x + \mu G^2,$$

$$B = \frac{x_i y_j - x_j y_i}{g_i + g_j} + \mu G.$$

One verifies that this system is compatible and is described by the Hamiltonian

$$H = -\frac{1}{2} \sum_{i < j} \frac{(x_i y_j - x_j y_i)^2}{g_i + g_j}.$$

This system is clearly related to those of Section 2 and agrees with (2.9) if we set

$$\begin{pmatrix} a & b \\ c & d \end{pmatrix} = \begin{pmatrix} 0 & 1 \\ -1 & 0 \end{pmatrix},$$

$$\alpha_j = \mu g_j^2, \qquad \beta_j = \mu g_j = \sqrt{\mu \alpha_j}\, .$$

Thus this system is integrable and its solutions are expressible in terms of hyperelliptic functions.

e. Hill's Equation (McKean and Trubowitz [9, 10])

We consider the Hill equation

$$-\left(\frac{d}{ds}\right)^2 y + q(s) y = \lambda y,$$

where $q(s + 1) = q(s)$. McKean and Trubowitz as well as Novikov et al. have studied the problem of recovering the potential $q = q(s)$ from various spectra. As a rule one needs two spectra, corresponding to two sets of different boundary conditions, to determine $q(s)$, whereas there exists a family of potentials giving rise to one spectrum determined by one set of boundary conditions. Of particular interest here is the boundary condition of periodicity, say

$$y(s + 1) = \pm y(s), \quad \text{or} \quad y(s + 2) = y(s).$$

The eigenvalues[7] $\lambda'_0 < \lambda'_1 \leqslant \lambda'_2 < \lambda'_3 \leqslant \lambda'_4 < \cdots$ form the endpoints of the band spectrum $[\lambda'_0, \lambda'_1] \cup [\lambda'_2, \lambda'_3] \cup \cdots$ of the continuous spectrum of Hill's equation considered on the line $-\infty < s < +\infty$. The other intervals $(-\infty, \lambda'_0)$, (λ'_1, λ'_2), (λ'_3, λ'_4), ... are called the instability intervals.

It is clear that the spectrum λ'_j cannot be prescribed arbitrarily but is subject to asymptotic restrictions for $j \to \infty$. Moreover, McKean and Trubowitz's work shows that the position of the λ'_{2j-1} is entirely determined by the λ'_{2j}. Of special interest is the case of a "finite-band" spectrum where all but a finite number of the eigenvalues are double, e.g.,

$$\lambda'_0 < \lambda'_1 < \cdots < \lambda'_{2N};$$
$$\lambda'_{2j-1} = \lambda'_{2j} \quad \text{for } j \geqslant N + 1. \tag{5.6}$$

In this case the entire spectrum is uniquely determined by $\lambda'_0, \lambda'_2, \ldots, \lambda'_{2N}$, and the corresponding potentials have been determined. They form an N-dimensional real torus which is the real part of the Jacobi variety of the hyperelliptic curve

$$w^2 = \prod_{j=0}^{2n} (z - \lambda'_j).$$

It is interesting that this problem is closely related to the above rank-2 perturbations. The following observation is due to E. Trubowitz (Moser [12]). Let $q(s)$ be any periodic potential with the spectrum (5.6), and $f_j(s)$ be the eigenfunctions normalized by

$$\int_0^1 f_j^2(s)\, ds = 1.$$

We consider the spectrum λ'_j fixed; then there exist positive numbers ϵ_j, $j = 0$, $1, \ldots, N$, just depending on the spectrum λ'_k, not on the particular potential, such that

$$\sum_{j=0}^{N} \epsilon_j f_{2j}^2(s) = 1, \qquad \sum_{j=0}^{N} \epsilon_j = 1.$$

The determination of the ϵ_j is irrelevant for us; it can be found in McKean and Trubowitz [9].

We observe that the point x with the coordinates

$$x_j = \sqrt{\epsilon_j}\, f_{2j}(s), \quad \alpha_j = \lambda'_{2j}, \quad j = 0, 1, \ldots, N, \tag{5.7}$$

is restricted to a sphere of radius 1. Moreover,

$$\frac{d^2 x_j}{ds^2} = \sqrt{\epsilon_j}\, f_{2j}''(s) = q x_j - \lambda_{2j} x_j.$$

---

[7] We supply the prime to avoid confusion with the eigenvalues λ_k of $L = L(x, y)$. Although the $\lambda'_0, \lambda'_1, \ldots$ are the eigenvalues for the two boundary conditions $y(s + 1) = \pm y(s)$, they are not sufficient to fix $q(s)$.

This equation can be viewed as constraining the motion $x_j'' + \lambda_{2j}' x_j = 0$ to a sphere, where $q = q(s)$ plays the role of the normal force,

$$q(s) = \sum_{j=0}^{N} \left(\lambda_{2j}' x_j^2(s) - x_j'^2(s) \right). \tag{5.8}$$

Thus this problem can be related to Neumann's constrained motion on the sphere if we set $n = N + 1$ and use (5.7). The potential $q(s)$ is then obtained from (5.8).

Now we know that all solutions of the Neumann problem are expressible as the inverses of hyperelliptic functions, i.e. are quasiperiodic functions associated to a torus of dimension $n - 1 = N$. Such a torus is characterized by prescribing the constant values of

$$G_k = \epsilon_k f_{2k}^2 + \sum_j{}' \frac{\epsilon_j \epsilon_k}{\lambda_{2k}' - \lambda_{2j}'} \left(f_{2j} f_{2k} - f_{2k} f_{2j}' \right)^2$$

or alternatively, by prescribing the function

$$\Phi_z(x, y) = \sum_{k=0}^{N} \frac{G_k}{z - \lambda_{2k}'} .$$

We will show that the periodicity condition $q(s + 1) = q(s)$ implies that

$$\Phi_z(x, y) = \frac{\displaystyle\prod_{k=1}^{N} (z - \lambda_{2k-1}')}{\displaystyle\prod_{k=0}^{N} (z - \lambda_{2k}')}, \tag{5.9}$$

so that Φ_z vanishes at $z = \lambda_{2k-1}'$ $(k = 1, 2, \ldots, N)$ and has poles at λ_0', $\lambda_2', \ldots, \lambda_{2N}'$.

Thus the N-gap potentials correspond to just one of the tori of the Neumann problem of Section 5(b). To prove (5.9) we consider the spectral problems given by

$$L = P_x (A - y \otimes y) P_x, \qquad M = P_x A P_x,$$

where the eigenvalues α_k of A are given by $\alpha_k = \lambda_{2k}'$, $k = 0, 1, \ldots, N$, and the eigenvalues λ_k of L are claimed to be λ_{2k-1}' $(k = 1, \ldots, N)$. By Section 4 the underlying hyperelliptic curve is given by

$$w^2 = P(z) = l^{(0)}(z) a(z)$$

$$= \prod_{1}^{N} (z - \lambda_k) \prod_{0}^{N} (z - \alpha_k).$$

On the other hand, in the theory of McKean and Trubowitz the hyperelliptic curve is given by

$$w^2 = \prod_{j=0}^{2N} (z - \lambda_j'),$$

which shows that the λ_k agree with the λ_{2k-1}', as we wanted to show.

The periodicity condition $q(s + 1) = q(s)$ forces that the $\lambda'_{2k-1} = \lambda_k$ are determined by the λ_{2k}, and it is natural to ask for those potentials belonging to the other tori, i.e. to tori for which the λ'_{2k-1} do not give rise to periodic potentials. From (5.8) and the above theory it is clear that for general choice of the λ'_{2k-1} the potential is a quasiperiodic function expressible in terms of θ-functions. In fact, such potentials have been considered by Dubrovin, Matveev, and Novikov [5]. We restricted ourselves to the finite-band case because it corresponds to a finite-dimensional mechanical system. Clearly this approach should be extended to the general case corresponding to the motion of a particle constrained to an infinite-dimensional sphere, or an isospectral flow in Hilbert space.

6. Appendix

We mention another integrable mechanical system whose integrals can be interpreted as eigenvalues of a matrix obtained by rank-two perturbation. The example is the motion of a particle on the sphere $|x| = 1$ in R^n under the influence of a potential

$$U = \frac{1}{2} \left(\langle Ax, x \rangle - \sum_{j=1}^{n} c_j^2 x_j^{-2} \right).$$

The corresponding differential equations are

$$\ddot{x}_j = -U_{x_j} = -\left[(Ax)_j + \frac{c_j^2}{x_j^3} \right] + \lambda x_j. \tag{6.1}$$

By separation of variables in the Hamilton–Jacobi equation Rosochatius [16] showed that this system is integrable. Here we want to show that this problem can be described as an isospectral flow of matrices of the form

$$L(x, y) = A + ax \otimes x + r(x \otimes y - y \otimes x)$$
$$+ x \otimes \frac{c}{x} + \frac{c}{x} \otimes x,$$

where c/x stands for the vector with components c_j/x_j. In particular, it follows that the eigenvalues of this matrix are in involution with respect to the standard symplectic structure. This matrix is again a rank-two perturbation, since

$$L = A + x \otimes \xi + \eta \otimes x$$

with

$$\xi = ax + ry + \frac{c}{x}, \qquad \eta = -ry + \frac{c}{x}.$$

Without giving the calculation, we give the generalization of Theorem 2 to this case.

With any fixed diagonal matrix $\beta = \mathrm{diag}(\beta_1, \beta_2, \ldots, \beta_n)$, define

$$B = r\beta + \left(\frac{\beta_i - \beta_j}{\alpha_i - \alpha_j} \left[r^2(x_i y_j - x_j y_i) + r\left(\frac{c_i}{x_i} x_j + \frac{c_j}{x_j} x_i \right) \right] \right) - (\delta_{ij} D_j),$$

where

$$D_j = r\sum_k{}' \frac{\beta_j - \beta_k}{\alpha_j - \alpha_k}\left(\frac{c_j}{x_j^2} + \frac{c_k}{x_k^2}\right)x_k^2.$$

Then the isospectral deformation

$$\frac{dL}{dt} = [B, L]$$

is equivalent to the Hamiltonian system

$$\dot{x} = H_y, \qquad \dot{y} = -H_x,$$

with

$$2H = a\langle \beta x, x\rangle - \sum_{i<j}\frac{\beta_i - \beta_j}{\alpha_i - \alpha_j}\left[r^2(x_iy_j - x_jy_i)^2 + x_i^2\frac{c_j^2}{x_j^2} + x_j^2\frac{c_i^2}{x_i^2}\right].$$

We write again

$$H = \sum_{j=1}^{n} \beta_j G_j$$

and obtain this way n rational functions in involution, and all the above X_H are integrable. In particular, for $\beta = A$ we obtain

$$2H = a\langle Ax, x\rangle - r^2(|x|^2|y|^2 - \langle x, y\rangle^2)$$

$$- \langle x, x\rangle \sum_{j=1}^{n}\frac{c_j^2}{x_j^2} + 2\sum_{j=1}^{n}c_j^2.$$

Restricting this system to the tangent bundle $\langle x, x\rangle = 1$, $\langle x, y\rangle = 0$ of the sphere gives the system (6.1) if we set $a = -1$, $r = 1$. Since also

$$2\sum_{j=1}^{n} G_j = a\langle x, x\rangle/2$$

belongs to the functions generated by G_j, the system (6.1) has the G_j as integrals.

The integration of the system can be carried out as above if one uses as auxiliary matrix

$$M = P_x A P_x.$$

The relevant hyperelliptic curve is

$$w^2 = P(z) = a^2(z)\left(1 - \sum_{j=1}^{n}\frac{c_j}{z - \alpha_j}\right)^2 - a(z)l(z).$$

Here $P(z)$ is a polynomial of degree $2n - 1$, and the curve of genus $n - 1$. But for $c_j \neq 0$ neither α_k nor the eigenvalues of λ_k of L are branch points.

P. Deift made the interesting observation that the system (6.1) can be derived from the Neumann system of Section 5(b) if one replaces n by $2n$, sets $\alpha_{k+n} = \alpha_k$ $(k = 1, 2, \ldots, n)$, and reduces the system by the n rotations in the $x_k - x_{k+n}$

planes. If one sets

$$x_k = r_k \cos \theta_k, \quad x_{k+n} = r_k \sin \theta_k; \quad y_j = x_j,$$

then the free Hamiltonian

$$\frac{1}{2} \sum_{k=1}^{n} \left(y_k^2 + y_{k+n}^2 + \alpha_k \left(x_k^2 + x_{k+n}^2 \right) \right)$$

becomes

$$\frac{1}{2} \sum_{k=1}^{n} \left(\dot{r}_k^2 + r_k^2 \dot{\theta}_k^2 + \alpha_k r_k^2 \right) = \frac{1}{2} \sum_{k=1}^{n} \left(\dot{r}_k^2 + \left(\frac{c_k}{r_k} \right)^2 + \alpha_k r_k^2 \right)$$

if $x_k y_{k+n} - x_{k+n} y_k = r_k^2 \dot{\theta}_k = c_k$ is the constant value of these integrals. Constraining this Hamiltonian to the unit tangent bundle gives the system (6.1).

This remark shows that matrices A with multiple eigenvalues are also of interest.

REFERENCES

[1] M. Adler, On a trace functional for formal pseudo-differential operators and the symplectic structure of the Korteweg–deVries equations. *Inv. Math.* **50** (3), 219–248 (1979).

[2] L. Bianchi, *Vorlesungen über Differentialgeometrie*, 2. Aufl. Teubner, Leipzig, Berlin, 1910.

[3] R. Devaney, Transversal homoclinic orbits in an integrable system. *Am. Math.* **100**, 631–648 (1978).

[4] L. A. Dikii, Hamiltonian systems connected with the rotation group. *Funct. Anal. and Its Appl.* **6** (4), 83–84 (1972).

[5] B. A. Dubrovin, V. B. Matveev, and S. P. Novikov, Nonlinear Equations of Korteweg–deVries type, finite-zone linear operators, and Abelian varieties. *Russ. Math. Survey* **31** (1), 59–146 (1976).

[6] C. G. J. Jacobi, Vorlesungen über Dynamik. In *Gesammelte Werke*, Supplementband, Berlin, 1884.

[7] T. Kato, *Perturbation Theory for Linear Operators*, 2nd ed., Springer, 1974, in particular pp. 244–250.

[8] S. V. Manakov, Remarks on the integrals of the Euler equations of the n-dimensional heavy top. *Func. Anal. and Its Appl.* **10** (4), 93–94 (1976).

[9] H. P. McKean and E. Trubowitz, Hill's operator and Hyperelliptic function theory in the presence of infinitely many branch points. *Comm. Pure Appl. Math.* **29**, 143–226 (1976).

[10] H. P. McKean and E. Trubowitz, Hill's surfaces and their theta functions. *Bull. Am. Math. Soc.* **84** (6), 1042–1085 (1979).

[11] A. S. Mischenko, Integral geodesics of a flow on a Lie group. *Funct. Anal. and Its Appl.* **4** (3), 73–78 (1970).

[12] J. Moser, Various aspects of integrable Hamiltonian systems. In *Proc. CIME Conference, Bressanone, Italy, June 1978. Prog. Math.* **8**, Birkhäuser (1980).

[13] C. Neumann, *Vorlesungen über Riemanns Theorie der Abelschen Integrale*, 2. Auflage. Teubner, Leipzig, 1884.

[14] C. Neumann, De problemate quodam mechanico, quod ad primam integralium ultraellipticorum classem revocatur. *Reine und Angew. Math.* **56**, 46–63 (1859).

[15] M. Reid, The complete intersection of two or more quadrics. Thesis, Cambridge Univ., 1972.

[16] E. Rosochatius, *Über die Bewegung eines Punktes.* (Inaugural Dissertation, Univ. Göttingen) Gebr. Unger, Berlin, 1877.

[17]   G. Salmon and W. Fiedler, *Analytische Geometrie des Raumes.* Teubner, Leipzig, 1863.

[18]   C. L. Siegel, *Topics in Complex Function Theory*, Vol. 2. Wiley-Interscience, New York, 1971.

[19]   P. Stäckel, Über die Integration der Hamilton–Jacobischen Differentialgleichung mittelst Separation der Variabeln. Habilitationsschrift, Univ. Halle-Wittenberg, 1891.

[20]   O. Staude, Geometrische Deutung der Additionstheoreme der hyperelliptishen Integrale und Functionen 1. Ordnung im System der confokalen Flächen 2. Grades. *Math. Ann.* **82**, 1–69, 145–176 (1883).

[21]   K. Uhlenbeck, Minimal 2-spheres and tori in S^k, preprint, 1975.

[22]   P. van Moerbeke, The spectrum of Jacobi matrices. *Inv. Math.* **37**, 45–81 (1976).

Remarks on Nonlinear Problems[1]

Louis Nirenberg*

Chern's work has covered all branches of the tree of differential geometry, including problems of hard analysis. Differential equations arising in geometric problems are often nonlinear, and this talk will be an informal presentation of some of the techniques that are used in proving existence of solutions of global nonlinear problems:

$$F(u) = y.$$

Here F is a mapping from some (function) space to another. There are two parts:

(1) Topological methods, in particular degree theory and generalizations.
(2) Variational methods: the solution is obtained as a stationary point of some functional.

In 1968 the AMS Global Analysis Conference was held here in Berkeley, and the Proceedings [2] give an excellent picture of the situation of these subjects at the time. This talk may be considered as a (rather biased) prolongation. It is not a comprehensive survey, but reflects my limited knowledge and taste. I will simply describe a few results with some indications of how they may be used in nonlinear partial differential equations. The talk is directed to nonexperts in the hope that they may be attracted to the subject. Many people are put off by an important analytic aspect of all these problems, that of finding estimates for the solutions. The estimates are unavoidable. How one obtains them varies from problem to problem, and we will not touch on that, but will try to avoid all technicalities.

I. Topological Methods

For convenience let us start by looking at a map F from one Banach space X to another Y, and consider the equation

$$F(u) = y. \qquad (1)$$

[1] This work was partially supported by NSF Grants MSC 76-07039 and INT-77-20878-A01 and by the Army Research Office Grant No. DDA 29-78-G-0127.

* Courant Institute of Mathematical Sciences, New York University, 251 Mercer St., New York, NY 10012, USA.

This might correspond to a partial differential equation in some function space. Let us suppose that F is a smooth mapping of X into Y in the sense of differential calculus.

A. Classical Things

A standard attack on the problem is to try to show that the range of the operator F is open and closed—it is then all of Y.

(i) To show openness one tries to use the implicit function theorem. For that one wants the linear operator $A = F'(u): X \to Y$ (i.e. the Jacobian matrix in finite dimensions) to be an invertible map from X onto Y. It suffices that it map onto Y.

(ii) To show closedness one usually tries to show that F is a *proper* map, i.e., the preimage of every compact set is compact. Here, estimates for the solution play a crucial role.

A more general method is the continuity method. Suppose F belongs (continuously in t) to a one-parameter family of maps F_t, $0 \leqslant t \leqslant 1$, with $F = F_1$. One tries to show that the set of values of t for which the equation

$$F_t(u) = y$$

has a solution is open and closed. The proofs involve the same ingredients as in the previous case.

Warning! Unfortunately, when applied to partial differential equations and function spaces X, Y, these methods seldom work for nonelliptic equations. The reason is that A^{-1}, if it exists, often maps Y not into X but into some larger function space. This usually expresses some loss of smoothness under the operation $A^{-1}F(u)$.

To continue with classical things, one of the most frequently used tools is the Leray–Schauder [15]:

Topological degree. Here $Y = X$ and

$$F = I - K \tag{2}$$

is a map defined on the closure $\overline{\Omega}$ of a bounded domain Ω in X. K is a compact operator, i.e., $K(\overline{\Omega})$ is compact. In case $F(\partial\Omega) \subset X \setminus \{y\}$, which is usually established via *a priori* estimates, the degree at y of the map F in Ω is well defined:

$$\deg(F, \Omega, y) = \nu, \quad \text{an integer.}$$

It is, roughly, the number of times y is covered—counted with orientation. This is obtained from the degree of finite dimensional maps of $\overline{\Omega} \subset R^n \to R^n$, by approximating K by maps into finite-dimensional space and using the degree of the map restricted to that space. Then one uses the fact that the degree doesn't change under suspension of a map, i.e. extension of a map to a product space with another space, by taking the product map with identity in the other space.

The degree has the following important properties:

(a) deg depends only on $F|_{\partial\Omega}$, and is invariant under homotopy within the class of maps of the form $I + $ (compact of $\partial\Omega) \to X \setminus \{y\}$.
(b) If deg $\neq 0$, then there exists a solution u of (1).

A solution u_0 of (1) is called regular if $F'(u_0)$ is invertible. It is then an isolated solution, and the local degree of F in a small neighbourhood of u_0, called the index of u_0, is ± 1. It is useful to know this when trying to determine the number of solutions of (1). Leray and Schauder gave a formula for computing it.

J. Sylvester [28] has recently derived an extension of their formula, which involves the geometry of a component of $F^{-1}(y)$. In bifurcation theory one sometimes encounters, as a connected component of solutions, a compact manifold M without boundary. It is natural to ask: what is the degree of the map in some neighborhood of M in the generic case that the map is "regular" on M?

Generic case. $\ker F'(u) = T_u(M) \ \forall u \in M$; here $T_u(M)$ is the tangent space to M at u. The local degree would then be called index M.

Assuming M is orientable, Sylvester proved:

(a) For dim M odd, index $M = 0$.
(b) For dim M even, index $M = \pm\chi(E) = \pm$ (the Euler characteristic of a finite-dimensional vector bundle E over M). The fibre E_u over u (with dimension $= \dim M$) is defined as

$$E_u = X/\text{Range}\, F'(u) \qquad \forall u \in M.$$

Furthermore, one can determine when to take $+$ or $-$. In particular, if $T_u M \cap \text{Range}\, F'(u) = 0, \ \forall u \in M$, then index $M = \pm\chi(M)$.

B. Generalizations

The classical degree theory has been extended in many directions, and we will just take up one or two.

1. Set-valued compact maps K. Here for every u, $K(u)$ is a convex set, and $\bigcup_{u \in \bar{\Omega}} K(u)$ is compact. Equation (1) now takes the form

$$u - y \in K(u).$$

Assumimg K is upper-semicontinuous and

$$u - y \notin K(u) \qquad \forall u \in \partial\Omega,$$

the degree deg(F, Ω, y) can be defined with properties similar to those above. See for example Lloyd [16]. Recently this has been used by K. C. Chang [9] in several problems, including so-called obstacle problems in which one seeks solutions of certain partial differential equations satisfying side conditions that they lie above an obstacle. References to other work may be found there.

2. Fredholm maps F between Banach manifolds X, Y of index zero. Here X and Y are infinite-dimensional manifolds with local covering charts which are diffeomorphic to fixed Banach spaces X_0, Y_0. F is a smooth map, and its derivative $F'(u)$ is a linear Fredholm map $\forall u$, i.e., $F'(u): X_0 \to Y$ has finite-dimensional kernel with dimension $d_1(u)$; Range $F'(u)$ is closed and has finite codimension $d_2(u)$. The index of F is defined as

$$\text{index } F = d_1(u) - d_2(u).$$

By standard properties of linear Fredholm maps one sees that for X connected, index F is independent of u.

Suppose index $F = 0$ and that the restriction of F to any closed bounded set on X (say X has a metric compatible with the local Banach-space structure) is *proper*. For various classes of Banach manifolds on which orientation can be introduced, Elworthy and Tromba [11, 12] defined the degree $\deg(F, \Omega, y)$ with properties as before. They made use of the Sard–Smale lemma. This has been done in a somewhat simpler way without the Sard–Smale lemma by using transversality more systematically. A very useful survey article is [7] by Borisovich, Zvyagin, and Sapronov. They also extend this degree to maps of the form

$$F = F_1 + K,$$

F_1 Fredholm, K compact.

Up to now, these generalizations have not been widely applied, but I believe more and more applications will arise. To illustrate what they can do that the Leray–Schauder theory cannot, let us consider an elliptic boundary-value problem. For simplicity, consider a second-order nonlinear elliptic equation for a real function u defined in a bounded domain G in R^n with smooth boundary, u satisfying the boundary condition $u = 0$ on ∂G. The Leray–Schauder theory is particularly suitable for studying quasilinear equations of the form (using summation convention)

$$a^{ij}(x, u, \text{grad } u) u_{x_i x_j} + f(x, u, \text{grad } u) = 0 \quad \text{in } G. \tag{3}$$

Define the map $K: u \to v$ as the solution v of

$$a^{ij}(x, u, \text{grad } u) v_{x_i x_j} + f(x, u, \text{grad } u) = 0 \quad \text{in } G,$$

$$v = 0 \quad \text{on } \partial G.$$

In suitable function spaces K is a smoothing operator and so compact. To solve the equation (3) we seek a fixed point of the map K, i.e., a solution of

$$F(u) = u - K(u) = 0.$$

If one has good *a priori* estimates for the solution, so that one can say the solution lies inside some fixed ball, then the degree at the origin of the map F in that ball is well defined, and one may try to compute it and show that it is nonzero.

Consider now the fully nonlinear elliptic equation

$$A(x, u, \text{grad } u, u_{x_i x_j}) = y(x) \quad \text{in } G,$$

$$u = 0 \quad \text{on } \partial G,$$

i.e.

$$A[u] = y,$$

and let us suppose that we have a good *a priori* estimate for solutions. The Leray–Schauder theory is not directly applicable, since it is not clear how to set this in the framework of (1) and (2). However, in suitable function spaces, for example Sobolev spaces with many derivatives in L^2, the theory of Fredholm maps may be applied. In particular

$$\deg(A, \text{ball in function space}, y) = \nu$$

is defined. If $\nu \neq 0$, a solution exists. For example, if A may be deformed to a linear elliptic operator via A_t, $0 \leqslant t \leqslant 1$, so that the *a priori* estimates hold for all solutions of $A_t[u] = y$, $0 \leqslant t \leqslant 1$, then $A[u] = y$ has a solution. As we said, up to now this has not been used in any significant way.

3. Fredholm maps with positive index. As before, the degree can be defined. It is no longer an integer but a Pontrjagin framed cobordism class (see Elworthy and Tromba [11, 12], Borisovich et al. [7], and references there). So far this is difficult to apply. One may also use stable homotopy classes, or cohomotopy. This has been applied in a theoretical way to elliptic equations (see Nirenberg [18]), but so far no *natural* example has come up in which it has been used. I believe some will arise.

Further work has been done, primarily by Böhme and Tromba. In particular, in their work on minimal surfaces spanning a curve in R^n they have used Fredholm maps of positive index [6]. Very recently Böhme has considered the following problem: Nitsche proved [20] that an analytic closed curve in R^3 with total curvature $\leqslant 4\pi$ admits a *unique* spanning minimal surface of the type of the disc. Böhme showed [5] that given any $\epsilon > 0$ and an integer $N > 0$, there exists a simple closed analytic curve in R^3 with total curvature $< 4\pi + \epsilon$ spanned by at least N different minimal surfaces of disc type. These curves are close to S^1 covered twice, and the minimal surfaces all have branch points.

Many people have contributed to this subject—just to mention a few more, M. A. Krasnoselskii, F. Browder, R. Nussbaum, and (for more general use of homotopy theory) K. Geba and A. Granas. The following are some useful references for recent developments and applications: Berger [4], Browder [8], Geba and Granas [14], Nirenberg [19], Sattinger [26], Schwartz [27], and Zeidler [29].

II. Variational Methods

In the 1960's Morse theory was extended to real functions f defined on infinite-dimensional spaces (Banach manifolds) by Palais, Smale, and others (see for example Schwartz [27]). This theory has found wide application. However, the functions are usually required to satisfy the Palais–Smale condition:

Condition P-S: If x_j is a sequence such that

$$a \leqslant f(x_j) \leqslant b$$

and
$$f'(x_j) \to 0 \quad \text{in norm,}$$
then a subsequence of the x_j converges.

To give an idea of what the condition means, here is an

ILLUSTRATION. Consider the boundary-value problem for a function u defined in a bounded domain G in R^n with smooth boundary:
$$\Delta u + u^\sigma = 0 \quad \text{in } G, \qquad u = 0 \quad \text{on } \partial G, \tag{4}$$
for which one seeks a positive solution. Here $\sigma > 1$ is a constant. A solution is a stationary point of the functional
$$f[u] = \int_G \left(\tfrac{1}{2} |\text{grad } u|^2 - \frac{u^{\sigma+1}}{\sigma+1} \right) dx. \tag{5}$$
It seems natural to work in the function space $W_0^{1,2}$, i.e. functions u in L^2 with grad u in L^2 and $u = 0$ on ∂G. When does condition P-S hold? The answer is
$$\text{P-S holds} \quad \Leftrightarrow \quad \sigma < \frac{n+2}{n-2}.$$
This is seen quite easily with the aid of the Sobolev embedding theorem. In fact for $\sigma < (n+2)/(n-2)$ the problem admits a nontrivial solution, while for $\sigma \geqslant (n+2)/(n-2)$ it does not if, for example, G is star-shaped.

For many problems of interest Condition P-S is too restrictive; it cannot however be dropped. The nonlinearities in the elliptic Yang–Mills equations are just on the borderline where Condition P-S fails.

Using Condition P-S, the Lyusternik–Schnirelman category theory has been extended and applied, as well as the related concept of genus (see Rabinowitz's lectures [22]). These approaches use min-max arguments to obtain critical, i.e. stationary, points of functions, which are neither minima nor maxima (see also Palais [21]). In recent years Rabinowitz has introduced a variety of variational arguments in connection with his work on Hamiltonian systems and on time-periodic solutions of a nonlinear string equation. Some of them may be regarded as extensions of the

Mountain-Pass Lemma. *Consider a point p in a valley completely ringed by a mountain range. Let q be a point outside the range such that any path going from q to p must rise above the vertical heights of p and q. If $f(x)$ represents the height of the land above any point $x = (x_1, x_2)$, then the height c of the lowest mountain pass going from q to p is a stationary value of f, i.e., c is the number*
$$c = \inf_h \max_t f(h(t)).$$

Here $h(t)$ represents a curve in the plane from q to p, and the infimum is taken with respect to all such curves.

This critical value will in general be much lower than sup f = height of the highest mountain. This simple lemma is valid also in Banach space for a function satisfying Condition P-S (see Ambrosetti and Rabinowitz [1]). It yields, in fact, a simple proof of the existence of a nontrivial solution of (4) in case $1 < \sigma < (n + 2)(n - 2)^{-1}$. In this case, f defined in (5) satisfies P-S in $W_0^{1,2}$, and by Sobolev's embedding theorem,

$$\int_G \frac{u^{\sigma+1}}{\sigma + 1}\, dx \leqslant C\left[\int_G \tfrac{1}{2}|\mathrm{grad}\, u|^2\, dx\right]^{(\sigma+1)/2} \qquad \forall u \in W_0^{1,2},$$

for some constant C. Consequently for small ϵ we see that

$$f[u] \geqslant \mathrm{const}(\epsilon^2 - \epsilon^{\sigma+1})$$

$$\geqslant c_0\epsilon^2, \qquad c_0 > 0,$$

$\forall u$ on the sphere $\|u\|_{W_0^{1,2}} = \epsilon$. On the other hand this sphere separates 0 and some point u_0 far out with $f(u_0) = f(0) = 0$. By the lemma there exists a nontrivial stationary point of f.

Rabinowitz has proved a useful extension of the mountain pass lemma and used it in [23] and [24]. Because of lack of time we will describe only one of the results in [24]—concerning Hamiltonian systems for two n-vectors $p(t)$, $q(t)$ which are functions of time and satisfy

$$\dot{p}(t) = -H_q \qquad (\text{i.e. } p_i(t) = -H_{q_i}, \ i = 1, \ldots, n), \qquad (6)$$

$$\dot{q}(t) = H_p.$$

Here the Hamiltonian $H(p,q)$ is a C^1 function defined on R^{2n}.

Theorem (*Rabinowitz* [24]): Assume

(i) $H \geqslant 0$ and $H = o(|p|^2 + |q|^2)$ *near the origin,*
(ii) $0 < H \leqslant \theta(p \cdot H_p + q \cdot H_q)$ *near infinity, θ fixed, $o < \theta < \tfrac{1}{2}$.*

Then $\forall T > 0$, there is a nontrivial periodic solution of (6) with period T.

Before touching on the proof we mention that Rabinowitz uses this via an elegant argument to prove the existence of a nontrivial periodic solution of (6) on any constant energy hypersurface $\Sigma: H = $ constant, provided grad $H \neq 0$ in Σ, and Σ is compact and strictly star-shaped about the origin. Recently, by a quite different argument, Clarke and Ekeland [10] have generalized this result in case H is convex and

$$\frac{H}{|p|^2 + |q|^2} \to \left\{ \begin{matrix} \infty \\ 0 \end{matrix} \right\} \quad \text{as } |p|^2 + |q|^2 \to \left\{ \begin{matrix} 0 \\ \infty \end{matrix} \right\}.$$

For any $T > 0$ they prove the existence of a periodic solution on $H = $ constant having T as *smallest* period.

Using a topological index due to himself and Fadell [13] and arguments related to those of Lyusternik and Schnirelman, Rabinowitz has proved [25] that the Theorem holds even if condition (i) is dropped. The Fadell–Rabinowitz index is rather complicated to describe.

To prove the theorem as stated (for convenience suppose $T = 2\pi$), Rabinowitz obtains the solution as a stationary point, in the space of 2π-periodic vector functions $p(t)$, $q(t)$, of the functional

$$f[p,q] = \int_0^{2\pi} \left[p \cdot \frac{dq}{dt} - H(p,q) \right] dt.$$

This functional is unbounded above and below, but he shows that f has a positive critical value.

First he reduces the problem by a suitable approximation to a finite-dimensional one in which he employs the extended mountain-pass lemma. We will conclude with this lemma, expressed in a more general form due to Ni [17].

Variational Lemma. *Let $f \in C^1(R^n)$ satisfy Condition P-S. For some nonnegative integer $k \leqslant n - 1$ let ϕ and ψ be continuous maps of S^k and S^{n-k-1}, respectively, into R^n, which have nontrivial linking. Assume that for some constants $c_0 < c_1$,*

$$f(\phi(x)) \leqslant c_0 < c_1 \leqslant f(\psi(y)) \qquad \forall x \in S^k, \quad \forall y \in S^{n-k-1}.$$

Then the following number $c \geqslant c_1$ is a critical value of f:

$$c = \inf_h \max_{x \in B^{k+1}} f(h(x)).$$

Here the infimum is taken with respect to all continuous extensions h of ϕ to the unit ball B^{k+1}.

This result seems far from obvious, even for $n = 3$, $k = 1$, but in fact the proof is not difficult. Ni [17] has also given an infinite-dimensional form of it. Recently Benci and Rabinowitz [3] have proved related variational lemmas in infinite dimensions and applied them to various problems.

REFERENCES

[1]   A. Ambrosetti, and P. H. Rabinowitz, Dual variational methods in critical point theory and applications, *J. Func. Anal.* **14**, 349–381 (1973).

[2]   American Mathematical Society, *Global Analysis, Proceedings of a Symposium on Pure Mathematics*, Providence, 1970, Vol. 15.

[3]   V. Benci and P. H. Rabinowitz, Critical point theorems for indefinite functionals, *Invent. Math.* **52**, 241–273 (1979).

[4]   M. S. Berger, *Nonlinearity and Functional Analysis*, Academic, New York, 1977.

[5]   R. Böhme, A plateau problem with many solutions, to appear.

[6]   R. Böhme and A. J. Tromba, The index theorem for classical minimal surfaces, Sonderforschungsbereich 72, Bonn, Reprint No. 165.

[7]   Yu. G. Borisovich, V. G. Zvyagin, and Yu. I. Sapronov, Nonlinear Fredholm maps and the Leray–Schauder theorem, *Uspekhi Mat. Nauk* **32**, 3–54 (1977); Engl. transl., *Russian Math. Surveys* **32**, 1–54 (1977).

[8]   F. Browder, Nonlinear operators in Banach spaces, In *Proc. Symp. Pure Math.* 18, Part 2, Amer. Math. Soc., Providence, R. I., 1976.

[9]   K. C. Chang, The obstacle problem and partial differential equations with discontinuous nonlinearities, *Comm. Pure Appl. Math.*, **33**, 117–146 (1980).

[10] F. H. Clarke and I. Ekeland, Hamiltonian trajectories having prescribed minimal period, *Comm. Pure Appl. Math.*, **33**, 103–116 (1980).

[11] K. D. Elworthy and A. J. Tromba, Differential structures and Fredholm maps. In *Proc. Symp. Global Anal. 15*, Amer. Math. Soc., Providence, R. I., 1970, pp. 45–94.

[12] K. D. Elworthy and A. J. Tromba, Degree theory on Banach manifolds. In *Proc. Symp. Pure Math. 18, Part 1, Nonlinear Functional Analysis*, Amer. Math. Soc., 1970, pp. 86–94.

[13] F. R. Fadell and P. H. Rabinowitz, Generalized cohomological index theories for Lie group actions with an application to bifurcation questions for Hamiltonian systems, *Inv. Math.*, **45**, 134–174 (1978).

[14] K. Geba and A. Granas, Infinite dimensional cohomology theories, *J. Math. Pure Appl.* **52**, 145–270 (1973).

[15] J. Leray and J. Schauder, Topologie at équations fonctionelles, *Ann. Sci. Ec. Norm. Sup.* **51**, 45–78 (1934).

[16] N. G. Lloyd, *Degree Theory*, Cambridge Tract in Math. 73, Cambridge U. P., London, 1978.

[17] W. M. Ni, Some minimax principles and their applications in nonlinear elliptic equations. *J. d'Analyse*, **37**, 248–275 (1980).

[18] L. Nirenberg, An application of generalized degree to a class of nonlinear problems, In *3rd Colloq. Anal. Fonct.*, Liège Centre Belge de Rech. Math., pp. 57–73 (1971).

[19] L. Nirenberg, *Topics in Nonlinear Functional Analysis*, Lecture Notes, Courant Inst., 1974.

[20] J. C. C. Nitsche, A new uniqueness theorem for minimal surfaces, *Arch. Rat. Mech. Anal.* **52**, 319–329 (1973).

[21] R. S. Palais, Critical point theory and the minimax principle, *Proc. Symp. Pure Math. 15*, Amer. Math. Soc., Providence, R. I., pp. 185–212 (1970).

[22] P. H. Rabinowitz, Variational methods for nonlinear eigenvalue problems (CIME), Verona (1974). In *Ediz. Cremonese*, Rome, 1974, pp. 141–195.

[23] P. H. Rabinowitz, Free vibrations for a semilinear equation, *Pure Appl. Math.* **31**, 31–68 (1978).

[24] P. H. Rabinowitz, Periodic solutions of Hamiltonian systems, *Comm. Pure Appl. Math.* **31**, 157–184 (1978).

[25] P. H. Rabinowitz, A variational method for finding periodic solutions of differential equations, M. R. C. Univ. of Wisconsin Report 1854, Mar. 1978.

[26] D. H. Sattinger, *Topics in Stability and Bifurcation Theory*. Springer Lecture Notes No. 309, 1973.

[27] J. Schwartz, *Nonlinear Functional Analysis*, Gordon and Breach, New York, 1969.

[28] J. Sylvester, Ph.D. Thesis, Courant Inst. Math. Sci., New York Univ., 1980.

[29] E. Zeidler, *Vorlesungen uber Nichtlineare Funktionalanalysis* I–III, Teubner Texte zur Math., Leipzig, 1976–1977.

Minimal Surfaces, Gauss Maps, Total Curvature, Eigenvalue Estimates, and Stability

Robert Osserman*

1. Introduction

The subject matter that I wish to discuss here is one that seems particularly appropriate to this occasion. It is in fact, as will become abundantly clear, one of the many parts of differential geometry where Chern's influence has been both fundamental and pervasive. What is more, a substantial portion of the body of recent work described here is closely related to, and in many cases directly inspired by, a single paper of Chern's [19]—his first paper devoted to the theory of minimal surfaces.

Before entering into a more detailed description of our subject, let me inject a note concerning the title. It was inspired by the title of a recent book by Farber [33]. What I hope to show, however, is that unlike Farber's book (which turns out to be a collection of separate essays), the topics in our title all end up as individual strands that intertwine to form a single integrated whole.

The principal objects of our study will be immersed manifolds. We shall use the notation

$$M \hookrightarrow \overline{M}$$

for a manifold M immersed in a manifold $\overline{M}$. It will also be convenient to have a notation for a manifold M to be immersed as a *minimal submanifold* in a Riemannian manifold $\overline{M}$. We shall adopt for that the notation

$$M \underset{\mathrm{MIN}}{\hookrightarrow} \overline{M}.$$

When relevant, the dimension of M will be denoted by a superscript: M^m.

2. The Classical Gauss Map

2.1

We begin by recalling the definition and some basic properties of the classical Gauss map for a two-dimensional oriented surface $M \hookrightarrow \mathbb{R}^3$.

To each point $p \in M$, we assign the unit normal $N(p)$ that together with the

*Department of Mathematics, Stanford University, Stanford, CA 94305, USA.

given orientation on M gives the correct orientation in $\mathbb{R}^3$. Considering $N(p)$ as a point of the unit sphere $\hat{S}$, we obtain the Gauss map

$$g : M \to \hat{S}$$

defined by

$$g : p \mapsto N(p).$$

Given a neighborhood $V(p)$ on M, Gauss considered the ratio

$$\frac{\text{Area } g(V(p))}{\text{Area } V(p)}$$

as a measure of the amount of curvature of the surface M near p. He defined the quantity $K(p)$, now known as the *Gauss curvature* of M at p, by setting

$$|K(p)| = \lim_{V(p)\downarrow\{p\}} \frac{\text{Area } g(V(p))}{\text{Area } V(p)} \tag{1}$$

and by considering $K(p)$ to be positive if orientation is preserved under g, negative if orientation is reversed. Thus, the area form $d\hat{A}$ of the image of M under g satisfies

$$d\hat{A} = |K| dA, \tag{2}$$

or in integral terms,

$$\hat{A}(g(M)) = \int_M |K| dA, \tag{3}$$

where, of course, the area on the left has to be considered with appropriate multiplicities, i.e., including the number of "sheets" of the image over each point. Gauss went on to show the basic distinction between *intrinsic* quantities associated with M—those that may be defined purely in terms of measurements along the surface itself—and *extrinsic* quantities, which depend on how the surface lies in space. He proved his celebrated *Theorema egregium*, asserting that the curvature K is in fact intrinsic, even though its definition would seem to be a strictly extrinsic one.[1]

Quite apart from its relation to Gauss curvature, the Gauss map has become a basic tool in the classical theory of surfaces. One result of particular interest to us is the following (see, for example, Spivak [68, Vol. 4, p. 385]).

Proposition 2.2. *Given $M \hookrightarrow \mathbb{R}^3$, the Gauss map g is conformal if and only if either M lies on a sphere or else M is a minimal surface.*

3. Minimal Surfaces

3.1

Note that minimal surfaces arise here in a totally different context from that of their origin as area-minimizing surfaces. Let us recall that if we start with a surface M in $\mathbb{R}^3$, we may construct a one-parameter family of surfaces M_t by

---

[1] For a description of the content of Gauss's fundamental paper, see Spivak [68, Vol. 2, Chapter 3] or Dombrowski [29]. In particular, Dombrowski includes Gauss's original Latin text, with a side-by-side English translation, as well as a beautiful discussion of its historical setting.

displacing each point p on M through the vector $tu(p)N(p)$, where u is a smooth function on M. Given a relatively compact domain D on M, to each t there corresponds a domain D_t on M_t whose area we denote by $A(t)$. Then the *first variation* of area is given by

$$\left.\frac{dA}{dt}\right|_{t=0} = -2\int_D uH\,dA, \tag{4}$$

where H is a function on M whose value at each point is defined to be the *mean curvature* of M at the point. M is a *minimal surface* if $H \equiv 0$. If D has least area among all surfaces with the same boundary, then $A(t)$ has a minimum at $t = 0$, so that $A'(0) = 0$ whatever the function u. It follows from (4) that $H \equiv 0$. Thus, area-minimizing surfaces are minimal, and in general, minimal surfaces are precisely the stationary solutions of the variational problem associated with minimizing area.

3.2

The connection between minimal surfaces and the conformality of the Gauss map indicated in Proposition 2.2 makes the use of complex variables most natural, and was exploited extensively in the 19th century by Riemann, Weierstrass, and Schwarz. An important 20th-century application was to the proof of Bernstein's theorem:

3.3. *Let $f(x, y)$ be defined for all $(x, y) \in \mathbb{R}^2$, and let the graph of f be a minimal surface. Then $f(x, y)$ is a linear function:* $ax + by + c$.

This theorem was originally construed as a Liouville-type theorem for the second-order nonlinear partial differential equation that $f(x, y)$ must satisfy in order for its graph to be a minimal surface. However, extracting the essential geometric content of the theorem, Nirenberg formulated a more general result as a conjecture:

3.4. *Let M be a complete minimal surface in $\mathbb{R}^3$. If the image of M under the Gauss map omits a neighborhood of some point, then M is a plane.*

Clearly, under the hypotheses of Bernstein's theorem, the image under the Gauss map omits an entire hemisphere.

Nirenberg's conjecture was proved in 1959 (Osserman [58]), and the conformality of the Gauss map plays a central role. Chern's paper [19] showed how to extend the result to $\mathbb{R}^n$, but for that purpose it was necessary to study first the appropriate generalization of the Gauss map.

4. The Generalized Gauss Map

4.1

We consider now an oriented manifold of arbitrary dimension and codimension immersed in $\mathbb{R}^n$:

$$M^m \hookrightarrow \mathbb{R}^n.$$

There are various ways one could go about generalizing the classical Gauss map. For example, if we consider g in the classical case to be essentially the map

$$(p, N(p)) \mapsto N(p),$$

then one could generalize it by using the same map, where in higher codimension there is not a unique unit normal at p, but a whole set of them: the unit sphere in the normal space to M at p. The resulting map has as domain a sphere bundle over M whose fibres are $(n - m)$-dimensional unit spheres and the range of the map is the unit sphere S^{n-1} in $\mathbb{R}^n$. One can then use a definition analogous to (1) to get a formula like (3). This is the route taken by Chern and Lashof [23, 24]. It leads to many interesting results, and to a whole theory of its own. (See, for example, the book of Ferus [34], and the paper by Kuiper in this volume. See also 9.8 and 9.9 below.) It turns out, however, not to be the appropriate route for the kind of problem we wish to treat here.

4.2

What we shall work with instead is the generalized Gauss map

$$g : p \mapsto T_p M \qquad\qquad (5a)$$

whose domain is M, and whose range is the Grassmannian $G_{m,n}$ of oriented m-planes in $\mathbb{R}^n$. Thus

$$g : M \to G_{m,n}. \qquad\qquad (5b)$$

The Grassmannians $G_{m,n}$ have been studied in great detail, and their structure, both from the geometric and from the topological point of view, is well known. (See, for example, Chern [17] and Leichtweiss [49], Wong [71], and Hiller [38].)

4.3

Let us start by examining an important special case: the Grassmannian $G_{2,4}$ of 2 planes in 4-space. (See for example, Eells and Lemaire [31, Section 6] and Wintgen [70, Section 2].[2]) Each oriented 2-plane corresponds to a wedge product $v \wedge w$, where (v, w) is an orthonormal basis of the plane. Thus $G_{2,4}$ can be identified with unit bivectors in the space $\Lambda = \Lambda^2(\mathbb{R}^4)$. The Hodge *-operator decomposes Λ into subspaces Λ_+, Λ_-, on which the star operator has eigenvalue $+1$, -1 respectively. The projection of $G_{2,4}$ onto each of these subspaces is the sphere of radius $1/\sqrt{2}$. Since $\Lambda^2(\mathbb{R}^4)$ is a 6-dimensional linear space, we have

$$G_{2,4} = S_1 \times S_2.$$

where $S_j = S^2(1/\sqrt{2}) \subset \mathbb{R}^3$. Let π_j be the projection of $G_{2,4}$ onto S_j.

4.4

Suppose now that we have

$$M^2 \hookrightarrow \mathbb{R}^4.$$

The Gauss map $g : M \to G_{2,4}$ decomposes into a pair of maps

$$g_i = \pi_j \circ g : M \to S_j. \qquad\qquad (6)$$

---

[2] See also Section 2 of Atiyah's paper in this volume

Various results from the theory of surfaces in $\mathbb{R}^3$ have been generalized to surfaces in $\mathbb{R}^4$ using this pair of maps. For example, we have the Gauss–Bonnet formula for a compact surface M:

$$\int_M K\, dA = 4\pi\chi \tag{7}$$

where χ is the Euler characteristic of M. If M is immersed in $\mathbb{R}^3$, the left-hand side of (7) can be interpreted, as we have seen in 2.1, as the total area of the image of M under the Gauss map, provided the area is counted algebraically—i. e., positively when orientation is preserved, and negatively when orientation is reversed.

Blaschke [7] gave a generalization to surfaces in $\mathbb{R}^4$:

Theorem 4.5. *Let M be a compact surface immersed in $\mathbb{R}^4$, and let χ be its Euler characteristic. Let A_j, $j = 1, 2$, be the algebraic area of the image of M under the map g_j defined in 4.4. Then*

$$A_1 + A_2 = 4\pi\chi.$$

A topological proof of this result was given by Chern and Spanier [26], who also proved that necessarily $A_1 = A_2$. An interesting application of Chern and Spanier's paper was made recently by Wintgen [70].

A generalization of 2.2 was given by Pinl [61]:

Theorem 4.6. *Given a minimal immersion*

$$M^2 \underset{\text{MIN}}{\hookrightarrow} \mathbb{R}^4,$$

the maps g_1, g_2 defined by (6) are both conformal.

Pinl's proof uses the quaternionic representation of $G_{2,4}$.

With all this as background, we come to the paper of Chern [19] referred to at the outset. We may state immediately one of his results, generalizing Nirenberg's conjecture 3.4.

Theorem 4.7. *Let M be a complete minimal surface in $\mathbb{R}^4$. If the image of M under each of the maps g_1, g_2 (defined by (6)) omits a neighborhood of some point, then M is a plane.*

Clearly, Chern's proof uses the conformality of the maps g_j. In order to extend the theorem to $\mathbb{R}^n$ it was necessary first to find a suitable substitute for the conformality of each map, and second to see how to generalize the hypothesis of omitting sets under each of the images. The answer in both cases was tied to a special feature of the Grassmannian $G_{2,n}$.

Lemma 4.8. *The Grassmannian $G_{2,n}$ can be identified with the complex quadric*

$$Q_{n-2} : z_1^2 + \cdots + z_n^2 = 0$$

in complex projective space $\mathbb{C}P^{n-1}$.

For various forms of this identification, see for example, Chern [19], Osserman [59], and Kobayashi and Nomizu [44, p. 278].

The first consequence of this fact is that $G_{2,n}$ has a natural complex structure. Using that, Chern showed [19]:

Lemma 4.9. *If M^2 is minimally immersed in $\mathbb{R}^n$, then the Gauss map g is antiholomorphic.*

This property of the Gauss map for minimal surfaces in $\mathbb{R}^n$ is absolutely fundamental, and has been a basic tool for all subsequent work.

As a first application, Chern gave the following generalization of 3.4:

Theorem 4.10. *Let M be a complete minimal surface in $\mathbb{R}^n$. Consider the Gauss map g as a map*

$$g : M \to Q_{n-2} \subset \mathbb{C}P^{n-1}.$$

If the image $g(M)$ does not intersect any hyperplane in a neighborhood of some fixed hyperplane of $\mathbb{C}P^{n-1}$, then M must be a plane.

Note an interesting feature of this theorem. Its statement depends on viewing the Grassmannian not intrinsically, but as a submanifold of projective space, and using extrinsic entities: the hyperplanes of the ambient space.

We next review briefly some additional results that have been obtained using Lemma 4.9.

5. Minimal Surfaces $M^2 \underset{\mathrm{MIN}}{\hookrightarrow} \mathbb{R}^n$

5.1

Throughout this section, M will denote a two-dimensional minimal surface in $\mathbb{R}^n$.

We start with a result (Osserman [59]) that is basically a finite version of Theorem 4.10.

Theorem 5.2. *Given $M \underset{\mathrm{MIN}}{\hookrightarrow} \mathbb{R}^n$ and $p \in M$, let d be the distance from p to the boundary of M (measured along M). Suppose that all normal vectors at all points of M make an angle of at least α with some fixed direction. Then the Gauss curvature K of M at p satisfies*

$$|K(p)| \leqslant \frac{1}{d^2} \frac{16(n-1)}{\sin^4 \alpha}. \tag{8}$$

Corollary 5.3. *If, under the same hypotheses, M is complete, then M is a plane.*

Indeed, M complete is equivalent to $d = \infty$, so that (8) implies $K(p) = 0$ for all $p \in M$. But a minimal surface in $\mathbb{R}^n$ with $K \equiv 0$ must be a plane.

The same paper also proves some results for minimal surfaces in $\mathbb{R}^3$ and

$\mathbb{R}^4$—in particular, a result which implies the following sharpening of Theorem 4.7:

Theorem 5.4. *If the image of a complete* $M \underset{\mathrm{MIN}}{\hookrightarrow} \mathbb{R}^4$ *omits sets of positive measure under each of the maps* g_1, g_2, *then* M *is a plane.*

Also a number of results on total curvature:

Theorem 5.5. *A complete* $M \underset{\mathrm{MIN}}{\hookrightarrow} \mathbb{R}^3$ *with finite total curvature is conformally equivalent to a compact Riemann surface with a finite number of points deleted. If the number of points deleted is* k, *and the Euler characteristic of* M *is* χ, *then*

$$\int_M K \, dA \leqslant 2\pi(\chi - k). \tag{9}$$

Theorem 5.6. *A complete* $M \underset{\mathrm{MIN}}{\hookrightarrow} \mathbb{R}^3$ *has total curvature either* $-\infty$, *or else an integral multiple of* 4π, *the integer being negative unless* M *is a plane.*

Theorem 5.7. *A complete* $M \underset{\mathrm{MIN}}{\hookrightarrow} \mathbb{R}^3$ *with minimal absolute total curvature*

$$\int_M K \, dA = -4\pi$$

must be either simply connected or doubly connected. In the former case, M *is Enneper's surface, and in the latter case,* M *is a catenoid.*

Finally, a result linking total curvature to the Gauss map.

Theorem 5.8. *Given a complete* $M \underset{\mathrm{MIN}}{\hookrightarrow} \mathbb{R}^3$ *with finite total curvature, if* M *is not a plane, then the image* $g(M)$ *under the Gauss map must cover the entire sphere except for at most 3 points.*

5.9

Theorems 5.5–5.8 turned out all to have extensions to $\mathbb{R}^n$ (Chern and Osserman [25]). The key observation that makes the extensions possible is that the basic formula (3) that holds for all surfaces in $\mathbb{R}^3$ continues to hold for minimal surfaces in $\mathbb{R}^n$; i.e., $M \underset{\mathrm{MIN}}{\hookrightarrow} \mathbb{R}^n \Rightarrow$

$$\hat{A}(g(M)) = -\int_M K \, dA. \tag{10}$$

Note that for minimal surfaces in $\mathbb{R}^n$, $K \leqslant 0$, so that the right-hand side of (10) is the absolute total curvature. The left-hand side of (10) represents the area of the image of M under the generalized Gauss map g, and for its definition, we need a metric in the Grassmannian $G_{2,n}$. Under the identification

$$G_{2,n} \leftrightarrow Q_{n-2} \subset \mathbb{C}P^{n-1} \tag{11}$$

the Grassmannian inherits a metric from the canonical Fubini–Study metric on $\mathbb{C}P^{n-1}$. The latter is defined up to a multiplicative constant as the unique metric invariant under unitary transformations. We have:

Proposition 5.10. *There exists a unique metric on complex projective space with constant holomorphic sectional curvature $\overline{K}$. If we choose the normalization*

$$\overline{K} = 2, \tag{12}$$

then the induced metric on Q_{n-2} defines a metric $d\hat{s}^2$ on the Grassmannian $G_{2,n}$ under the identification (11), satisfying

$$g^*(d\hat{s}^2) = -K\,ds^2 \tag{13}$$

for any minimal surface M in $\mathbb{R}^n$, where ds^2 is the metric on M, K the Gauss curvature of M, and $g: M \to G_{2,n}$ is the generalized Gauss map.

Equation (13) implies in particular that the Gauss map is conformal for a minimal surface, and that

$$g_*(d\hat{A}) = -K\,dA, \tag{14}$$

which is equivalent to (10).

Using Equation (10), one is able to translate statements about the total curvature of minimal surfaces in $\mathbb{R}^n$ into corresponding statements about the area of holomorphic curves in $\mathbb{C}P^{n-1}$, which in turn may be treated by integral-geometric methods involving the measure of the set of hyperplanes intersecting the curve. In that way, it was possible to extend all the results of Theorems 5.5–5.8. In particular, Theorem 5.5 goes over without change, while the "quantization" result, Theorem 5.6, has to be modified in that the total curvature of a complete $M \underset{\text{MIN}}{\hookrightarrow} \mathbb{R}^n$ is an integral multiple of 2π, instead of 4π. Only recently have appropriate generalizations of Theorem 5.7 been made (C. C. Chen [14–16], Hoffman and Osserman [40]). In particular, Chen proved [16]:

Theorem 5.11. *A complete $M \underset{\text{MIN}}{\hookrightarrow} \mathbb{R}^n$ with total curvature*

$$\int_M K\,dA = -4\pi \tag{15}$$

must be either simply connected or doubly connected. In the former case it lies in some affine $\mathbb{R}^6 \subset \mathbb{R}^n$, and in the latter case, in some $\mathbb{R}^5 \subset \mathbb{R}^n$.

We note that Theorem 5.11 can also be deduced from work of Gackstatter [36].

One can, in fact, give a complete description of complete minimal surfaces satisfying (15) (Hoffman and Osserman [40]) and one finds that the dimensions "5" and "6" in Chen's theorem are sharp. It turns out that the doubly connected surfaces are all a kind of "skew catenoid" generated by a one-parameter family of ellipses.

Chen also proved [14]

Theorem 5.12. *A complete $M \underset{\text{MIN}}{\hookrightarrow} \mathbb{R}^n$ with total curvature*

$$\int_M K\,dA = -2\pi \tag{16}$$

lies in an affine $\mathbb{R}^4 \subset \mathbb{R}^n$, and with respect to a suitable complex structure on $\mathbb{R}^4$, M is a holomorphic curve in $\mathbb{C}^2$.

In fact, M is congruent to the graph of the function

$$w = cz^2$$

for some real $c > 0$.

5.13

Holomorphic curves in $\mathbb{C}^m$, considered as real surfaces in $\mathbb{R}^{2m}$, are minimal surfaces, and play an important role in the theory. We owe our basic understanding of that role to work of Calabi [11] and Lawson [47]. In particular, Lawson made effective use of the generalized Gauss map. He showed in fact that one could characterize completely by means of the Gauss map those minimal surfaces in $\mathbb{R}^n$ that arise as the real form of holomorphic curves:

Theorem 5.14. *Given $M \underset{\text{MIN}}{\hookrightarrow} \mathbb{R}^{2m}$, there exists a complex structure on $\mathbb{R}^{2m}$ with respect to which M is a holomorphic curve in $\mathbb{C}^m$ if and only if the image of M under the Gauss map lies in some $\mathbb{C}P^k \subset Q_{2m-2}$.*

5.15

Note that for $M \underset{\text{MIN}}{\hookrightarrow} \mathbb{R}^n$, if $\hat{M} = g(M)$, then (13) implies that $\hat{M}$ is regular wherever M is regular and $K \neq 0$. At all nonsingular points of $\hat{M}$, we have a well-defined Gauss curvature $\hat{K}$. It follows from the normalization (12) and the fact that $\hat{M}$ lies minimally in $\mathbb{C}P^{n-1}$ that

$$\hat{K} \leqslant 2, \qquad (17)$$

an important observation in recent work on stability, as we shall see later. One has the following result (Hoffman and Osserman [40]) closely related to Theorem 5.14.

Theorem 5.16. *Given $M \underset{\text{MIN}}{\hookrightarrow} \mathbb{R}^n$, let $\hat{M} = g(M)$. Then*

$$\hat{K} \equiv 2 \qquad (18)$$

if and only if M lies in some affine $\mathbb{R}^4 \subset \mathbb{R}^n$, and is a holomorphic curve in $\mathbb{C}^2$ with respect to a suitable complex structure on $\mathbb{R}^4$.

One also has:

Lemma 5.17. *Given $M \underset{\text{MIN}}{\hookrightarrow} \mathbb{R}^n$, let $\hat{M} = g(M)$. Then*

$$\hat{K} \equiv 1 \qquad (19)$$

if and only if M is locally isometric to a minimal surface in $\mathbb{R}^3$.

Using Lemma 5.17 and results of Calabi [9], Lawson was able to give a complete answer to a question posed earlier by Pinl [62]: to characterize those $M \to \mathbb{R}^n$ that are isometric to minimal surfaces in $\mathbb{R}^3$. We refer to Lawson [47, 48] for details on this, and for a wealth of further results.

Still more applications of the generalized Gauss map to the study of minimal M^2 in $\mathbb{R}^n$ will appear later in our discussion of stability. We now turn to the study of arbitrary submanifolds of $\mathbb{R}^n$, and the role of the Gauss map.

6. The Gauss Map for Arbitrary Submanifolds of $\mathbb{R}^n$

6.1

Throughout this section we consider immersed submanifolds

$$M^m \hookrightarrow \mathbb{R}^n$$

and their associated Gauss maps

$$g : M \to G_{m,n},$$

as described in 4.2. There is a canonical metric on the Grassmannian $G_{m,n}$, which induces a metric $d\hat{s}^2$ on the Gauss image

$$\hat{M} = g(M) \subset G_{m,n}.$$

The first study of the metric $d\hat{s}^2$ for arbitrary submanifolds of $\mathbb{R}^n$ appears to have been made in 1874 by Jordan [43].[3] For a more modern approach, see Chern [18], which provides, incidentally, a beautiful overview of many topics closely related to those we are considering here.

For our present purposes, the key result is a formula due to Obata [56] relating the metric $d\hat{s}^2$ to the fundamental quantities associated with the submanifold M:

$$g^*(d\hat{s}^2) = mH \cdot B - \mathrm{Ric}(M). \tag{20}$$

Here B is the vector-valued second fundamental form of M; H denotes the mean curvature vector of M, taken as an *average* (so that mH corresponds to the mean curvature vector defined as the *trace* of B); and $\mathrm{Ric}(M)$ is the Ricci form, whose value on any unit vector is the Ricci curvature of M in that direction. For two-dimensional surfaces we have

$$\mathrm{Ric}(M) = K \, ds^2, \qquad m = 2. \tag{21}$$

In particular, when $m = 2$, $n = 3$, Equation (20) reduces to the classical equation connecting the three fundamental forms of surfaces in $\mathbb{R}^3$:

$$\mathrm{III} = 2H\,\mathrm{II} - K\,\mathrm{I}. \tag{22}$$

(See for example Spivak [68, Vol. 3, p. 88].)

Equation (22) is one of the basic equations of surface theory. It may be used, for example, in the proof of Proposition 2.2. It seems likely that Obata's equation, (20), will play an important role in future work on submanifolds. The

---

[3] We owe this reference to a paper by Leichtweiss [50], who was one of the first to study problems related to those considered in this section.

remainder of this section includes a number of consequences of (20), mostly taken from the paper of Hoffman and Osserman [41].

First we note some immediate corollaries of (20).

Corollary 6.2. (Obata [56]). *Given* $M^m \hookrightarrow \mathbb{R}^n$, *consider the following properties*:

(a) *The Gauss map* $g: M \to G_{m,n}$ *is conformal*,
(b) M *is an Einstein manifold*,
(c) M *is either minimal or pseudoumbilic*.

Then any two of these properties imply the third.

In fact,

(a) $\Leftrightarrow g^*(d\hat{s}^2) = \rho^2 ds^2$,
(b) $\Leftrightarrow \text{Ric}(M) = \mu^2 ds^2$,
(c) $\Leftrightarrow B \cdot H = \lambda^2 ds^2$.

(The last equation is the definition of pseudoumbilic when $H \neq 0$.)

Corollary 6.3. *For minimal* $M^m \underset{\text{MIN}}{\hookrightarrow} \mathbb{R}^n$, *the* ($m$-*dimensional*) *area* $\hat{A}$ *of* $g(M)$ *is given by*

$$\hat{A} = \int_M \sqrt{\det(-\text{Ric}(M))} \, dA. \tag{23}$$

Note in particular that the right-hand side of (23) is an *intrinsic* quantity, so that Corollary 6.3 is a kind of *theorema egregium* for arbitrary minimal submanifolds. In fact, when $H \equiv 0$, (20) reduces to

$$g^*(d\hat{s}^2) = -\text{Ric}(M), \tag{24}$$

so that not only the area ratio is intrinsic, but also the metric pullback itself. Note also that (24) implies the (elementary) fact that on a minimal submanifold the Ricci form is negative semidefinite. That in turn implies that the determinant appearing in the integrand of (23) is nonnegative.

We may note in passing that the determinant of the Ricci form is a quantity that does not seem to occur in other contexts. However, it arises completely naturally here.

We note also the following question that arises in this connection. For $M^m \underset{\text{MIN}}{\hookrightarrow} \mathbb{R}^n$, consider the negative of the Ricci form as defining a new metric on M. What can one say about the properties of this metric (such as curvatures)? For example, when $m = 2$, we would be considering the metric

$$d\hat{s}^2 = -K \, ds^2,$$

and as we have seen earlier, the Gauss curvature of this metric satisfies $\hat{K} \leq 2$, with $\hat{K} \equiv 1$ when $n = 3$. Barbosa and doCarmo [6] have studied the metric

$$d\hat{s}^2 = -R \, ds^2$$

for minimal $M^{n-1} \underset{\text{MIN}}{\hookrightarrow} \mathbb{R}^n$, where R is the scalar curvature of M, and have shown that the scalar curvature $\hat{R}$ of the metric $d\hat{s}^2$ satisfies $\hat{R} \leqslant 2n - 5$. Since both $-R\,ds^2$ and $-\text{Ric}(M)$ reduce to $K\,ds^2$ when $m = 2$, either one may be considered as a suitable generalization. However, in view of (24), the latter seems the more natural choice. Whatever properties of the new metric one finds may then be interpreted as providing necessary conditions for a given metric to be realizable on a minimal submanifold of $\mathbb{R}^n$.

We turn next to an important result of Ruh and Vilms [63].

6.4. *Given $M^m \hookrightarrow \mathbb{R}^n$, the tension field τ of the Gauss map $g : M \to G_{m,n}$ satisfies*

$$\tau = \nabla H \tag{25}$$

where ∇ denotes the covariant derivative associated with the normal bundle of M.

Corollary 6.5. *The Gauss map is harmonic if and only if M has parallel mean curvature.*

The subject of harmonic maps has been extensively studied in recent years (see for example Eells and Lemaire [30]) as has the class of submanifolds with parallel mean curvature[4] (see for example the book by B. -Y. Chen [13].) The link between the two via the Gauss map was both beautiful and unexpected. However, it appears most natural if we note (following Eells and Lemaire [30, p. 30]) that after suitable identifications, the differential of the Gauss map may be interpreted as being precisely the second fundamental form B. But the tension field τ of a map f is just

$$\tau(f) = \text{tr}\,\nabla(df).$$

Hence

$$\tau(g) = \text{tr}\,\nabla(dg) = \text{tr}\,\nabla(B) = \nabla\,\text{tr}(B) = \nabla H,$$

which is (25). As for Corollary 6.5, it follows immediately from (25), since a map is harmonic if and only if its tension field vanishes.

Another proof of the Ruh–Vilms Theorem can be found in Chern–Goldberg [22]. A striking application of the theorem has been made by Hildebrandt, Jost and Widman [37], who generalize Bernstein's Theorem 3.3 to minimal graphs of arbitrary dimension and codimension. Their paper contains excellent comprehensive discussions of harmonic mappings into Riemannian manifolds and of the geometry of the Grassmannian, en route to combining them via the Gauss map to obtain their result.

There is one fact in particular concerning harmonic maps that will be of importance to us:

---

[4]To the best of my knowledge, it was Chern who first suggested, in the mid sixties, that the notion of parallel mean curvature would be a fruitful one, as the natural extension of constant mean curvature for hypersurfaces.

Lemma 6.6. *Let* $f: M^m \to \overline{M}^n$ *be a conformal map of one Riemannian manifold into another; i.e.,* $|f_*(X)| = \rho(p)|X|$ *for* $X \in T_p M$, *where* ρ *is a smooth positive function on* M. *Then when* $m = 2$, f *is harmonic if and only if* $f(M)$ *is minimal in* $\overline{M}$; *when* $m > 2$, f *is harmonic if and only if* $f(M)$ *is minimal in* $\overline{M}$ *and* f *is homothetic.*

The lemma follows immediately from the following formula for the tension field of a conformal map (see Hoffman and Osserman [41, p. 3]):

$$\tau = m\rho^2 H - (m - 2)f_*(\nabla \log \rho),$$

where H is the mean curvature vector of $f(M)$ in $\overline{M}$.

Another useful fact is the following (see, for example, U. Simon [66], Theorem 4.3 and the references given there):

Lemma 6.7. *Given* $M^m \hookrightarrow \mathbb{R}^n$, *the following are equivalent:*

(i) *M is pseudoumbilic and has parallel mean curvature;*
(ii) *M lies in some hypersphere* $S^{n-1}(r)$, *and* M *is a minimal submanifold of* $S^{n-1}(r)$.

We now specialize to the case of surfaces: $m = 2$. In view of (21), (20) reduces to

$$g^*(d\hat{s}^2) = 2H \cdot B - K \, ds^2 \qquad (m = 2). \tag{26}$$

This is the equation for a general surface, with the basic equation (13) for minimal surfaces appearing as a special case. From (26), we see the appropriate generalization of Proposition 2.2:

Lemma 6.8. *Given* $M^2 \hookrightarrow \mathbb{R}^n$, *the Gauss map is conformal if and only if at each point of* M *either* $H = 0$ *or* M *is pseudoumbilic.*

Also, a very useful formula, part way between (13) and (26):

Lemma 6.9. *Given* $M^2 \hookrightarrow \mathbb{R}^n$, *if the Gauss map* g *is conformal, then the metric* $d\hat{s}^2$ *on* $g(M)$ *satisfies*

$$g^*(d\hat{s}^2) = (2|H|^2 - K) \, ds^2. \tag{27}$$

Finally, combining Corollary 6.5 with Lemmas 6.6, 6.7, and 6.8, we come to the following conclusions.

Theorem 6.10. *Given* $M^2 \hookrightarrow \mathbb{R}^n$, *with Gauss map* g, *the following are equivalent:*

(a) g *is conformal and harmonic,*
(b) g *is conformal and* $g(M)$ *is minimal in* $G_{2,n}$,
(c) *M is either a minimal surface in* $\mathbb{R}^n$, *or else a minimal surface in some hypersphere* $S^{n-1}(r)$.

Corollary 6.11. *Given* $M^2 \underset{\text{MIN}}{\hookrightarrow} S^{n-1}(r) \hookrightarrow \mathbb{R}^n$, *let* $g : M \to G_{2,n}$ *be the Gauss map of* M *considered as a surface in* $\mathbb{R}^n$. *Then*

$$g^*(d\hat{s}^2) = \left(\frac{2}{r^2} - K \right) ds^2. \tag{28}$$

Corollary 6.12. *Under the same hypotheses, the Gauss curvature* $\hat{K}$ *of* $\hat{M} = g(M)$ *satisfies*

$$\hat{K} \leqslant 2. \tag{29}$$

Corollary 6.11 follows immediately from (27) and the fact that if M is minimal in a hypersphere of $\mathbb{R}^n$, then the mean curvature vector H of M considered as a surface in $\mathbb{R}^n$ coincides with the mean curvature vector of the sphere whose length is the reciprocal of the radius.

Corollary 6.12 uses the fact that under the normalization (12) of the Fubini–Study metric on $\mathbb{C}P^{n-1}$, the quadric Q_{n-2} has sectional curvature $\tilde{K}$ satisfying $0 \leqslant \tilde{K} \leqslant 2$. (See, for example, Ogiue [57, Proposition 2.2a].) Since, by Theorem 6.10, $\hat{M}$ is a minimal surface in Q_{n-2}, the Gauss curvature of $\hat{M}$ at each point is no greater than the sectional curvature of Q_{n-2} at the point, so that (29) follows.

6.13

Minimal surfaces in spheres have been studied extensively, starting with basic papers of Calabi [10], Simons [67], Chern [20, 21], and Lawson [45, 46]. We shall see in Section 8 below how Theorem 6.10 and its corollaries can provide further insight into the theory.

7. Other Gauss Maps

7.1

Let S^{n-1} denote the unit sphere in $\mathbb{R}^n$. Given an immersed $M^m \hookrightarrow S^{n-1}$, we may of course consider M as immersed in $\mathbb{R}^n$ and apply all of the theory discussed above. However, we may also operate intrinsically on S^{n-1} as a constant-curvature manifold, without reference to any enveloping space. Thus, at each point $p \in M$, we may consider the totality of great circles in S^{n-1} tangent to M at p. They will span a great m-sphere, which we denote by $g_s(p)$. Thus we define a *spherical Gauss map* g_s whose image space is the set of all great m-spheres in S^{n-1}. Since each great m-sphere is uniquely determined by the linear $(m + 1)$-space of $\mathbb{R}^n$ containing it, we may identify the image space with the standard Grassmannian $G_{m+1,n}$. Thus the spherical Gauss map is of the form

$$g_s : M \to G_{m+1,n}. \tag{30}$$

Using the canonical metric on $G_{m+1,n}$, one can make the same sort of computation as we discussed in 6.1 for Euclidean space. Obata [56] carried out the computation, and derived, in addition to his formula (20) for the standard Gauss map, the corresponding one for g_s:

$$g_s^*(d\hat{s}^2) = mH_s \cdot B_s - \mathrm{Ric}(M) + (m-1)\,ds^2, \tag{31}$$

where B_s is the second fundamental form of M with respect to S^{n-1}, and H_s the corresponding mean curvature vector of M.

It follows that Corollary 6.2 holds without change for the spherical Gauss map, and that for minimal submanifolds of a sphere (i.e., $H_s \equiv 0$) one has instead of (24)

$$g_s^*(d\hat{s}^2) = -\mathrm{Ric}(M) + (m-1)\,ds^2. \tag{32}$$

In particular, the metric is again intrinsic.

For a minimal $M^2 \xrightarrow[\mathrm{MIN}]{\hookrightarrow} S^{n-1}$, (32) becomes

$$g_s^*(d\hat{s}^2) = (1 - K)\,ds^2. \tag{33}$$

In his fundamental paper on minimal surfaces in S^3 [46], Lawson introduces the spherical Gauss map in a somewhat different fashion. Given an oriented $M^2 \hookrightarrow S^3$, at each point $p \in M$ there is a unique unit vector $N(p)$ normal to M and tangent to S^3 at p, with the proper orientation. Lawson thus defines the spherical Gauss map

$$g_s : M \to S^3 \tag{34}$$

by

$$g_s : p \mapsto N(p).$$

In fact, since the unit tangent vectors to S^3 are clearly in one–one correspondence with the oriented great 2-spheres in S^3, Lawson's and Obata's Gauss maps are equivalent. Said differently, the Grassmannian $G_{3,4}$ is just the unit sphere S^3 in $\mathbb{R}^4$, the normalization being the same, since Lawson derived the same equation (33) for the pullback of the metric on the unit sphere under his spherical Gauss map g_s, when M is a minimal surface in S^3. Among the other results he obtains are the following.

Theorem 7.2. *Given a minimal* $M^2 \xrightarrow[\mathrm{MIN}]{\hookrightarrow} S^3$, *the image* $\hat{M}$ *of* M *under* g_s *is a branched minimal surface in* S^3 *with branch points where* $K = 1$. *The Gauss curvature* $\hat{K}$ *of* $\hat{M}$ *satisfies*

$$\hat{K} = -\frac{K}{1-K}. \tag{35}$$

Theorem 7.3. *Given* $M^2 \xrightarrow[\mathrm{MIN}]{\hookrightarrow} S^3$, *define a map*

$$\tilde{g} : M^2 \to \mathbb{R}^6 \tag{36}$$

by

$$\tilde{g} : x \mapsto x \wedge g_s(x) \tag{37}$$

where both the point $x \in M^2$ *and its Gauss image* $g_s(x)$ *are considered as points of* $\mathbb{R}^4$. *Then the map* $\tilde{g}$ *is conformal, the image lies in the unit sphere* $S^5 \subset \mathbb{R}^6$, *and*

$\tilde{M} = \tilde{g}(M)$ *is a regular minimal surface in* S^5. *Further, if* $d\tilde{s}^2$ *is the induced metric from* S^5 *on* $\tilde{M}$, *then*

$$\tilde{g}^*(d\tilde{s}^2) = (2 - K)\,ds^2. \tag{38}$$

Lawson calls $\tilde{M}$ the *bipolar surface* associated with M.

These are the only results that we shall need in the next section. However, let us conclude by noting that Obata [56] also considers the Gauss map for submanifolds of hyperbolic space. In fact, if $\overline{M}^n(c)$ is the n-dimensional space form of constant sectional curvature c, and if $M^m \hookrightarrow \overline{M}^n(c)$, then the Gauss map g assigns to each point $p \in M$ the totally geodesic m-submanifold of $\overline{M}$ tangent to M at p. When $c = 0$ or 1 this reduces to the cases already considered. For $c < 0$ there is again a canonical metric on the corresponding Grassmannian, and if $d\hat{s}$ is the induced metric on $\hat{M} = g(M)$, one has the formula

$$g^*(d\hat{s}^2) = mH \cdot B - \text{Ric}(M) + (m-1)c\,ds^2 \tag{39}$$

that holds for all values of c, regardless of sign.

8. Stability and Eigenvalues

8.1

As we noted in 3.1, minimal surfaces have their origin in the variational approach to the problem of minimizing area. We have seen that they also arise in other contexts involving properties of the Gauss map, and not related to minimizing area. However, it is important to relate the general theory back to the original problem, and to ask when a given minimal surface does indeed minimize area. For that purpose, we consider the *second variation* of area. With the same notation as the first variation formula (4), we find

$$\frac{d^2A}{dt^2}\bigg|_{t=0} = \int_D \left(|\nabla u|^2 + 2Ku^2\right)dA. \tag{40}$$

We say that D is *stable* if the second variation is positive for all variations that keep the boundary fixed. In other words, by (40), stability is equivalent to

$$\int_D |\nabla u|^2\,dA > -2\int_D Ku^2\,dA \tag{41}$$

for all smooth u in $\overline{D}$ vanishing on the boundary. It is convenient to rewrite (41) using a new metric

$$d\hat{s} = -K\,ds^2. \tag{42}$$

Then

$$d\hat{A} = -K\,dA \tag{43}$$

and

$$|\nabla u|^2 = -K|\hat{\nabla} u|^2, \tag{44}$$

where $\hat{\nabla}$ denotes the gradient in the new metric. We may then rewrite (41) as

$$\int_D |\hat{\nabla} u|^2 \, d\hat{A} > 2 \int_D u^2 \, d\hat{A} . \tag{45}$$

We recall now that the ratio

$$Q(u) = \frac{\displaystyle\int_D |\nabla u|^2 \, dA}{\displaystyle\int_D u^2 \, dA}$$

is called the *Rayleigh Quotient*, and the quantity

$$\lambda_1(D) = \inf Q(u) \tag{46}$$

represents the first eigenvalue of the problem

$$\begin{aligned} \Delta u + \lambda u &= 0 \quad \text{in } D, \\ u &= 0 \quad \text{on } \partial D. \end{aligned} \tag{47}$$

The "inf" in (46) may be taken over all piecewise smooth functions in $\overline{D}$ that vanish on the boundary, while the Δ in (47) represents the Laplace operator with respect to a given metric on D. If D has reasonably smooth boundary, then (47) has a solution u_1 corresponding to the eigenvalue λ_1, and the "inf" in (46) is actually attained when $u = u_1$.

Putting all this together, we find that the stability condition (41) (or (45)) is simply the statement

$$\lambda_1(D) > 2, \tag{48}$$

where we use the metric (42) in Equation (47).

But for a minimal surface, the metric (42) is exactly the pullback under the Gauss map of the metric on the unit sphere. This leads to

Theorem 8.2. *Let D be a relatively compact domain on a minimal $M \overset{\hookrightarrow}{\text{MIN}} \mathbb{R}^3$. Suppose that the Gauss map g maps D one–one onto a domain $\hat{D}$ on the unit sphere. If $\lambda_1(\hat{D}) < 2$, then D cannot be area-minimizing with respect to its boundary.*

In fact, retracing the above argument, we see that the condition $\lambda_1(\hat{D}) < 2$ implies that there exists a function u on $\overline{D}$, vanishing on the boundary, which makes the second variation (30) negative. One thus gets a one-parameter family of surfaces with smaller area than D.

It is a well-known and elementary fact that if D' is a hemisphere on the unit sphere, then $\lambda_1(D') = 2$. We therefore have an immediate consequence of Theorem 8.2:

Corollary 8.3. *Let D be a relatively compact domain in a minimal $M \overset{\hookrightarrow}{\text{MIN}} \mathbb{R}^3$. If the Gauss map g maps D one–one onto a domain containing a hemisphere, then D cannot be area-minimizing.*

This result is a classical one due to H. A. Schwarz. The above discussion is taken from a paper of Barbosa and doCarmo [3], who go on to prove an important result in the other direction.[5]

Theorem 8.4. *Let D be a domain on a minimal $M \underset{\mathrm{MIN}}{\hookrightarrow} \mathbb{R}^3$. Let $\hat{D}$ be the image of M under the Gauss map. If*

$$\mathrm{Area}(\hat{D}) < 2\pi, \tag{49}$$

then D is stable.

Corollary 8.5. *The same conclusion holds if (49) is replaced by the condition*

$$\int_D |K|\, dA < 2\pi. \tag{50}$$

In fact, the left-hand side of (50) represents the area of $\hat{D}$ *counting multiplicities.* For a one–one map the two quantities coincide; otherwise Area $(\hat{D})$ is smaller.

Now if the Gauss map is one–one, then Theorem 8.4 follows directly from a result of Peetre [60]:

Theorem 8.6. *Let D be a domain on the unit sphere, and let $\tilde{D}$ be the geodesic disk on the sphere having the same area as D. Then*

$$\lambda_1(D) \geqslant \lambda_1(\tilde{D}). \tag{51}$$

Applying this to a domain $\hat{D}$ satisfying (49), the disk $\tilde{D}$ will lie in a hemisphere D', and the monotonicity of λ_1 would yield

$$\lambda_1(\hat{D}) \geqslant \lambda_1(\tilde{D}) > \lambda_1(D') = 2.$$

Thus (48) holds, and that is equivalent to stability.

In the general case, where the Gauss map takes g onto a branched covering of the sphere, the argument is far more delicate. However, it is still based on the above line of reasoning.

As an example of the importance of Theorem 8.4, it was immediately used by Nitsche [55] to give a new uniqueness theorem for Plateau's problem:

Theorem 8.7. *Let γ be a real analytic Jordan curve in $\mathbb{R}^3$ whose total absolute curvature is at most 4π. Then there exists a unique simple connected minimal surface spanning γ.*

At the end of Barbosa and doCarmo's paper they suggest that by using the generalized Gauss map for an $M^2 \underset{\mathrm{MIN}}{\hookrightarrow} \mathbb{R}^n$, one might be able to obtain in similar fashion a sufficient condition for stability. That proved to be the case, and the argument was provided by Spruck [69].

---

[5] To our knowledge, it was Chern who first proposed that the condition (49) should be sufficient for stability.

Theorem 8.8. *For all* $n \geqslant 3$, *there exists a constant* $\epsilon(n) > 0$ *such that if* D *is a relatively compact domain on a minimal surface* $M^2 \xhookrightarrow{} \mathbb{R}^n$, *and if*

$$\int_D |K| \, dA < \epsilon(n), \tag{52}$$

then D *is stable.*

Spruck shows in fact that for an $M^2 \xhookrightarrow{} \mathbb{R}^n$, one has a second-variation formula generalizing (40) that implies the inequality

$$\left. \frac{d^2 A}{dt^2} \right|_{t=0} \geqslant \int_D (|\nabla u|^2 + 2Ku^2) \, dA. \tag{53}$$

Thus (41) is again a sufficient condition for stability, and introducing the metric (42) reduces the problem as before to (45). But by virtue of (13), the metric (42) is that induced on the image $\hat{D}$ of D under the generalized Gauss map. Chern's basic Lemma 4.9 implies that $\hat{D}$ is a minimal surface in Q_{n-2}. Spruck then applies the Sobolev inequality of Hoffman and Spruck [42] that holds for minimal submanifolds of a Riemannian manifold to prove (45) and hence the theorem.

Barbosa and doCarmo returned to the same problem in a recent paper [4] and gave a considerable improvement of Spruck's theorem. They showed that Theorem 8.8 holds with (52) replaced by

$$\int_D |K| \, dA < \tfrac{4}{3}\pi, \tag{54}$$

provided that D is simply connected.

Their proof is based on the observation that since the Gauss image $\hat{D}$ of D is the conjugate of a holomorphic curve, its Gauss curvature $\hat{K}$ at any point is at most equal to the holomorphic curvature $\overline{K}$ of $\mathbb{C}P^{n-1}$, which by the normalization (12) is equal to 2. Thus, with respect to the metric (42), the Gauss curvature at any point of D is at most 2. But we may now invoke a theorem of Bandle [1] (see also Chavel and Feldman [12]) generalizing Peetre's theorem 8.6:

Theorem 8.9. *If* D *is a simply connected domain whose Gauss curvature* K *satisfies* $K \leqslant K_0$, *and if the area* A *of* D *satisfies* $K_0 A < 4\pi$, *then*

$$\lambda_1(D) \geqslant \lambda_1(\tilde{D}) \tag{55}$$

where $\tilde{D}$ *is the geodesic disc of the area* A *on the simply connected surface of constant curvature* K_0.

Finally, Barbosa and doCarmo prove that if $\tilde{D}$ is a geodesic disc on $S^2(1/\sqrt{2})$ with area less than $\tfrac{4}{3}\pi$, then $\lambda_1(\tilde{D}) > 2$. Combining this estimate with (55) for the case $K_0 = 2$ and with the arguments preceding Theorem 8.9 gives Theorem 8.8 with the improved assumption (54).

As was the case in $\mathbb{R}^3$, the above argument is complete only if the Gauss curvature of M never vanishes, so that the metric (42) is everywhere regular. In the general case, more elaborate arguments are necessary.

In Spruck's paper [69] he also gave conditions for the stability of higher-dimensional minimal submanifolds in $\mathbb{R}^n$. H. Mori [51, 52] and D. Hoffman [39] independently showed how Spruck's results could be generalized to minimal submanifolds of a Riemannian manifold. Using the general formula for the second variation of area, they find:

Lemma 8.10. *Let* $\overline{M}$ *be a Riemannian manifold whose sectional curvature is bounded above by a constant* $\overline{K}$. *Given* $M^m \xrightarrow{\subset}_{\text{MIN}} \overline{M}$, *and* $D \subset\subset M$, *the second variation of area satisfies*

$$\frac{d^2 A}{dt^2}\bigg|_{t=0} \geqslant \int_D \left[|\nabla u|^2 - (m\overline{K} + \|B\|^2)u^2 \right] dA \tag{56}$$

where B *is the vector-valued second fundamental form of* M *in* $\overline{M}$, *and the variation vector field is of the form* uv: v *a unit normal vector field to* M, *and* u *a smooth function vanishing on the boundary of* D. *In the case of surfaces,* $m = 2$, (56) *becomes*

$$\frac{d^2 A}{dt^2} \geqslant \int_D \left[|\nabla u|^2 - 2(2\overline{K} - K)u^2 \right] dA. \tag{57}$$

Corollary 8.11. D *is stable if*

$$\int_D |\nabla u|^2 \, dA > 2 \int_D (2\overline{K} - K)u^2 \, dA \tag{58}$$

for all u *vanishing on the boundary of* D.

The relation between (56) and (57) depends on the Gauss equation, which for a two-dimensional surface M relates the mean curvature vector H and the Gauss curvature K of M at a point p to the sectional curvature $\overline{K}_\sigma$ of $\overline{M}$ at $T_p M$ by

$$\|B\|^2 = 2\left[2|H|^2 + \overline{K}_\sigma - K \right].$$

Using this in (56) together with the inequality $\overline{K}_\sigma \leqslant \overline{K}$ gives (57).

For the special case of a minimal surface in the unit 3-sphere, Mori [52] used Corollary 8.11 to obtain a very explicit estimate:

Theorem 8.12. *Let* D *be a relatively compact domain on an* $M \xrightarrow{\subset}_{\text{MIN}} S^3(1)$. *Suppose that*

$$\operatorname*{Sup}_D K = K_0 < 1 \tag{59}$$

and

$$\int_D (1 - K) \, dA < \frac{1}{54\pi} \frac{1 - K_0}{2 - K_0}. \tag{60}$$

Then D *is stable.*

Mori's idea is to introduce on D the new metric

$$d\hat{s}^2 = (1 - K) \, ds^2, \tag{61}$$

which, as we saw in (33), is exactly the metric on the image of M under the *spherical* Gauss map. But by Lawson's Theorem 7.2, the image of a minimal M in $S^3(1)$ under the spherical Gauss map is again minimal. Thus the Sobolev inequality of Hoffman and Spruck can be applied, as in Spruck's originial argument, to give the desired result.

Mori's theorem was much improved by Barbosa and doCarmo [4], who gave a sharp result for this case:

Theorem 8.13. *Let D be a simply connected relatively compact domain on an* $M \xrightarrow[\text{MIN}]{\hookrightarrow} S^3(1)$. *If*

$$\int_D (2 - K)\, dA < 2\pi, \tag{62}$$

then D is stable.

Their argument involves a computation to show that the metric

$$d\hat{s}^2 = (2 - K)\, ds^2 \tag{63}$$

has Gauss curvature $\hat{K}$ satisfying

$$\hat{K} \leqslant 1. \tag{64}$$

However, we can conclude (64) directly from the fact that by (38), the metric (63) is precisely the one on the bipolar surface $\tilde{M}$ associated with M, and since $\tilde{M}$ is a minimal surface in the unit sphere S^5, its Gauss curvature satisfies (64). The rest of the proof is exactly as in Barbosa and doCarmo's improved version of Spruck's Theorem 8.8, but even simpler, since the metric (63) is everywhere regular, and one does not have the added complications of singular points.

Finally, Barbosa and doCarmo considered stability for minimal immersions in arbitrary constant-curvature spaces [5]. In particular, they proved:

Theorem 8.14. *Given a minimal surface in a sphere of radius r*:

$$M \xrightarrow[\text{MIN}]{\hookrightarrow} S^{n-1}(r) \subset \mathbb{R}^n,$$

let D be a simply connected relatively compact domain in M. If

$$\int_D \left(\frac{2}{r^2} - K \right) dA < \frac{2n-6}{2n-7}\pi, \tag{65}$$

then D is stable.

Their proof is as follows:
Step 1. Introduce a new metric

$$d\tilde{s}^2 = \left(\frac{2}{r^2} - K \right) ds^2. \tag{66}$$

Then since $\bar{K} = 1/r^2$ in this case, (58) takes the form

$$\int_D |\tilde{\nabla} u|^2\, d\tilde{A} > 2\int_D u^2\, d\tilde{A} \tag{67}$$

which, as before, is equivalent to

$$\tilde{\lambda}_1(D) > 2, \tag{68}$$

where $\tilde{\lambda}_1$ is the first eigenvalue of the Laplacian in D with respect to the metric (66).

Step 2. Use Bandle's Theorem 8.9 to show that if the Gauss curvature $\tilde{K}$ of D in the metric (66) satisfies

$$\tilde{K} \leqslant K_0 = \frac{1}{r_0^2}, \tag{69}$$

then

$$\tilde{\lambda}_1(D) \geqslant \lambda_1(D_0), \tag{70}$$

where D_0 is the geodesic disc of area $\tilde{A}$ on the sphere of radius r_0, and $\tilde{A}$ is the area of D in the metric (66).

Step 3. Show that if D_0 is a geodesic disc of area A on a sphere of radius r_0, and if

$$A = \frac{4\pi}{1 + K_0}, \qquad K_0 = \frac{1}{r_0^2}, \tag{71}$$

then

$$\lambda_1(D_0) \geqslant 2. \tag{72}$$

Step 4. Show that if M is a minimal surface in $S^{n-1}(r)$, then (69) holds with

$$K_0 = 3 - \frac{3}{n-3}. \tag{73}$$

The theorem follows by combining these four steps with the observation that the left-hand side of (65) is precisely the area of D in the metric (66), while the right-hand side is obtained by substituting (73) in (71).

With Theorem 8.14 we come very close to fulfilling the promise made in our introductory remarks: the statement and proof of this one theorem involve all the separate parts of our title with the single exception of the Gauss map. It only remains to note that by rectifying this omission we quickly arrive at a simpler proof of a stronger theorem (Hoffman and Osserman [41]).

Theorem 8.15. *The conclusion of Theorem* 8.14 *holds if* (65) *is replaced by*

$$\int_D \left(\frac{2}{r^2} - K \right) dA < \tfrac{4}{3}\pi. \tag{74}$$

The key observation in the proof is that by (28), the metric (66) is precisely the pullback of the metric on $G_{2,n}$ under the Gauss map g. But then, by Corollary 6.12, we have $\tilde{K} \leqslant 2$. We may therefore replace (73) by $K_0 = 2$, and inserting this in (71), we find that (74) implies (68) and hence stability.

8.16

It may not be inappropriate at this point, where we are in danger of getting bogged down in details, to step back momentarily and insert a note of a more philosophic nature.

Ever since Gauss, in connection with his *theorema egregium*, drew attention to the distinction between intrinsic and extrinsic quantities, there has been a decided split in the field of differential geometry. Riemannian geometers in general favored the intrinsic, while submanifold specialists studied the extrinsic. Insofar as one can see a trend, it would certainly appear to have been toward the intrinsic, the underlying philosophy being that one should wherever possible strip away extraneous considerations. However, in recent years there may have been some movement in the opposite direction, as when one uses an embedding theorem to put a manifold in Euclidean space in order to use properties of the ambient space. Some of the proofs given above carry this process one step further, using extrinsic properties of the embedded manifold to derive additional information. Thus, in the proof of Theorem 8.15, we get a better understanding of the metric (66) by not treating it intrinsically, in the manner of Barbosa and doCarmo, but considering it as the induced metric on a surface lying minimally in the Grassmannian, and then in turn considering the Grassmannian as a Kähler submanifold of complex projective space. Similarly, the problem with Mori's approach to minimal surfaces in S^3 (Theorem 8.12) is that he considers (naturally enough) the intrinsic (spherical) Gauss map, where the induced metric is $(1 - K) ds^2$, which leads to the unnecessary assumption (59) and the strange form of the right-hand side of (60). It turns out that for this particular problem, the best approach is to embed S^3 in $\mathbb{R}^4$, consider the surface M as lying in $\mathbb{R}^4$, and use the bipolar surface that is constructed from that embedding.

8.17

There is another recent result on stability that is not quite as central to our main theme, but that seems worth noting here.

We recall that Bernstein's Theorem 3.3 involves nonparametric minimal surfaces represented as a graph $z = f(x, y)$. One would like to replace the assumption that the surface is representable as a graph by something more geometric. Nirenberg's conjecture 3.4 uses the property of a graph that its Gauss map is restricted to a hemisphere. Another property that holds for minimal graphs is that they are *globally stable*; in fact, every domain D on a minimal graph is area-minimizing for its boundary. That property leads naturally to the question of generalizing Bernstein's theorem by using stability.

Theorem 8.18. *Given $M \underset{\text{MIN}}{\hookrightarrow} \mathbb{R}^3$, assume that M is complete and globally stable. Then M is a plane.*

This theorem was proved independently by Fischer–Colbrie and Schoen [35] and by doCarmo and Peng [27], after the same result had been proved by doCarmo and daSilveira [28] under the additional assumption that the total curvature of M is finite.

8.19

There is a whole other theory of the stability of minimal surfaces for the case of compact surfaces in a Riemannian manifold. We content ourselves here with mentioning that compact stable minimal surfaces play a crucial role in the important recent work of Schoen and Yau. We refer to the papers of Schoen and Yau [64, 65] and to Yau's paper in the present volume.

9. Total Curvature and the Gauss Map; Summary

We return now for a last look at the Gauss map, and focus on two basic questions:

(1) How much of the geometry of a surface M is encoded in properties of its Gauss map?
(2) Which of the basic properties of the classical Gauss map remain valid for the generalized Gauss map?

We shall restrict ourselves in this section to two-dimensional surfaces in $\mathbb{R}^n$. The following theorems are in part a synthesis of results already referred to, but also include some additional facts.

Theorem 9.1. *Given* $M^2 \hookrightarrow \mathbb{R}^n$, *let* $g : M \to Q_{n-2}$ *be the associated Gauss map. Then*

(a) g holomorphic $\Leftrightarrow$ M is a Euclidean 2-sphere in some affine
                                        $\mathbb{R}^3 \subset \mathbb{R}^n$,
(b) g antiholomorphic $\Leftrightarrow$ M is minimal,
(c) g conformal $\Leftrightarrow$ M is minimal or pseudoumbilic,
(d) g conformal and harmonic $\Leftrightarrow$ M minimal in $\mathbb{R}^n$ or M minimal in $S^{n-1} \subset \mathbb{R}^n$.

Theorem 9.2. *Given* $M^2 \xrightarrow{\text{MIN}} \mathbb{R}^n$, *let* $\hat{K}$ *be the Gauss curvature of* $\hat{M} = g(M)$. *Then*

(a) $\hat{K} \equiv 1 \Leftrightarrow M \subset \mathbb{R}^6$ *and* M *locally isometric to a minimal surface in* $\mathbb{R}^3$,
(b) $\hat{K} \equiv 2 \Leftrightarrow M \subset \mathbb{R}^4$ *and* M *a holomorphic curve in* $\mathbb{C}^2$.

Theorem 9.3. *Given* $M \xrightarrow{\text{MIN}} \mathbb{R}^4$, *the Gauss curvature* $\hat{K}$ *of* $\hat{M} = g(M)$ *is constant* $\Leftrightarrow$ $\hat{K} \equiv 1$ *or* $\hat{K} \equiv 2$ *and* M *is as described in Theorem 9.2.*

Theorems 9.1(a) and 9.3 are proved in Hoffman and Osserman [40]. The remaining parts of the above theorems summarize the results of 4.9, 5.16, 5.17, 6.8, and 6.10. We should also recall at this point Ruh and Vilms's result (Corollary 6.5) that for $M^m \hookrightarrow \mathbb{R}^n$,

$$g \text{ harmonic} \quad \Leftrightarrow \quad M \text{ has parallel mean curvature.}$$

If we now review our discussion of the classical Gauss map in Section 2, we find that Proposition 2.2, describing which surfaces have conformal Gauss map, has no single generalization to surfaces in $\mathbb{R}^n$, but tends to fragment into the different parts of Theorem 9.1. In fact, for surfaces in $\mathbb{R}^3$, the various parts of Theorem 9.1 are related by

$$(a) \text{ or } (b) \quad \Leftrightarrow \quad (c) \quad \Leftrightarrow \quad (d).$$

Certainly the most fundamental property of the classical Gauss map is the one expressed in Equation (3) and used by Gauss to define Gauss curvature:

$$\text{Area}(g(M)) = \int_M |K| \, dA.$$

A most natural and basic question, posed by C. C. Chen, is: *for which surfaces in* $\mathbb{R}^n$ *does this equation continue to hold?*—i.e., for which surfaces does the total absolute curvature equal the area of the image under the Gauss map? We saw in (5.9) that minimal surfaces in $\mathbb{R}^n$ have this property, and that it plays a key role in their study. Thus, Chen's question is whether this property is peculiar to minimal surfaces, or whether it holds more widely. The answer is provided by the following theorem (Hoffman and Osserman [41]):

Theorem 9.4. *Given* $M^2 \hookrightarrow \mathbb{R}^n$, *denote by* H *the mean curvature vector of* M, *by* B *the vector-valued second fundamental form, and by* $\tilde{B}$ *the projection of* B *orthogonal to* H (*we may set* $\tilde{B} = B$ *where* $H = 0$); *let* $\hat{A}$ *be the area of the image of* M *under the Gauss map. Then*

$$\hat{A} = \int_M \sqrt{K^2 + 2|H|^2 \|\tilde{B}\|^2}\, dA. \tag{75}$$

Corollary 9.5. *Given an arbitrary* $M \hookrightarrow \mathbb{R}^n$,

$$\hat{A} \geqslant \int_M |K|\, dA. \tag{76}$$

Equality holds in (76) *if and only if at each point of* M *either* $H = 0$ *or* $\tilde{B} = 0$.

Thus we see one more case where minimal surfaces arise in connection with a question that seems initially to have no relevance to them. The answer to Chen's question is that the basic equation (3) holds in $\mathbb{R}^n$ only for minimal surfaces and for those surfaces having the property that $\tilde{B} = 0$ at all points where $H \neq 0$. Note that for an arbitrary surface in $\mathbb{R}^3$, there is a single normal direction, so that H and B are collinear and $\tilde{B} = 0$ where $H \neq 0$. Thus (75) implies that (3) holds in the two cases where it was previously known: minimal surfaces in $\mathbb{R}^n$ and arbitrary surfaces in $\mathbb{R}^3$. Furthermore, one can deduce from (75) that "generically", those two cases are the only ones where (3) holds. For that we need the following result, derived by Yau (oral communcation):

Lemma 9.6. *In the notation of Theorem* 9.4, *if* $H \neq 0$ *and* $\tilde{B} \equiv 0$, *then on any domain where* $K \neq 0$, *the mean curvature vector* H *is parallel.*

Corollary 9.7. *Given* $M \hookrightarrow \mathbb{R}^n$, *if* $H \neq 0$, $K \neq 0$, *and* $\tilde{B} \equiv 0$, *then* M *lies in an affine* $\mathbb{R}^3 \subset \mathbb{R}^n$.

To derive the corollary from the lemma, one can use, for example, a theorem of Erbacher [32] on reduction of codimension.

The only case left to consider is therefore when $K = 0$. In fact, if one chooses a curve C in $\mathbb{R}^{n-1}$ with nonzero curvature such that C does not lie in any affine subspace, and forms the surface

$$M = C \otimes \mathbb{R},$$

then on M, one has $H \neq 0$, $\tilde{B} \equiv 0$, and $K \equiv 0$. For $n > 3$, the surface M will not lie in any 3-dimensional subspace and will not be minimal, but by (75), Equation (3) will hold.

9.8

Finally we note that another way to answer Chen's question, and another way to prove Corollary 9.5, is by making use of the notion of total absolute curvature due to Chern and Lashof [23]: given $M^2 \hookrightarrow \mathbb{R}^n$, at any point $p \in M$ let S^{n-3} be the unit sphere in the normal space to M at p, and let c_{n-3} be the area of S^{n-3}. For each $\nu \in S^{n-3}$, $B \cdot \nu$ is the second fundamental form of M in the direction ν. Set

$$K^*(p) = \frac{n-2}{c_{n-3}} \int_{S^{n-3}} |\det(B \cdot \nu)|.$$

Then the *total absolute curvature* of M in the sense of Chern and Lashof is the quantity

$$\int_M K^* \, dA$$

(up to a multiplicative normalizing constant which is chosen differently in different places.) The relation between this quantity and the two appearing in the inequality (76) is easily stated (see Hoffman and Osserman [41, Section 5]):

Theorem 9.9. *For arbitrary $M^2 \hookrightarrow \mathbb{R}^n$,*

$$\int_M |K| \, dA \leqslant \int_M K^* \, dA \leqslant \hat{A}(g(M)). \tag{77}$$

Note that for surfaces in $\mathbb{R}^3$, the three expressions in (77) all coincide. For surfaces in $\mathbb{R}^4$, the right-hand inequality in (77) was proved by Wintgen [70]. By way of conclusion, let us note what seems to be an interesting open question.

Conjecture 9.10 *The right-hand inequality in (77) is valid for submanifolds M of arbitrary dimension in $\mathbb{R}^n$; i.e., the (suitably normalized) Chern–Lashof total curvature of M is bounded above by the area of the image of M under the generalized Gauss map.*

If this conjecture is true, it will, for example, imply a recent result of Mutō [54] on critical points of the generalized Gauss map.

REFERENCES

[1] C. Bandle, Konstruktion isoperimetrischer Ungleichungen der mathematischen Physik aus solchen der Geometrie, *Comment. Math. Helv.* **46**, 182–213 (1971).

[2] J. L. Barbosa, On minimal immersions of S^2 in S^{2m}. *Trans. Amer. Math. Soc.* **210**, 75–106 (1975).

[3] J. L. Barbosa, and M. doCarmo, On the size of a stable minimal surface in R^3. *Amer. J. Math.* **98**, 515–528 (1976).

[4] ———, Stability of minimal surfaces and eigenvalues of the Laplacian, *Math. Z.*, to appear.

[5] ———, Stability of minimal surfaces in spaces of constant curvature. Preprint.

[6] ———, A necessary condition for a metric in M^n to be minimally immersed in $\mathbb{R}^{n+1}$. Preprint.

[7] W. Blaschke, Sulla geometria differenziale delle superficie S_2 nello spazio euclideo S_4. *Ann. Mat. Pura Appl* (4) **28** 205–209 (1949).

[8]   O. Borůvka, Sur les surfaces représentées par les fonctions sphériques de première espèce. *J. Math. Pures Appl.* **12**, 336–383 (1933).

[9]   E. Calabi, Isometric imbedding of complex manifolds, *Ann. of Math.* **58**, 1–23 (1953).

[10] ———, Minimal immersions of surfaces in euclidean spheres, *J. Diff. Geom.* **1**, 111–125 (1967).

[11] ———, Quelques applications de l'analyse complexe aux surfaces d'aire minima, in *Topics in Complex Manifolds*, Presses de l'Université de Montréal, 1968, pp. 58–81.

[12] I. Chavel and E. A. Feldman, Isoperimetric inequalities on curved surfaces. *Adv. Math.*, to appear.

[13] B. -Y. Chen, *Geometry of Submanifolds*. Marcel Dekker, New York, 1973.

[14] C. C. Chen, Complete minimal surfaces with total curvature -2π. *Boletim da Soc Mat. Brasil.* **10**, 71–76 (1979).

[15] ———, A characterization of the catenoid, *An. Acad. Brasil Ciênc.* **51** 1–3 (1979).

[16] ———, Elliptic functions and non-existence of complete minimal surfaces of certain type. *Proc. Amer. Math. Soc.*, **79**, 289–293 (1980).

[17] S. -S. Chern, On the multiplication in the characteristic ring of a sphere bundle, *Ann. of Math.* **49**, 362–372 (1948).

[18] ———, La géométrie des sous-variétés d'un espace euclidien à plusieurs dimensions, *L'Enseignement Math.* **40**, 26–46 (1951–1954).

[19] ———, Minimal surfaces in an euclidean space of N dimensions. In *Differential and Combinatorial Topology*, Princeton U. P., 1965, pp. 187–198.

[20] ———, On the minimal immersions of the two sphere in a space of constant curvature. In *Problems in Analysis*, Princeton U. P. 1970, pp. 27–40.

[21] ———, On minimal spheres in the four-sphere. In *Studies and Essays Presented to Y. W. Chen*, Taiwan, 1970, pp. 137–150.

[22] S. -S. Chern and S. I. Goldberg, On the volume-decreasing property of a class of real harmonic mappings. *Amer. J. Math.* **97**, 133–147 (1975).

[23] S. -S. Chern and R. K. Lashof, On the total curvature of immersed manifolds. *Amer. J. Math.*, **79**, 306–318 (1957).

[24] ———, On the total curvature of immersed manifolds II. *Mich. Math. J.* **5**, 5–12 (1958).

[25] S. -S. Chern and R. Osserman, Complete minimal surfaces in Euclidean n-space. *J. d'Anal. Math.* **19**, 15–34 (1967).

[26] S. -S. Chern and E. Spanier, A theorem on orientable surfaces in four-dimensional space. *Comment. Math. Helv.* **25**, 1–5 (1951).

[27] M doCarmo and C. N. Peng, Stable complete minimal surfaces in $\mathbb{R}^3$ are planes. Preprint.

[28] M. doCarmo and A. M. daSilveira, Globally stable complete minimal surfaces in $\mathbb{R}^3$. *Proc. Amer. Math. Soc.*, to appear.

[29] P. Dombrowski, 150 Years after Gauss' "Disquisitiones generales circa superficies curvas." *Astérisque* **62**, 1–153 (1979).

[30] J. Eells and L. Lemaire, A report on harmonic maps. *Bull. London Math. Soc.* **10**, 1–68 (1978).

[31] ———, On the construction of harmonic and holomorphic maps between surfaces. Preprint.

[32] J. A. Erbacher, Reduction of the codimension of an isometric immersion. *J. Diff. Geom.* **5**, 333–340 (1971).

[33] L H. Farber *Lying, Despair, Jealousy, Envy, Sex, Suicide, Drugs and the Good Life*. Harper and Row, New York, 1978.

[34] D. Ferus, *Totale Absolutkrümmung in Differentialgeometrie und Topologie*. Springer, Berlin, 1968.

[35] D. Fischer-Colbrie and R. Schoen, The structure of complete stable minimal surfaces in 3-manifolds of nonnegative scalar curvature. Preprint.

[36] F. Gackstatter, Über die Dimension einer Minimalfläche und zur Ungleichung von St. Cohn–Vossen. *Arch. Rat. Mech. Anal.* **61**, 141–152 (1976).

[37] S. Hildebrandt, J. Jost, and K.-O. Widman, Harmonic mappings and minimal submanifolds. Preprint.

[38] H. L. Hiller, On the cohomology of real Grassmannians. *Trans. Amer. Math. Soc.* **257**, 521–533 (1980).

[39] D. Hoffman, Lower bounds on the first eigenvalue of the Laplacian of Riemannian manifolds. In *Minimal Submanifolds and Geodesics*. (Proceedings of the Japan–United States Seminar on Minimal Submanifolds, including Geodesics, Toyko, 1977), Kaigai, Tokyo, 1978 pp. 61–72.

[40] D. Hoffman and R. Osserman, The geometry of the generalized Gauss map. Memoirs *Amer. Math. Soc.*, to appear.

[41] ———, The area of the generalized Gaussian image and the stability of minimal surfaces in S^n and $\mathbf{R}^n$. Preprint.

[42] D. Hoffman and J. Spruck, Sobolev and isoperimetric inequalities for Riemannian submanifolds. *Communications Pure Appl. Math.* **27**, 715–727 (1974), correction *ibid.* **28**, 765–766 (1975).

[43] C. Jordan, Généralisation du théorème d'Euler sur la courbure des surfaces. *C. R. Acad. Sci. Paris* **79**, 909–912 (1874).

[44] S. Kobayashi and K. Nomizu, *Foundations of Differential Geometry*. Vol II, Interscience, New York, 1969.

[45] H. B. Lawson, Jr., The global behavior of minimal surfaces in S^n. *Ann. of Math.* **92**, 224–237 (1970).

[46] ———, Complete minimal surfaces in S^3. *Ann. of Math.* **92** 335–374 (1970).

[47] ———, Lectures on Minimal Surfaces. IMPA, Rio de Janeiro 1970.

[48] ———, Some intrinsic characterizations of minimal surfaces. *J. d'Anal. Math.* **24**, 151–161 (1971).

[49] K. Leichtweiss, Zur Riemannschen Geometrie in Grassmannschen Manningfaltigkeiten. *Math. Z.* **76**, 334–366 (1961).

[50] ———, Uber eine Art von Krümmungsinvarianten beliebiger Untermannigfaltigkeiten des n-dimensionalen euklidischen Raums. *Abh. Math. Seminar Univ. Hamburg* **26**, 155–190 (1963).

[51] H. Mori, Notes on the stability of minimal submanifolds of Riemannian manifolds. *Yokohama Math. J.* **25**, 9–15 (1977).

[52] ———, A note on the stability of minimal surfaces in the 3-dimensional unit sphere. *Indiana Univ. Math. J.* **26**, 977–980 (1977).

[53] Y. Mutō, The Gauss map of a submanifold in a Euclidean space. *J. Math. Soc. Japan* **30**, 85–100 (1978).

[54] ———, Index of some Gauss-critical submanifolds. *Tôhoku Math. J.* **30**, 561–573 (1978).

[55] J. C. C. Nitsche, A new uniquesness theorem for minimal surfaces. *Arch. Rat. Mech. Anal.* **52**, 319–329 (1973).

[56] M. Obata, The Gauss map of immersions of Riemannian manifolds in spaces of constant curvature. *J. Diff. Geom.* **2**, 217–223 (1968).

[57] K. Ogiue, Differential geometry of Kaehler submanifolds. *Advances in Math.* **13**, 73–114 (1974).

[58] R. Osserman, Proof of a conjecture of Nirenberg. *Comm. Pure Appl. Math.* **12**, 229–232 (1959).

[59] ———, Global properties of minimal surfaces in E^3 and E^n. *Ann. of Math.* **80**, 340–364 (1964).

[60] J. Peetre, A generalization of Courant's nodal line theorem. *Math. Scand.* **5**, 15–20 (1959).

[61] M. Pinl, B-Kugelbilder reeler Minimalflächen in R^4. *Math. Z.* **59**, 290–295 (1953).

[62] ———, Über einen Satz von G. Ricci-Curbastro und die Gaussche Krümmung der Minimalflächen I, II. *Arch. Math.* **4**, 369–373 (1953), **15**, 232–240 (1964).

[63] E. A. Ruh and J. Vilms, The tension field of the Gauss map. *Trans. AMS* **149**, 569–573 (1970).

[64] R. Schoen and S.-T. Yau, Existence of incompressible minimal surfaces and the topology of three dimensional manifolds with non-negative scalar curvature. *Ann. of Math.* **110**, 127–142 (1979).

[65] ———, On the proof of the positive mass conjecture in general relativity. *Comm. Math. Phys.* **65**, 45–76 (1979).

[66] U. Simon, A further method in global geometry. *Abh. Math. Sem. Univ. Hamburg* **44**, 52–69 (1975).

[67] J. Simons, Minimal varieties in Riemannian manifolds. *Ann. of Math.* **88**, 62–105 (1968).

[68] M. Spivak, *A Comprehensive Introduction to Differential Geometry*. 2nd Ed. Vols. 1–5, Publish or Perish, Inc., Berkeley, 1979.

[69] J. Spruck, Remarks on the stability of minimal submanifolds of $\mathbb{R}^n$. *Math. Z.* **144**, 169–174 (1975).

[70] P. Wintgen, On the total curvature of surfaces in E^4. *Colloq. Math.* **39**, 289–296 (1978).

[71] Y. C. Wong, Differential geometry of Grassmann manifolds. *Proc. Nat. Acad. Sci. U.S.A.* **57**, 589–594 (1967).

[72] ———, Sectional curvatures of Grassmanian manifolds. *Proc. Natl. Acad. Sci.* **60**, 75–79 (1968).

de Rham–Sullivan Measure of Spaces and Its Calculability

Wu Wen-tsün*

1. de Rham–Sullivan Theorem for Complexes

In the first paper on *L'Analysis Situs*, dated 1895, Poincaré introduced funda-mental notions which are nowadays called differential manifolds, complexes, Betti numbers, fundamental groups, etc., thus laying down the foundations of modern algebraic topology. In addition Poincaré posed the problem of determin-ing the Betti numbers of differential manifolds by means of exterior differential forms; see Section 9 of that paper. The problem was clarified by E. Cartan, and only in 1931 was it completely solved by de Rham. The result, now known as the de Rham theorem, may be stated as follows.

Let $A^*(V)$ be the differential graded anticommutative algebra (abbreviated DGA) of exterior differential forms on a differential manifold V. Then the homology of $A^*(V)$ is algebraically (i.e., additively as well as multiplicatively) isomorphic to the real cohomology ring of V as a topological space, viz.

$$H(A^*(V)) \underset{\text{alg}}{\approx} H_R^{**}(V). \qquad (1.1)$$

To avoid the difficulties arising from the nonmanipulable notion of "hom-ologous" in the case of differential manifolds, Poincaré developed in the second and third supplements to *L'Analysis Situs* the combinatorial topology with *complex* instead of differential manifolds as the basic subject to be studied. It is natural to ask whether the theory of Cartan and de Rham may be carried over to the much more general case of complexes. However, it was only in the early 1970s that the question was completely settled by Sullivan in the following manner [5, 6].

For a simplicial complex K let us associate to each simplex σ an exterior differential form (abbreviated EDF) $\omega(\sigma)$ on $|\sigma|$ such that for any face τ of σ the restriction of $\omega(\sigma)$ to τ is just $\omega(\tau)$. Call such a compatible set of EDFs $\omega(\sigma)$ simply an EDF on K. Then under natural operations such EDFs on K become a DGA, called the *de Rham–Sullivan algebra* of K and denoted by $A^*(K)$. The Sullivan extension of the de Rham theorem then states that the homology of

*Institute of Systems-Science, Academia Sinica, Peking, China.

$A^*(K)$ is algebraically isomorphic to the real cohomology ring of the space of K, viz.

$$H(A^*(K)) \underset{\text{alg}}{\approx} H_R^{**}(K). \tag{1.2}$$

The original de Rham theorem has been much studied and proved in various ways, and quite recently several different proofs of the extended de Rham–Sullivan theorem have also appeared. We remark here that one of the proofs of the original de Rham theorem, due to A. Weil in 1952, is extremely instructive in that it is simple in principle, is naturally extendable to the general case of complexes, and moreover is constructive in the sense that it gives an explicit determination of the isomorphisms (1.1) and (1.2) in question. For this reason we give the following

Sketch of Weil's proof (for differential manifolds V). Construct a locally finite simple open covering $\mathfrak{U} = (U_i)_{i \in I}$ of V with associated partition of unity (f_i) and nerve N. Define a differential coelement of degree (m, p) as any system $\Omega = (\omega_H)$ $= (\omega_{i_0, \ldots, i_p})$ of EDFs of degree m in $U_{|H|} = \bigcap_{0 \leqslant \nu \leqslant p} U_{i_\nu} \neq \varnothing$ attached to the sequences $H = (i_0, \ldots, i_p) \subset I$, which depend alternatively on the indices $i_0, \ldots, i_p$. To Ω we have naturally two operators d and δ with $d\Omega$ and $\delta\Omega$ coelements of bidegrees $(m + 1, p)$ and $(m, p + 1)$ respectively. The retraction of each $U_H \neq \varnothing$ defines an operator I_H in U_H such that $I\Omega = (I_H\omega_H)$ is a coelement of bidegree $(m - 1, p)$ satisfying the relation $\Omega = Id\Omega + dI\Omega$ for $m > 0$, with a similar relation for $m = 0$. Again, for any set $J' = J \cup \{i\} \subset I$ with $U_{J'} \neq \varnothing$ and ω_J an EDF in U_J, let us denote by $f_i\omega$ the EDF in U_J equal to $f_i\omega$ in $U_{J'}$, and to 0 in $U_J - U_{J'}$. The partition of unity $\{f_i\}$ defines now an operator K which associates to each coelement $\Omega = (\omega_H)$ of bidegree (m, p) with $p > 0$ the coelement $K\Omega = (\zeta_{i_0 \ldots i_{p-1}})$ of bidegree $(m, p - 1)$ with

$$\zeta_{i_0 \ldots i_{p-1}} = \sum f_k \omega_{k i_0 \ldots i_{p-1}}, \tag{1.3}$$

the $\sum$ being extended over such $k \in I$ with $U_{(k i_0 \ldots i_{p-1})} \neq \varnothing$. Similarly $K\Omega$ is defined for Ω of bidegree $(m, 0)$. We have then $\Omega = K\delta\Omega + \delta K\Omega$.

It may then easily be seen that for a closed EDF ω on V of degree m, $\Xi = \delta(I\delta)^m\omega$ will be a cocycle of dimension m in N, and conversely for a cocycle $\Xi = (\zeta_{i_0 \ldots i_m})$ of dimension m in N, $\omega = K(dK)^m\Xi$ will be a closed EDF of degree m on V. The correspondence between ω and Ξ will establish additive isomorphism between $H(A^*(V))$ and $H_R^{**}(V)$. It is easy to verify that the isomorphism is also multiplicative as asserted. (See also Bott [2].) $\square$

The above proof can be easily modified to the general case of complexes:

Proof of extended de Rham–Sullivan theorem for complexes. The complex K has a natural simple covering $\mathfrak{U}$ consisting of open stars U_i of vertices v_i of K, so that the nerve of $\mathfrak{U}$ coincides with K. We may then define coelements of bidegree (m, p) as well as operator I as before. To define the operator K, we may replace the partition of unity $\{f_i\}$ in the following manner. For each vertex v_i let t_i be the function in U_i which takes on the value $t_i = x_i$ for any point in barycentric coordinates $x_i v_i + \sum_{k=1}^p x_{jk} v_{jk}$ in a p-simplex of vertices $v_i, v_{j_1}, \ldots, v_{j_p}$. K is then again defined by (1.3) with f_k replaced by t_k. The proof then runs as before. $\square$

Remark. The above proof shows that the de Rham–Sullivan theorem remains true if we consider only EDFs $\sum \alpha_{i_0 \ldots i_p} dx^{i_0} \cdots dx^{i_p}$ in a simplex for which $\alpha_{i_0 \ldots i_p}$ are polynomials in the barycentric coordinates x^i with *rational* coefficients.

2. I^*-Measure of Spaces

In what follows we shall consider only DGAs A on R with

$$H_0(A) \approx R, \qquad H_1(A) = 0.$$

For such DGAs Sullivan has shown how to attach a *minimal model* $M = \min A$, unique up to DGA isomorphism, which is characterized by the following two conditions

(M$_1$) M is free as an algebra and is decomposable, i.e., $dx \in M^+ \cdot M^+$ for any $x \in M$ (M^+ = set of elements of positive degree in M).

(M$_2$) There exist *canonical* DGA-morphisms $\rho : M \to A$, unique up to DGA homotopy, with induced algebraic isomorphism

$$\rho_* : H(M) \approx H(A).$$

For a DGA morphism

$$f : A \to B$$

it is also proved that there will be induced morphisms $\tilde{f}$ of minimal models to make the following diagram homotopically commutative ($M = \min A$, $N = \min B$, ρ_A, ρ_B) the corresponding canonical morphisms):

$$
\begin{array}{ccc}
A & \xrightarrow{\;f\;} & B \\[4pt]
{\scriptstyle \rho_A} \uparrow & & \uparrow {\scriptstyle \rho_B} \\[4pt]
M & \xrightarrow{\;\tilde{f}\;} & N_B
\end{array}
$$

The following complement is sometimes very useful:

Lemma. *If $f : A \to B$ is an epimorphism, then with ρ_B given, $\tilde{f}$ and ρ_A can be chosen to make the above diagram* strictly *commutative.*

Proof. Straightforward, following the usual steps for the proof of existence of minimal model and the corresponding canonical DGA morphism. $\square$

Consider now connected simplicial complex K for which $H_R^1(K) = 0$ (to be called *c-complexes* for simplicity) so that the minimal model of the de Rham–

Sullivan algebra $A^*(K)$ is well defined. Denote it by $I^*(K)$; then it is easily seen that $I^*(K)$ depends only on the space of K. By the very definition there exist canonical morphisms

$$\rho : I^*(K) \rightarrow A^*(K)$$

inducing algebraic isomorphisms

$$\rho_* : H(I^*(K)) \approx H(A^*(K)) \approx H_R^{**}(K),$$

which shows that $I^*(K)$ contains complete information about real cohomology of K. In addition, Sullivan has also proved that the graded module associated to $I^*(K)$ is isomorphic to $\pi_*(K) \otimes R$, so that $I^*(K)$ contains also at least partial information about homotopy groups of K.

It goes without saying that the definition of $I^*(X)$ may be easily extended to arbitrary spaces X that are connected and such that $H_R^1(X) = 0$, to be called c-spaces in what follows.

The above connection of I^* with H_R^{**} permits us to determine in many cases the I^* of simple c-spaces from the knowledge of H_R^{**}. We have thus:

Theorem 2.1. *Consider $H_R^{**}(X)$ as a DGA with trivial differential; then we have*

$$I^*(X) \approx \min H_R^{**}(X)$$

for c-spaces X in the following cases: compact Lie groups G, their classifying spaces B_G, G/U with U a closed connected subgroup of maximum rank in G, and Riemannian symmetric spaces. It is also true for c-spaces X with H_R^{**} free or H_R^{**} a subalgebra in $A^*(X)$.

Let $M = G/U$ be a homogeneous space with G a compact connected Lie group and U a closed connected subgroup (always with $H_R^1 = 0$). It is known that $H_R^{**}(G)$ is an exterior algebra on transgressive elements $x_1, \ldots, x_N$ of odd degree and that $H_R^{**}(B_G)$ is a polynomial algebra on generators y_i of even degree $= \deg x_i + 1$, which are transgressives of x_i. Similarly we have $H_R^{**}(U)$ as an exterior algebra on generators $u_1, \ldots, u_L$ and $H_R^{**}(B_U)$ a polynomial algebra on generators $v_1, \ldots, v_L$ with v_j as transgressives of u_j. Let the canonical homomorphism

$$\rho^* : H_R^{**}(B_G) \rightarrow H_R^{**}(B_U)$$

be given by

$$\rho^* y_i = P_i(v_1, \ldots, v_L)$$

with P_i some polynomials. Introduce the following Cartan algebra:

$$C_M^* = H_R^{**}(G) \otimes H_R^{**}(B_U)$$

with twisted differential

$$dx_i = P_i(v_1, \ldots, v_L),$$
$$dv_i = 0.$$

Then we have

Theorem 2.2. $I^*(M) \approx \min C_M^*$, *so that on taking homology on both sides, we get Cartan's theorem*

$$H_R^{**}(M) \approx H(C_M^*).$$

Proof. See Wu [9].

3. Determination of H^{**} of a Space by Means of I^*

The determination of cohomology (or homology) groups and rings is one of the oldest problems in algebraic topology and is far from trivial in its appearance. Besides the direct cell-subdivision method based on the very definition of topology, there are developed since methods of differential forms (for manifolds only) due to E. Cartan, of spectral sequences due to the French school, and of twisted products due to E. H. Brown.

The introduction of I^*-measure furnishes a new method which seems to be a much more powerful one for the determination of cohomology ring in the case of *real* coefficients. It is based on the following principle. The I^*-measures of relatively simple spaces can usually be determined from the very definition as shown in the last section. Now I^*-measure contains information on the real cohomology as well as the real homotopy of a space, so that the knowledge of the I^*-measure of simpler constitutents of a complicated space may be sufficient to determine completely the real cohomology of that complicated space, while the mere knowledge of the real cohomologies of the various constituents is insufficient to do so. It turns out that this is usually in fact the case, and we shall illustrate the applications of this principle in some concrete cases below.

EXAMPLE 3.1 (Cone Construction). Let L be a c-subcomplex of a c-complex K with injection $j: L \subset K$. Let C_L be a cone over L and $\Delta = \Delta_L(K) = K \cup C_L$ which is homotopically the space K/L in shrinking L to a point. Simple examples show that $H_R^{**}(\Delta)$ as a ring is by no means determined by the knowledge of

$$j^* : H_R^{**}(K) \to H_R^{**}(L).$$

However, as $I^*(K), I^*(L)$ contain more information than mere $H_R^{**}(K), H_R^{**}(L)$, it turns out that $H_R^{**}(\Delta)$ can be completely determined in the present case by the knowledge of

$$j^I : I^*(K) \to I^*(L) \tag{3.1}$$

induced by $j: L \subset K$ in the following manner.

For any morphism of DGA algebras

$$j : A \to B$$

let C be the algebraic map cone of j considered as morphism of modules. For the induced morphism of homology

$$j_H : H(A) \to H(B)$$

we have then the exact sequence

$$0 \to \operatorname{Coker} j_H \overset{\delta_H}{\to} H(C) \overset{i_H}{\to} \operatorname{Ker} j_H \to 0.$$

We may take then module basis

$$H(A) = (X_1, \ldots, X_n, Y_1, \ldots, Y_s),$$
$$H(B) = (j_H Y_1, \ldots, j_H Y_s, Z_1, \ldots, Z_t),$$
$$H(C) = (X_1^c, \ldots, X_n^c, Z_1^c, \ldots, Z_t^c)$$

with $i_H X_i^c = X_i$, $\delta_H Z_j = Z_j^c$.

Take now $\xi_i \in X_i$ and $(\xi_i, \beta_i) \in X_i^c$ with

$$d_A \xi_i = 0,$$
$$d_B \beta_i = -j\xi_i.$$

Suppose that

$$X_i X_j = \sum \lambda_{ij}^k X_k$$

(λ_{ij}^k redundant in incorrect degrees), so that

$$\xi_i \xi_j = \sum \lambda_{ij}^k \xi_k + d_A \alpha_{ij}$$

for some $\alpha_{ij} \in A$. Set

$$\zeta_{ij} = -\sum \lambda_{ij}^k \beta_k - \beta_i d_B \beta_j + j\alpha_{ij} \in B.$$

Then it is easily verified that $(0, \zeta_{ij})$ is a cycle of C with class Z_{ij}^c depending only on the X's but not on the choice of the ξ's, β's and α's.

Introduce now multiplications and a trivial differential in $H(C)$ to turn it into a DGA, to be denoted by $J(j)$, as follows:

$$X_i^c X_j^c = \sum \lambda_{ij}^k X_k^c + Z_j^c,$$
$$X_i^c Z_j^c = 0, \qquad Z_i^c Z_j^c = 0.$$

We have then

Theorem 3.1. $H_R^{**}(\Delta) = J(j')$.

In the same manner, for the union K of two c-complexes K_1, K_2 along a common c-subcomplex L we can determine $H_R^{**}(K)$ in terms of the DGA morphisms

$$I^*(K_1) \overset{j_1^I}{\to} I^*(L) \overset{j_2^I}{\leftarrow} I^*(K_2), \tag{3.2}$$

but not so in terms of the morphisms

$$H_R^{**}(K_1) \overset{j_1^H}{\to} H_R^{**}(L) \overset{j_2^H}{\leftarrow} H_R^{**}(K_2).$$

The construction, however, is too complicated to give here.

EXAMPLE 3.2 (Eilenberg–Moore spectral sequence). The cohomology of a space is quite often determined by means of spectral sequences of fibrations. For example, suppose that $X \times_B Y$ is the fibre space induced by a map $f : X \to B$ from a fibration $F \subset Y \xrightarrow{j} B$ with B simply connected, so that we have a fibre square

$$
\begin{array}{ccc}
F \subset X \underset{B}{\times} Y & \longrightarrow & Y \\
\downarrow & & \downarrow {\scriptstyle j} \\
X & \xrightarrow{\;f\;} & B
\end{array}
\tag{3.3}
$$

Then there exists a spectral sequence convergent to $H_k^{**}(X \underset{B}{\times} Y)$ over a coefficient field k for which the E_2 term is determined by the morphisms

$$
H_k^{**}(X) \xleftarrow{f^*} H_k^{**}(B) \xrightarrow{j^*} M_k^{**}(Y)
\tag{3.4}
$$

in the following manner:

$$
E_2 \approx \operatorname{Tor}_{H_k^{**}(B)}(H_k^{**}(X), H_k^{**}(Y)) \Rightarrow H_k^{**}\!\left(X \underset{B}{\times} Y\right).
\tag{3.5}
$$

In particular we have for the fibration $F \subset Y \xrightarrow{j} B$ the spectral sequence

$$
E_2 \approx \operatorname{Tor}_{H_k^{**}(B)}(k, H_k^{**}(Y)) \Rightarrow H_k^{**}(F).
\tag{3.6}
$$

These spectral sequences due to Eilenberg and Moore furnish some information about cohomology of $X \times_B Y$ or F from the knowledge of these of X, Y, B. However, only in very rare cases will these spectral sequences collapse to give a complete determination of those of $X \times_B Y$ or F with exact isomorphisms

$$
\operatorname{Tor}_{H_k^{**}(B)}(H_k^{**}(X), H_k^{**}(Y)) \approx H_k^{**}\!\left(X \underset{B}{\times} Y\right)
\tag{3.7}
$$

or

$$
\operatorname{Tor}_{H_k^{**}(B)}(k, H_k^{**}(Y)) \approx H_k^{**}(F).
\tag{3.8}
$$

Such cases do arise, but only under very strong conditions, say with $Y \to B$ the canonical fibration $B_U \to B_G$ and $X = $ Riemannian symmetric space, as considered by Baum and Smith [1], or $X = G'/U'$ with U' of maximum rank in G', as considered by Wolf [8] ($k = $ real field, G, G' compact connected Lie groups, U, U' closed connected subgroups).

Now in case $k = R$, if instead of the cohomology H_R^{**} we use the I^*-measures, which contain much more information than H_R^{**}'s, then it turns out that the induced morphisms

$$
I^*(X) \xleftarrow{f^I} I^*(B) \xrightarrow{j^I} I^*(Y)
\tag{3.9}
$$

are enough for the complete determination of $H_R^{**}(X \times_B Y)$ (or $H_R^{**}(F)$), viz.

Theorem 3.2. *For the fibre square* (3.3) *of c-spaces we have*

$$
H_R^{**}\!\left(X \underset{B}{\times} Y\right) \approx \operatorname{Tor}_{I^*(B)}(I^*(X), I^*(Y)).
\tag{3.10}
$$

In particular we have for the fibration $F \subset Y \to B$ of c-spaces

$$H_R^{**}(F) \approx \text{Tor}_{I^*(B)}(R, I^*(Y)). \tag{3.11}$$

For proofs see Wu [10]. See also the corresponding theorems in Section 4. We remark only that in the cases considered by Baum and Wolf I^* coincides with H_R^{**} for the various spaces X, Y, B involved, and the above isomorphisms become those in (3.7), (3.8).

EXAMPLE 3.3 (Twisted product for fibre space). For a fibration $F \subset Y \to B$ Brown has shown that

$$H^{**}(Y) \approx H(S^*(B) \otimes S^*(F)),$$

in which S^* denotes singular cohomology and H is the homology with respect to a certain twisted differential d_τ in the tensor product. As S^* is transfinite, it is desirable to replace if possible S^* by say H^{**}, which is of course rarely possible. However, if we consider in the case of a real coefficient field I^* instead of H_R^{**}, then it turns out again that I^* will give enough information to determine $H_R^{**}(Y)$ completely, as in the following

Theorem 3.3. *For a fibration $F \subset Y \to B$ of c-spaces we have*

$$H_R^{**}(Y) \approx H(I^*(B) \otimes I^*(F)),$$

with some twisted differential d_τ in the tensor product $I^*(B) \otimes I^*(F)$.

For a proof see Wu and Wang [12]. See also the corresponding theorem in Section 4.

In certain cases the twisted differential d_τ may be explicitly determined. Thus for F a sphere S^n $(n > 1)$ we have

Theorem 3.4. *For the sphere bundle*

$$S^n \subset Y \to B$$

the twisted differential d_τ in $I^*(B) \otimes I^*(S^n)$ is given as follows:
For n odd, so that $I^*(S^n) = \Lambda(x)$ with $\deg x = n$, $dx = 0$, we have

$$d_\tau x = e,$$

where $e \in I^*(B)$ corresponds to the Euler class of the bundle on passing to the homology.
For n even, so that

$$I^*(S^n) = \Lambda(x, y)$$

with $\deg x = n$, $\deg y = 2n - 1$, $dx = 0$, $dy = x^2$, we have

$$d_\tau x = 0,$$

$$d_\tau y = x^2 - p_n,$$

in which p_n corresponds to the nth real Pontrjagin class of the bundle on passing to homology.

Sketch of Proof. d_τ is determined from consideration of universal bundles over Grassmannian manifolds whose I^*-measures are easily determined. □

4. Calculability of I^*-Measure

In algebraic topology one usually associates various algebraic structures (or numbers) to spaces reflecting their topological properties. We may mention thus dimensions, Betti numbers, homology and cohomology groups on rings, homotopy groups, etc. In comparison with the notions length, area, volume, or content associated to elementary geometrical figures reflecting their metrical properties, we shall call such algebraic structures associated to spaces also their "measures". Accordingly, we shall denote by $H_A^*(X)$ $[H_A^{**}(X)]$ the cohomology group [ring] of a space X on a coefficient group [resp. ring] A, instead of the usual notation $H^*(X, A)$, to make clear the character of H_A^* or H_A^{**} being a "measure". Now for a certain measure of a topological space to be fruitful, the measure in question should satisfy, besides being invariant and of finite type in character, the further condition of being "constructive" or "calculable" in the following sense.

In algebraic topology we frequently have to construct new spaces from given ones (as in forming space products, shrinking part of a space to a point, forming a suspension or loop space, etc.) and then determine or "calculate" for the new space the measure in question from those of given ones. In view of this, we shall introduce the notion of "calculability" of a measure M on certain category of spaces as follows:

Definition. Let X_α be a set of spaces and G a certain geometric procedure producing some space X from X_α. Then we shall say that M is *calculable* vis-à-vis the geometrical construction G if $M(X)$ is completely determined by means of some algebraic construction from $M(X_\alpha)$ together with the inherent interrelations of the latter ones arising from their mutual geometrical relations and the geometric construction G.

In this sense, the measures H_Z^* and H_R^{**} are both calculable vis-à-vis the space-product construction, and the Künneth formulae give the precise manner of the corresponding algebraic determination. However, it turns out that even the simplest measures in algebraic topology are often noncalculable in the above sense vis-à-vis quite simple geometrical constructions. Thus simple examples show that:

(a) The integral cohomology ring measure H_Z^{**} is not calculable vis-à-vis the space-product construction, though the integral group measure H_Z^* is.
(b) The real cohomology ring measure H_R^{**} is not calculable vis-à-vis the geometrical cone construction, and neither is the integral cohomology group measure H_Z^*, though the real cohomology group measure H_R^* is.

The above cases show that even so simple a measure as H_A^* is not an appropriate one from the point of view of calculability. On the other hand, the

I^*-measure introduced in Section 2, based on theory of Sullivan, besides being invariantive and of finite type, is calculable vis-à-vis practically all geometrical constructions usually met in algebraic topology. We shall cite below a few examples.

EXAMPLE 4.1 (Space-union construction). For any diagram of DGA morphisms

$$A_1 \xrightarrow{f_1} B \xleftarrow{f_2} A_2,$$

let $J(f_1, f_2)$ be the DGA consisting of all elements (a_1, a_2) such that $a_i \in A_i$, $\deg a_1 = \deg a_2$, $f_1 a_1 = f_2 a_2$, with an evident DGA structure. For any minimal DGA N with free generators x_i, let N^{tr} be the DGA with free generators u_i, v_i such that

$$\deg u_i = \deg x_i, \qquad \deg v_i = \deg x_i + 1,$$
$$du_i = v_i, \qquad dv_i = 0.$$

We shall also note by

$$\tau_N : N^{\mathrm{tr}} \to N \tag{4.1}$$

the DGA morphism defined by

$$\tau_N(u_i) = x_i, \qquad \tau_N(v_i) = dx_i. \tag{4.2}$$

It is clear that $H(N^{\mathrm{tr}}) = 0$ and τ_N is onto, and accordingly Λ^{tr} will be called the *trivialization* of N.

With these notions we are now in a position to prove that I^*-measure is calculable vis-à-vis the space-union construction with an explicit algebraic determination:

Theorem 4.1. *Let K be the union of two c-complexes K_1, K_2 with a c-subcomplex L in common. Let the injection of L in K_1, K_2 induce the DGA morphisms*

$$M_1 \xrightarrow{\varphi_1} N \xleftarrow{\varphi_2} M_2, \tag{4.3}$$

in which $M_i = I^*(K_i)$, $N = I^*(L)$. Set $\tilde{M}_i = M_i \otimes N^{\mathrm{tr}}$, $\tilde{\varphi}_i = \varphi_i \otimes \tau_N$, and form the diagram

$$\tilde{M}_1 \xrightarrow{\tilde{\varphi}_1} N \xleftarrow{\tilde{\varphi}_2} \tilde{M}_2. \tag{4.4}$$

Then we have

$$I^*(K) \approx \min J(\tilde{\varphi}_1, \tilde{\varphi}_2). \tag{4.5}$$

In particular, for $L =$ a point, so that $K = K_1 \vee K_2$, the J on the right-hand side of (4.5) is simply the direct sum of $I^*(K_1)$ and $I^*(K_2)$ with evident DGA structure, so that (4.5) becomes

$$I^*(K) \approx \min(I^*(K_1) \oplus I^*(K_2)). \tag{4.6}$$

Sketch of Proof. Let $A_i = A^*(K_i)$, $B = A^*(L)$ with DGA morphisms induced by injections

$$A_1 \xrightarrow{f_1} B \xleftarrow{f_2} A_2.$$

The morphisms f_i are epimorphisms and it is clear that

$$A^*(K) \approx J(f_1, f_2)$$

so that

$$I^*(K) \approx \min J(f_1, f_2). \tag{4.7}$$

Now by the Lemma of Section 2 we see that the canonical morphism $\rho_B : N \to B$ and the epimorphisms f_i above can be completed to a *strictly* commutative diagram of DGA morphisms with ρ_i canonical ones:

$$
\begin{array}{ccccc}
A_1 & \xrightarrow{f_1} & B & \xleftarrow{f_2} & A_2 \\
\rho_1 \uparrow & & \uparrow \rho_B & & \uparrow \rho_2 \\
M_1 & \xrightarrow{\psi_1} & N & \xleftarrow{\psi_2} & M_2
\end{array}
$$

Setting $\tilde{M}_i = M_i \otimes N^{\mathrm{tr}}$, $\tilde{\psi}_i = \psi_i \otimes \tau_N$, then, as f_i are epimorphisms and N^{tr} is free, we can extend the above diagram further into a strictly commutative one as shown below:

$$
\begin{array}{ccccc}
A_1 & \xrightarrow{f_1} & B & \xleftarrow{f_2} & A_2 \\
\tilde{\rho}_1 \uparrow & & \uparrow \rho_B & & \uparrow \tilde{\rho}_2 \\
\tilde{M}_1 & \xrightarrow{\tilde{\psi}_1} & N & \xleftarrow{\tilde{\psi}_2} & \tilde{M}_2
\end{array} \tag{4.8}
$$

By diagram-chasing arguments we prove easily that the natural DGA morphism

$$J(\tilde{\psi}_1, \tilde{\psi}_2) \to J(f_1, f_2)$$

defined by the pair $(\tilde{\rho}_1, \tilde{\rho}_2)$ will induce an algebraic isomorphism in homology:

$$H(J(\tilde{\psi}_1, \tilde{\psi}_2)) \approx H(J(f_1, f_2)).$$

It follows that

$$\min J(\tilde{\psi}_1, \tilde{\psi}_2) \approx \min J(f_1, f_2).$$

As it is easy to verify that $\min J(\tilde{\psi}_1, \tilde{\psi}_2) \approx \min J(\tilde{\varphi}_1, \tilde{\varphi}_2)$, we get the theorem from (4.7).   □

EXAMPLE 4.2 (Cone construction). As a special case of the preceding one we see that the I^*-measure is also calculable vis-à-vis the cone-construction. As this case is rather of particular importance (cf. Section 5) we give an alternative explicit algebraic construction as follows:

Theorem 4.2. *Let* $\Delta = \Delta_L(K)$ *be the cone construction over a c-subcomplex L of a c-complex K. The injection $L \subset K$ will induce a DGA morphism*

$$\varphi : M \to N$$

in which $M = I^*(K)$, $N = I^*(L)$. *Setting* $\tilde{M} = M \otimes N^{\mathrm{tr}}$, $\tilde{\varphi} = \varphi \otimes \tau_N$, *then we have*

$$I^*(\Delta) \approx \min \operatorname{Ker} \tilde{\varphi}. \tag{4.9}$$

Moreover, we have a canonical DGA morphism

$$\gamma(\varphi) : I^*(\Delta) \to I^*(K) = M \tag{4.10}$$

corresponding to the injection $K \subset \Delta$ *which is defined as the composition*

$$I^*(\Delta) \to \operatorname{Ker} \tilde{\varphi} \to \tilde{M} \to M,$$

with the last DGA *morphism the natural projection.*

Sketch of Proof. Form the exact sequence

$$0 \to \operatorname{Ker} f \to A^*(K) \xrightarrow{f} A^*(L) \to 0, \tag{4.11}$$

in which the DGA morphism f is induced from the injection $L \subset K$. It is easy to verify that

$$I^*(\Delta) \approx \min \operatorname{Ker} f. \tag{4.12}$$

As f is onto, we can complete (4.11) to a strictly commutative diagram

$$\begin{array}{ccccccccc}
0 & \longrightarrow & \operatorname{Ker} f & \longrightarrow & A^*(K) & \xrightarrow{\ f\ } & A^*(L) & \to & 0 \\
& & \rho \uparrow & & \rho_1 \uparrow & & \uparrow \rho_2 & & \\
0 & \longrightarrow & \operatorname{Ker} \tilde{\psi} & \longrightarrow & \tilde{M} & \xrightarrow{\ \tilde{\psi}\ } & N & \to & 0
\end{array}$$

in which ρ_1, ρ_2 induce algebraic isomorphisms in homology. It follows that ρ also induces algebraic isomorphism in homology, so that the theorem follows immediately from (4.12) and the relation $\min \operatorname{Ker} \tilde{\psi} \approx \min \operatorname{Ker} \tilde{\varphi}$, easily verified. $\square$

As a consequence of Theorem 4.2 we get for the suspension construction:

Theorem 4.3. *Let ΣK be the suspension of a finite c-complex K. Let $N = I^*(K)$, then*

$$I^*(\Sigma K) \approx \min \operatorname{Ker} \tau_N.$$

EXAMPLE 4.3 (Holing construction). As the reverse operation for a particular cone construction, we have the following holing construction: Let K be a complex of dimension n and σ a simplex of highest dimension n of K. Let K^σ be the complex obtained from K by removing the interior of σ. Then the I^*-measure is again calculable vis-à-vis such a holing construction of K^σ from K. For this let us first explain some terminology.

For any minimal DGA M we can take generators in each dimension $m \geqslant 2$ to be of the form

$$x_1^m, \ldots, x_{h_m}^m, y_1^m, \ldots, y_{k_m}^m,$$

in which $dx_i^m = 0$ while $dy_1^m, \ldots, dy_{k_m}^m$ are linearly independent. For any cycle $z = z_n$ of degree n of M, let us form a DGA M^z in the following manner. The M^z will possess generators

$$x_1^m, \ldots, x_{h_m}^m, y_1^m, \ldots, y_{k_m}^m \qquad \text{in degree } m \leqslant n - 2,$$

$$x_1^{n-1}, \ldots, x_{h_{n-1}}^{n-1}, y_1^{n-1}, \ldots, y_{k_{n-1}}^{n-1}, u_{n-1} \qquad \text{in degree } n - 1,$$

$$x_1^n, \ldots, x_{h_n}^n \qquad \text{in degree } n,$$

while M^z has no generators in degrees $> n$. The differential will be the same as in M with the further relation $du_{n-1} = z_n$. The multipications in M^z are subject to the sole condition that any product in degree $> n$ of generators of M^z is 0. We have now the following:

Theorem 4.4. *Let K^σ be the holing construction of a complex K of finite dimension n with an n-simplex σ removed. Consider the exact sequence*

$$H_R^{n-1}(K^\sigma) \xrightarrow{j} H_R^{n-1}(\dot\sigma) \xrightarrow{\delta} H_R^n(K).$$

If j is onto, then let us take $z = z_n$ to be 0 in $M = I^*(K)$. If $j = 0$, then take z to be a cycle of degree n in M whose class corresponds to the δ-image of a generator of $H_R^{n-1}(\dot\sigma)$. Then we have

$$I^*(K^\sigma) \approx \min M^z. \tag{4.13}$$

The holing construction was suggested by Professor W. Y. Hsiang, and the theorem above will be used in his study of fixed points under group actions. The proof of the theorem is based on the direct consideration of $A^*(K) \to A^*(K^\sigma)$ and is elementary in character.

EXAMPLE 4.4 (Fibre-product construction). For a fibre square of c-spaces

$$
\begin{array}{ccc}
F \subset X \times Y & \longrightarrow & Y \\
{\scriptstyle B} \downarrow & & \downarrow {\scriptstyle g} \\
X & \xrightarrow{\ f\ } & B
\end{array}
\tag{4.14}
$$

we have

Theorem 4.5. *Consider $I^*(X)$ and $I^*(Y)$ respectively as a right and a left $I^*(B)$-DGA via induced morphisms f^I and g^I. Take the corresponding bar construction $\operatorname{Bar} I^*(X)$. Then $I^*(X \times_B Y)$ is completely determined from the diagram*

$$I^*(X) \xleftarrow{f^I} I^*(B) \xrightarrow{g^I} I^*(Y) \tag{4.15}$$

as

$$I^*\!\left(X \underset{B}{\times} Y\right) \approx \min\!\left(\operatorname{Bar} I^*(X) \underset{I^*(B)}{\otimes} I^*(Y)\right). \tag{4.16}$$

In particular we have

$$I^*(F) \approx \min\!\left(\operatorname{Bar} R \underset{I^*(B)}{\otimes} I^*(Y)\right), \tag{4.17}$$

in which $\operatorname{Bar} R$ is the bar construction of R considered as a trivial right $I^*(B)$-DGA.

Sketch of proof. Let

$$A^*(X) \xleftarrow{f^A} A^*(B) \xrightarrow{g^A} A^*(Y)$$

be the diagram of induced DGA morphisms of the corresponding de Rham–
Sullivan algebras. Using the usual spectral-sequence arguments, we prove first
that the homology of $X \times_B Y$ is isomorphic to the homology of $\mathrm{Bar}\, A^*$
$(X) \otimes_{A^*(B)} A^*(Y)$ or $\mathrm{Tor}_{A^*(B)}(A^*(X), A^*(Y))$, so that

$$I^*\left(X \underset{B}{\times} Y\right) \approx \min\left(\mathrm{Bar}\, A^*(X) \underset{A^*(B)}{\otimes} A^*(Y)\right).$$

The next step consists in replacing all A^* by the corresponding I^*. For details see
Wu [10]. □

EXAMPLE 4.5 (Fibre-space construction). From Theorem 4.5 we deduce the
following theorem. For the details of the proof we refer again to Wu and Wang
[12]. We remark that Theorems 3.2 and 3.3 are just consequences of Theorems
4.5 and 4.6 on merely taking the homology of both sides in the respective DGA
isomorphisms.

Theorem 4.6. *For a fibration $F \subset Y \to B$ of e-spaces there exists some twisted
differential d_τ in $I^*(B) \otimes I^*(F)$ such that*

$$I^*(Y) \approx \min(I^*(B) \otimes I^*(F)). \tag{4.18}$$

5. Effective Computation and Axiomatic System of
I*-Measure on the Category of Simply Connected Finite
Polytopes

As any finite complex can be obtained from its 2-dimensional skeleton by adding
successively higher-dimensional simplexes each step of which is equivalent to a
cone construction, Theorem 4.2 furnishes us a means to compute effectively the
I^*-measure of any simply connected finite complex once we know how to
compute the I^*-measure of a simply connected 2-dimensional complex. Now any
simply connected 2-dimensional complex has the same homotopy type as a
budget of 2-spheres joined at a single point. Hence the I^*-measure of a simply
connected finite complex can be effectively computed on account of Theorems
4.1–4.3 and the well-known fact that I^*-measure is a homotopy invariant.

Now consider any connected and simply connected finite complex K. Let us
represent such a complex by

$$K = K_m \supset K_{m-1} \supset \cdots \supset K_1 \supset K_0,$$

in which K_0 is the 2-dimensional squelette, and K_r is the union of K_{r-1} with an
additional simplex Δ_r, the boundary of which is

$$\dot{\Delta}_r = L_{r-1} \subset K_{r-1},$$

so that

$$K_r = K_{r-1} \cup \Delta_r \simeq K_{r-1}/L_{r-1}.$$

Let

$$j_{r-1} : L_{r-1} \subset K_{r-1}$$

be inclusion map, and

$$j^I_{r-1} : I^*(K_{r-1}) \rightarrow I^*(L_{r-1})$$

any of the associated morphisms of I^*-measures.

Now we have the following

Lemma. *If we know how to compute* $I^*(L'_{r-1}), I^*(L''_{r-1})$ *and the associated* DGA *morphisms*

$$k^I_{r-1} : I^*(L'_{r-1}) \rightarrow I^*(L''_{r-1})$$

for any c-subcomplexes $L''_{r-1} \subset L'_{r-1}$ *of* K_{r-1}, *then we know also how to compute* $I^*(L'_r), I^*(L''_r)$, *and the associated* DGA *morphisms*

$$k^I_r : I^*(L'_r) \rightarrow I^*(L''_r)$$

for any c-subcomplexes $L''_r \subset L'_r$ *of* K_r.

Proof. We consider three cases separately.

Case 1. $L''_r \subset L'_r \subset K_{r-1}$. In this case $I^*(L'_r)$, $I^*(L''_r)$, and $k^I_r : I^*(L'_r) \rightarrow I^*(L''_r)$ are already known by hypothesis.

Case 2. $L'_r \subset K_{r-1}$ but $L'_r \not\subset K_{r-1}$. In that case L'_r contains Δ_r as a subsimplex and may be considered as the cone construction over L_{r-1} of some subcomplex $\bar{L}'_r$ of K_{r-1} ($L'_r = \Delta_r \cup \bar{L}'_r$), while L''_r is a subcomplex of $\bar{L}'_r$. It follows from Theorem 4.2 that $I^*(L'_r)$ is determined by the DGA morphism

$$j^I_r : I^*(\bar{L}'_r) \rightarrow I^*(L_{r-1}),$$

already known by hypothesis. The associated DGA morphisms

$$I^*(L'_r) \rightarrow I^*(L''_r)$$

are then the composition

$$I^*(L'_r) \overset{\gamma(j^I_r)}{\rightarrow} I^*(\bar{L}'_r) \rightarrow I^*(L''_r),$$

of which the morphism on the right is again known by hypothesis.

Case 3. $L''_r \not\subset K_{r-1}$. In this case both L'_r, L''_r contain Δ_r as a subsimplex and can be considered as cone constructions over L_{r-1}:

$$L''_r = \Delta_r \cup \bar{L}''_r, \qquad L'_r = \Delta_r \cup \bar{L}'_r.$$

Now $\bar{L}''_r \subset \bar{L}'_r \subset K_{r-1}$, so that $I^*(\bar{L}'_r)\, I^*(\bar{L}''_r)$, and $j^I_r : I^*(\bar{L}'_r) \rightarrow I^*(\bar{L}''_r)$ are known by hypothesis; thus

$$I^*(L'_r), \quad I^*(L''_r), \quad \text{and} \quad k^I_r : I^*(L'_r) \rightarrow I^*(L''_r)$$

can be determined by Theorem 4.2. $\square$

In view of Theorems 4.1–4.3 and the above Lemma we thus arrive at the following

Theorem 5.1. *There is an algorithmic procedure permitting us to compute effectively the I^*-measure of any connected and simply connected finite complex up to any prescribed degree.*

The same considerations also furnish us immediately an axiomatic system for the I^*-measure over the category of connected and simply connected finite complexes (or polytopes) as follows.

To each complex K in the category is associated a minimal $I^*(K)$, and to each pair of complex K and subcomplex K' is associated a set of DGA morphisms determined up to homotopy:

$$k^I : I^*(K) \to I^*(K').$$

These DGA algebras and morphisms are characterized by the following axiomatic system (Wu [11]):

Axiom 1. I^* is a homotopy measure, in other words, for K, K' having same homotopy type in the category, we have $I^*(K) \approx I^*(K')$.

Axiom 2. If $K'' \subset K' \subset K$ are complexes all in the above category and $k^I : I^*(K) \to I^*(K')$, $k'^I : I^*(K') \to I^*(K'')$ are associated DGA morphisms for the pairs $K' \subset K$ and $K'' \subset K'$, then $k'^I k^I : I^*(K) \to I^*(K'')$ are also associated DGA morphisms for the pair $K'' \subset K$.

Axiom 3. If $\Delta = \Delta_L(K)$ is the cone construction over a subcomplex L of K, both being in the above category, then $I^*(\Delta)$ is given by

$$I^*(\Delta) \approx \min \operatorname{Ker} \tilde{j}^I,$$

where

$$j^I : I^*(K) \to I^*(L)$$

in any of the associated DGA morphisms of the pair $L \subset K$, and $\tilde{j}^I : I^*(K) \otimes I^*(L)^{\mathrm{tr}} \to I^*(L)$ is determined from j^I.

Axiom 4. For the cone construction $\Delta = \Delta_L(K)$ as in Axiom 3, the associated DGA morphisms of the pair $K \subset \Delta$ are given by the natural DGA morphisms

$$\gamma(j^I) : I^*(\Delta) \to I^*(K).$$

Axiom 5. The I^*-measure $I^*(\Delta)$ and the morphism $\gamma(j^I)$ in Axioms 3 and 4 are functorial in character. In other words, if we have pairs $L \subset K$ and $L' \subset K'$ is the category with L', K' as subcomplexes of L and K respectively, then we have the following homotopically commutative diagram with evident DGA morphisms:

$$
\begin{array}{ccc}
I^*(\Delta) & \to & I^*(K) \\
\downarrow & & \downarrow \\
I^*(\Delta') & \to & I^*(K')
\end{array}
$$

Axiom 6. For any two complexes K', K'' in the category we have

$$I^*(K' \vee K'') \approx \min(I^*(K') \oplus I'(K'')),$$

and the associated DGA morphisms

$$I^*(K' \vee K'') \to I^*(K')$$

are determined by the composition

$$I^*(K' \vee K'') \to I^*(K') \oplus I^*(K'') \to I^*(K').$$

Axiom 7. For a 2-sphere s^2 the I^*-measure is determined as in Theorem 4.5: $I^*(s) = \Lambda(x, y)$ with $\deg x = 2$, $\deg y = 3$, $dx = 0$, $dy = x^2$.

REFERENCES

[1] P. Baum and L. Smith, The real cohomology of differentiable fiber bundles. *Comm. Math. Helv.* **42**, 171–179 (1967).

[2] R. Bott, Lectures on characteristic classes and foliations. In *Lectures on Algebraic and Differential Topology* (R. Bott, S. Gitler, and I. M. James, Eds.), Springer, Berlin, Heidelberg, New York, 1972.

[3] E. Friedlander, P. A. Griffiths, and J. Morgan, Homotopy theory and differential forms. Mimeo. Notes, 1972.

[4] H. Poincaré, *Analysis Situs.* t. 1 J. de L'Ecole Polytechnique, 1895.

[5] D. Sullivan, Differential forms and topology of manifolds. *Symp. Tokyo on Topology*, 1973.

[6] ———, *Infinitesimal Computations in Topology*. Publ. Math. No. 47, Inst. de Hautes Etudes Scientifiques, 1978.

[7] A. Weil, Sur les theoremes de de Rham. *Comm. Math. Helv.* **20**, 119–145 (1952).

[8] J. Wolf, The real and rational cohomology of differential fiber bundles. *Trans. Amer. Math. Soc.* **245**, 211–220 (1975).

[9] Wu Wen-tsün, Theory of I^*-functions in algebraic topology—Real topology of homogeneous space. (In Chinese) *Acta Math, Sinica* **18**, 162–172 (1975).

[10] ———, Real topology of fiber squares. *Scientia Sinica* **18**, 464–482 (1975).

[11] ———, Effective calculation and axiomatization of I^*-functors on complexes. *Scientia Sinica* **10**, 647–664 (1976).

[12] Wu Wen-tsün and Wang Qi-ming, I^*-functor of a fibre space. *Scientia Sinica* **21**, 1–18 (1978).

[13] Wu Wen-tsün, *Calculability of I^*-Measure vis-à-vis Space-Union and Related Constructions.* Kuxue Tong Pao, 1979.

Fibre Bundles and the Physics of the Magnetic Monopole

Chen Ning Yang*

It is a privilege on my part to speak at this symposium honoring Professor Chern. I think I can claim, among all the participants here, to have known him for the longest time, perhaps even longer than Mrs. Chern has know him.

I shall first discuss some recent developments in the theory of the magnetic monopole, using ideas borrowed from fibre-bundle theory. I shall then make some remarks about the relationship between mathematics and physics.

1. Magnetic Monopole and Nontrivial Bundles

The magnetic monopole is the magnetic charge. While the idea of magnetic monopoles must have been discussed in classical physics early in the history of electricity and magnetism, modern discussions date back to 1931 and the important paper of Dirac [5], in which he pointed out that magnetic monopoles in quantum mechanics exhibit some extra and subtle features. In particular, with the existence of a magnetic monopole of strength g, electric charges and magnetic charges must necessarily be quantized, in quantum mechanics. We shall give a new derivation of this result in a few minutes.

If one wants to describe the wave function of an electron in the field of a magnetic monopole, it is necessary to find the vector potential $\vec{A}$ around the monopole. Dirac chose a vector potential which has a string of singularities. The necessity of such a string of singularities is obvious if we prove the following theorem. (Wu and Yang [8]).

Theorem. *Consider a magnetic monopole of strength $g \neq 0$ at the origin, and consider a sphere of radius R around the origin. There does not exist a vector potential $\vec{A}$ for the monopole magnetic field which is singularity-free on the sphere.*

This theorem can be easily proved in the following way. Suppose there were a singularity-free $\vec{A}$, and consider the loop integral

$$\oint A_\mu \, dx^\mu$$

* Institute of Theoretical Physics, State University of New York, Stony Brook, NY 11794, USA.

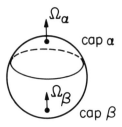

Figure 1

around a parallel on the sphere as indicated in Figure 1. By Stokes's theorem this loop integral is equal to the total magnetic flux through cap α:

$$\oint A_\mu\, dx^\mu = \Omega_\alpha. \tag{1}$$

Similarly we can apply Stokes's theorem to cap β, obtaining

$$\oint A_\mu\, dx^\mu = \Omega_\beta. \tag{2}$$

Here Ω_α and Ω_β are the total upward magnetic flux through caps α and β, both of which are bordered by the parallel. Subtracting these two equations, we obtain

$$0 = \Omega_\alpha - \Omega_\beta, \tag{3}$$

which is equal to the total flux *out* of the sphere, which in turn is equal to $4\pi g \neq 0$. We have thus reached a contradiction.

Having proved this theorem, we observe that R is arbitrary. Thus one concludes that there must be a string of singularities or strings of singularities in the vector potential to describe the monopole field. Yet we know that the magnetic field around the monopole is singularity-free. This suggests that the string of singularities is not a real physical difficulty.

The problem here is of course well known to the mathematicians. The string of singularities is an "obstruction". It can be removed if we allow a piecing together of different vector potential $\vec{A}$ in different regions of space. In other words, we use the idea used for the description of a manifold.

To be more specific, divide space outside the monopole into two overlapping regions. Call the region above the lower cone in Figure 2, region R_a, and the region below the upper cone, region R_b. The union of the two gives all points

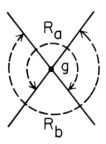

Figure 2

outside of the origin where the monopole is situated. In R_a we shall choose a vector potential for which there is only one nonvanishing component of A, the azimuthal component:

$$(A_r)^a = (A_\theta)^a = 0, \qquad (A_\phi)^a = \frac{g}{r \sin \theta} (1 - \cos \theta). \tag{4}$$

It is important to notice that this vector potential has no singularities anywhere in R_a. Similarly in R_b we choose the vector potential

$$(A_r)^b = (A_\theta)^b = 0, \qquad (A_\phi)^b = \frac{-g}{r \sin \theta} (1 + \cos \theta), \tag{5}$$

which has no singularities in R_b. It is simple to prove that the curl of either of these two potential gives correctly the magnetic field of the monopole.

In the region of overlap, since both of the two sets of vector potentials share the same curl, the different between them must be curl-free and therefore must be a gradient. Indeed a simple calculation shows

$$(A_\mu)^a - (A_\mu)^b = \partial_\mu \alpha, \qquad \text{where } \alpha = 2g\phi, \tag{6}$$

where ϕ is the azimuthal angle. The Schrödinger equation for an electron in the monopole field is thus

$$\left[\frac{1}{2m} (p - eA_a)^2 + V \right] \psi_a = E\psi_a \quad \text{in } R_a,$$

$$\left[\frac{1}{2m} (p - eA_b)^2 + V \right] \psi_b = E\psi_b \quad \text{in } R_b,$$

where ψ_a and ψ_b are respectively the wave functions in the two regions. The fact that the two vector potentials in these two equations differ by a gradient tells us, by the well-known gauge principle, that ψ_a and ψ_b are related by a phase-factor transformation

$$\psi_a = S\psi_b, \qquad S = e^{ie\alpha}, \tag{7}$$

or

$$\psi_a = e^{2iq\phi} \psi_b, \qquad q = eg. \tag{8}$$

Around the equator, which is entirely in R_a, ψ_a is single-valued. Similarly, since the equator is also entirely in R_b, ψ_b is single-valued around the equator. Therefore, S must return to its original value when one goes around the equator. That implies Dirac's quantization condition:

$$2q = 2eg = \text{integer.} \tag{9}$$

Two ψ's, ψ_a and ψ_b, in R_a and R_b respectively, that satisfy the *condition of transition* (8) in the overlap region, are called a *section* by the mathematicians. We see that around a monopole the *electron wave function is a section and not an ordinary function.* We shall call these wave sections. Different wave sections (belonging to different energies, for example) clearly satisfy the same condition of transition (8) with the same q. Thus we need to develop (Wu and Yang [9]) the concept of a Hilbert space of sections. This can be done in a natural fashion, and the physics of the magnetic monopole based on these ideas has been developed. (Wu and Yang [9, 10]).

The section defined by (8) is a section on a nontrivial bundle. The quantization condition, (9) above, already given [5] in 1931 by Dirac, is the simplest case of the deep Chern–Weil Theorem [4] applied to a U_1 bundle over S_2.

The application of the concept of nontrivial bundles to the theory of magnetic monopoles thus removes the string singularity which has impeded progress in this field ever since Dirac's paper.

2. Mathematics and Physics

The development of physics in the twentieth century is characterized by the repeated borrowing from mathematics at the fundamental conceptual level:

Physics	Mathematics
Special relativity	4-dimensional space-time
General relativity	Reimannian geometry
Quantum mechanics	Hilbert space
Electromagnetism and nonabelian gauge fields	Fibre bundles

Yet it should be emphasized that in each of these cases, the conceptual origin of the physical development was rooted in physics, and not in mathematics. There was, in fact, often a certain amount of resistance among physicists to the mathematization of physics. That this is the case is not surprising, since the fundamental value judgement in physics resides in relevance to the physical universe around us. Physicists are therefore by training more pragmatic, than mathematicians, if that is the right way to describe it. A good example of the resistance to the mathematization of physics can be found in a letter Faraday wrote to Maxwell in 1857. Faraday was a great experimental physicist with deep intuition, but not much mathematical training. He was the one who originated the concept of the lines of force. Maxwell, forty years Faraday's junior, set about to express Faraday's ideas in mathematical language. Faraday was suspicious of such efforts, as a true experimentalist should be, since from the viewpoint of physics, most mathematical formalisms do not lead to real understanding, but at best, to pointless decorations, and at worst, to clutterings that impede progress. Faraday's attitude was vividly revealed in the following passage in his letter (March 25, 1857):

> . . . My Dear Sir—I received your paper, and thank you very much for it. I do not say I venture to thank you for what you have said abont "Lines of Force," because I know you have done it for the interests of philosophical truth; but you must suppose it is work grateful to me, and gives me much encouragement to think on. I was at first almost frightened when I saw such mathematical force made to bear upon the subject, and then wondered to see that the subject stood it so well.

Faraday's reaction is all the more understandable if we take a look at the

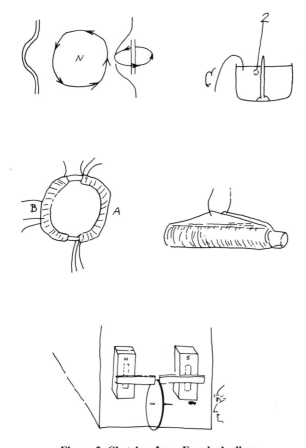

Figure 3. Sketches from Faraday's diary

sketches he made of his experiments (Figure 3). It is with such mundane instruments that he grappled with the secrets of nature. No wonder he had little use for a discipline apparently far removed from reality.

Maxwell's equations turned out to be the greatest achievement of the physics of the nineteenth century. In the hundred years since Maxwell, physicists have found that his equations give amazingly accurate and complete descriptions of much of the physical world. And the study of Maxwell equations played essential roles in the conceptual development of all the great revolutions in physics in the early twentieth century: special relativity, general relativity, and quantum mechanics. It continues to do so now.

But why did nature choose Maxwell's equations and not, for example, scalar equations? Today we know the answer to this question, though the full meaning and full implication of the answer remain to be worked out. Maxwell equations —indeed, the equations describing *all* fundamental forces of nature—are gauge field equations which are based on the geometrical concept of connections on fibre bundles.

The beauty and profundity of the geometry of fibre bundles were to a large extent brought forth by the work [4] of the man we are here to honor today. I must admit, however, that the appreciation of this beauty came to physicists only in recent years. Let me use myself as an example. I was impressed as a graduate student in the forties by the gauge principle for electromagnetism, since it was, besides general relativity, the only principle known for choosing interactions. In 1954 Mills and I generalized this principle to isotopic-spin symmetry. In so doing we were interested in the equations (see Yang [11] for a short historical discussion) and not the geometrical meaning of the equations. Around 1968 I realized that gauge fields, nonabelian as well as abelian ones, can be formulated in terms of nonintegrable phase factors, i.e. path-dependent group elements. I asked my colleague Jim Simons about the mathematical meaing of these nonintegrable phase factors, and he told me they are related to connections on fibre bundles. But I did not then appreciate that the fibre bundle was a deep mathematical concept. In 1975 I invited Jim Simons to give to the theoretical physicists at Stony Brook a series of lectures on differential forms and fibre bundles. I am grateful to him that he accepted the invitation and I was among the beneficiaries. Through these lectures T. T. Wu and I finally understood the concept of nontrivial bundles and the Chern–Weil theorem, and realized how beautiful and general the theorem is. We were thrilled to appreciate that the nontrivial bundle was exactly the concept with which to remove, in monopole theory, the string difficulty which had been bothersome for over forty years.

Later in 1975, Belavin, Polyakov, Schwartz, and Tyupkin [3] found the instanton solution, which in mathematical language is a connection on a nontriv-

Mathematics   Physics

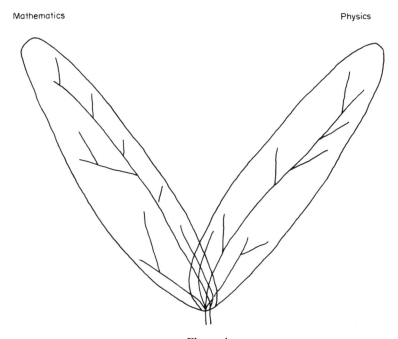

Figure 4

ial SU_2 bundle over S_4. It became clear that nontrivial bundles will play an important part in elementary-particle theory. Powerful mathematics (Atiyah et al. [1, 2]) using the Atiyha–Singer theorem and the methods of algebraic geometry have been brought to bear on the instanton problem.

While we all rejoice at the renewed common interest of the mathematicians and physicists, it would be wrong, however, to think that the two disciplines overlap that much. They do not (Figure 4). And they have their separate aims and tastes. They have distinctly different value judgements, and they have different traditions. At the fundamental conceptual level they amazingly share some concepts, but even there, the life force of each discipline runs along its own veins.

REFERENCES

[1]   M. F. Atiyah, N. J. Hitchin, and I. M. Singer, *Proc. Nat. Acad. Sci. U. S. A.* **74** 2662 (1977).

[2]   M. F. Atiyah, N. J. Hitchin, V. G. Drinfeld, and Yu. I. Manin, *Phys. Lett.* **65A** 185 (1978).

[3]   A. Belavin, A. Polyakov, A. Schwartz, and Y. Tyupkin, *Phys. Lett.* **59B**, 85 (1975).

[4]   Shiing-Shen Chern, *Ann. Math.* **45** 747 (1944); *Bull. Am. Math. Soc.* **52** 1 (1946); *Topics in Differential Geometry*, Institute for Advance Study, Princeton, N. J., (1951).

[5]   P. A. M. Dirac, *Proc. Roy. Soc.* **A113**, 60 (1931).

[6]   Yoichi Kazama, Chen Ning Yang, and A. S. Goldhaber, *Phys. Rev. D* **15**, 2287 (1977).

[7]   Tung Sheng Tu, Tai Tsun Wu, and Chen Ning Yang, *Scientia Sinica* **21** 317 (1978).

[8]   Tai Tsun Wu and Chen Ning Yang, *Phys. Rev. D* **12** 3845 (1975).

[9]   Tai Tsun Wu and Chen Ning Yang, *Nucl. Phys. B* **107** 365 (1976).

[10] Tai Tsun Wu and Chen Ning Yang, *Phys. Rev. D* **14** 437 (1976); **16** 1018 (1977).

[11] Chen Ning Yang, *Ann. N. Y. Acad. Sci.* **294** 86 (1977).

The Total Mass and the Topology of an Asymptotically Flat Space-Time

Shing-Tung Yau*

Exactly ten years ago, I met Professor Chern in Hong Kong for the first time. Since then, his teaching and his encouragement have had a tremendous influence on me. I dedicate this article to him in appreciation of his generous help.

In this talk we report on joint work with R. Schoen on general relativity. Our research began in the fall of 1977, and our aim is to try to understand the subject from the point of view of differential geometry. Here I will focus my attention on the positive-mass conjecture which we solved.

Roughly speaking, the positive-mass conjecture states that for an isolated physical system, the positivity of the local energy density implies the positivity of the total energy. The total energy has contributions from both the matter and the gravitational energy. It is defined by a nonlinear process, which makes the problem nontrivial.

In order to explain what is going on mathematically, we recall that in general relativity, the gravitational potential is described by a Lorentz metric g_{ij} on a four-dimensional manifold. The Einstein field equation says that

$$R_{ij} - \frac{R}{2} g_{ij} = T_{ij}, \qquad (1)$$

where R_{ij} is the Ricci tensor defined by g_{ij}, $R = \sum_{i,j} g^{ij} R_{ij}$ is the scalar curvature, and T_{ij} is the energy stress tensor.

Hence the Einstein field equation (1) relates the geometry of the space-time to the physics. In order for the global theory of space-time to be meaningful, it is necessary to impose some conditions on T_{ij} and some boundary conditions on the space-time.

The condition that one wants to impose on T_{ij} is usually called *the dominated energy condition* (see Hawking and Ellis [3]). It says that for any future-oriented nonspacelike vector n^i, $T_{ij} n^j$ is also future-oriented nonspacelike. This is considered to be physically reasonable, because most fields satisfy this condition.

The boundary condition that we want to impose on the space-time is that it should be asymptotically flat. Physically this means that the system is isolated. We shall describe this physical system by a three-dimensional spacelike hypersurface N in a space-time. By the existence and the uniqueness of the Cauchy problem (see Choquet-Bruhat [2], Hawking and Ellis [3]) for the Einstein equa-

*Department of Mathematics, Stanford University, Stanford, CA 94305, USA.

tion, a small neighborhood of N in the space-time is uniquely determined by the metric tensor g_{ij} of N and the second fundamental form p_{ij} of N in the space-time. We shall say that N is asymptotically flat if for some compact set C of N, $N \backslash C$ consists of k components $N_1, \ldots, N_k$ such that each N_l is diffeomorphic to the complement of a compact set in R^3 and if under this diffeomorphism, asymptotically, we have

$$g_{ij} = \delta_{ij} + \frac{2M_l}{r} \delta_{ij} + 0\left(\frac{2}{r^2}\right),$$

$$P_{ij} = 0\left(\frac{1}{r^2}\right),$$

$$\tag{2}$$

where r is the Euclidean distance and $0(1/r^2)$ is a term whose supremum norm is bounded by $1/r^2$ and whose first three derivatives decay correspondingly.

The constant M_l is called the total mass of the end N_l. There are other expressions of M_l. This formulation is due to Arnowitt, Desner, and Misner [1]. It is also called the ADM mass. The physical reason for calling it the total mass can be found there.

The dominant energy condition is a condition on the energy stress tensor T_{ij} which is defined on the space-time. In order to understand its implications on N, one applies the Einstein field equation (1), the Gauss curvature equation, and the Codazzi equation. The last two equations relate the geometry of space-time to the geometry of N. More specifically, let e_1, e_2, e_3, e_4 be an orthonormal Lorentz frame on N such that e_4 is the future normal to N. Then we can express T_{ij} in this frame. The dominated energy condition implies that

$$T_{44} \geqslant \sqrt{\sum_{i=1}^{3} T_{4i}^2} \ . \tag{3}$$

In physics literature, one denotes T_{44} by μ and T_{4i} by J_i. The former is interpreted as the local mass density, and the latter is interpreted as the local angular momentum. The Gauss curvature equation for N to be a hypersurface in the spacetime implies that

$$\mu = \frac{1}{2} \left\{ R - \sum_{i,j} p_{ij}^2 + \left(\sum_i P_{ii}\right)^2 \right\}, \tag{4}$$

where R is the scalar curvature of N.

The Codazzi equation for a hypersurface in the space-time says that

$$J_i = \sum_i p_{ji,j} - \sum_j P_{jj,i}, \tag{5}$$

where covariant derivatives on N are taken.

We can now state the positive-mass conjecture mathematically as a problem on the three-dimensional manifold N as follows:

Let N be a three-dimensional Riemannian manifold with metric tensor g_{ij} and another symmetric tensor p_{ij} which satisfy Equation (2). Suppose the inequality (3) holds on M, where the T_{4i}'s are related to g_{ij} and p_{ij} by (4) and (5). Then for each end of N, the total mass M_l is nonnegative. It can be zero only if the initial data set on N is trivial, i.e., the space-time obtained by evolving from the initial data on N according to the Einstein equation is the (flat) Minkowski space-time. More precisely, the last sentence is equivalent to saying that N can be embedded as a hypersurface in the Minkowski space-time so that g_{ij} is the induced metric and p_{ij} is the second fundamental form of the embedding.

Our proof of the positive-mass conjecture can be divided into two steps. In the first step, we replace the dominant energy condition (3) by the assumption that the scalar curvature R of the metric g_{ij} is nonnegative. In this step, the second fundamental form p_{ij} does not play a role. If the mean curvature of N is zero, i.e., if the trace $p = \sum_i p_{ii}$ of the second fundamental form p_{ij} with respect to the metric g_{ij} is zero, then the dominant energy condition (3) does imply that $R \geqslant 0$. We proved that under the assumption $R \geqslant 0$, the total mass $M_l \geqslant 0$, and it is zero only if the Riemannian manifold is isometric to the flat Euclidean space. The detailed proof of the first step has appeared already in [5].

The second step is to reduce the positive-mass conjecture to the first step. At this point we need an idea of Jang [4]. The idea can be explained as follows.

One considers the case when the initial data on N is "trivial" and tries to study the equation that governs the trivial case. Thus let N be a complete spacelike hypersurface in the (flat) Minkowski space-time with second fundamental form p_{ij}. Then N is the graph of a function w defined on the standard linear hypersurface $x_4 = 0$ whose gradient has length less than one. Jang considered the standard linear space as a graph over N which can then be considered as a hypersurface in $N \times R$. It is a direct computation that the metric of the linear space can be written as $g_{ij} + D_i w D_j w$, where $|\nabla w|$ is the length of the gradient of w taken with respect to g_{ij}. Also the second fundamental form p_{ij} is given by $D_i D_j w / (1 + |\nabla w|^2)^{1/2}$. Conversely, it was also observed by Jang that if on the given initial data set N, we can find a smooth function w such that $g_{ij} + D_j w D_j w$ is flat and $p_{ij} = D_i D_j w / (1 + |\nabla w|^2)^{1/2}$, then N is embeddable in Minkowski space. As a result, Jang suggested studying the equation

$$\sum_{i,j} \left(g_{ij} - \frac{D_i w D_j w}{(1 + |\nabla w|^2)} \right) \left[p_{ij} - \frac{D_i D_j w}{(1 + |\nabla w|^2)^{1/2}} \right] = 0 \qquad (6)$$

and the metric $g_{ij} + D_i w D_j w$ defined on N.

Jang did not find the solution of Equation (6), and the demonstration of its existence is in fact the most difficult part of step two. In order for the solution to be useful, one should impose the boundary condition $|\nabla w| = 0(1/r)$ so that the total mass corresponding to the new metric $g_{ij} + D_i w D_j w$ is the same as the total mass of g_{ij}. One can also prove uniqueness if one assumes further that w tends to zero at infinity. Hence one tries to solve (6) under the boundary condition $w = 0(1/r)$ and $|\nabla w| = 0(1/r^2)$. It turns out that two spheres Σ in N, whose mean curvature is equal to the trace of p_{ij} restricted to Σ, form obstructions to the existence of (6). However, for the purpose of proving the mass conjecture, it is enough to solve (6) in the complement of these spheres with suitable boundary conditions on them. For simplicity, we assume no two such spheres exist.

Thus in the following, we shall assume that we have found a smooth solution

w to equation (6) which satisfies the boundary condition $|\nabla w| = 0(1/r)$. Then as was mentioned above, the metric $\bar{g}_{ij} = g_{ij} + D_i w D_j w$ has the same total mass as g_{ij}. It suffices, therefore, to demonstrate the nonnegativity of the total mass of $\bar{g}_{ij}$.

In order to make use of equation (6), we interpret it more geometrically as follows. The function w defines a graph H in $N \times R$. If we assign the product metric on $N \times R$, the graph H has the induced metric given by $\bar{g}_{ij}$. We extend the tensor p_{ij} and the vector field J ($J_i = T_{4i}$) to the whole space $N \times R$, by translation vertically. Let e_1, e_2, e_3, e_4 be a local orthonormal frame field with respect to the product metric in a neighborhood of a point in H, such that e_4 is orthogonal to H. Then restricting the tensor p_{ij} to H and taking its trace, we obtain a scalar $\sum_i p_{ii}$ defined on H. Equation (6) can then be reinterpreted as

$$\sum_i h_{ii} = \sum_i p_{ii}, \tag{7}$$

where h_{ij} is the second fundamental form of H in $N \times R$.

We can translate the vector e_4 vertically and obtain a vector field. If ∇ is the connection of $N \times R$, we define

$$h_{i4} = \langle \nabla_{e_4} e_4, e_i \rangle. \tag{8}$$

By direct computation, we find

$$2(\mu - \langle J, e_4 \rangle) = \bar{R} - \sum_{i,j=1}^{3} (h_{ij} - p_{ij})^2$$
$$- 2 \sum_{i=1}^{3} (h_{i4} - p_{i4})^2$$
$$+ 2 \sum_{i=1}^{3} D_i(h_{i4} - p_{i4}), \tag{9}$$

where D_i is the induced connection on H, and $\bar{R}$ is the scalar curvature of H.

By the dominated energy condition, the left-hand side of (9) is nonnegative. Hence $\bar{R}$ is nonnegative up to a divergence term.

From the nonnegativity of the right-hand side of (9), we show that it is possible to solve the equation

$$\Delta \phi - \frac{\bar{R}}{8} \phi = 0 \tag{10}$$

on H with $\phi > 0$ and asymptotically

$$\phi = 1 + \frac{A_l}{2r} + 0\left(\frac{1}{r^2}\right), \tag{11}$$

where A_l is a constant associated to each end N_l.

The effect of solving Equation (10) is that the metric $\phi^4 \bar{g}_{ij}$ has zero scalar curvature, so that we can apply the result in Schoen and Yau [5], which is step one mentioned above. Therefore the total mass of the metric $\phi^4 \bar{g}_{ij}$ is nonnegative and is zero only if $\phi^4 \bar{g}_{ij}$ is flat.

However, the total mass of $\phi^4 \bar{g}_{ij}$ is $A_l + M_l$, and so it is sufficient to prove $A_l \leqslant 0$. This inequality can also be verified by using Equation (9). It also follows that in case $A_l = 0$, (9) becomes an equality. This can happen only if $\mu = J = 0$, because H is a graph and $\mu \geqslant |J|$. From the way that we use Equation (9), we can demonstrate that $A_l = 0$ implies $h_{ij} = p_{ij}$, $h_{i4} = p_{i4}$ for $i, j \leqslant 3$. Hence $\bar{R} = 0$, and we can apply the result in [5] to prove that $\bar{g}_{ij}$ is flat. Since the equalities $h_{ij} = p_{ij}$ and $h_{i4} = p_{i4}$ imply that $p_{ij} = D_i D_j w / (1 + |\bar{\nabla}w|^2)^{1/2}$, the initial data on N are in fact trivial in the sense that the space-time evolved from it is the flat Minkowski spacetime.

While the details of solving Equation (6) will be reported elsewhere, we would like to point out that its solution gives extra information about the space-time besides the positivity of the total mass. For example, we are able to determine the topology of any asymptotically flat space-time which satisfies the dominated energy condition. This comes from the observation that solvability of (6) guarantees that an asymptotically flat space-time which satisfies the dominated energy condition admits an asymptotically flat metric with zero scalar curvature. Our previous method [5] then tells us that the three-dimensional asymptotically flat space should, modulo the Walhausen conjecture in topology, have the following form. It is the complement of finite number of points of the following space:

$$S^3/\Gamma_1 \# S^3/\Gamma_2 \# \cdots \# S^3/\Gamma_k \# S^2 \times S^1 \# S^2 \times S^1 \# \cdots \# S^2 \times S^1,$$

where Γ_i are finite groups acting freely on S^3, and $\#$ means connected sum.

More precisely, we have proved that after compactifying the asymptotically flat space by adding a finite number of points, the fundamental group does not contain any surface group of genus $\geqslant 1$.

Most of our results in this paper have their counterparts for the compact universe. We shall report on this elsewhere.

References

[1]   Arnowitt, S. Deser, and C. Misner, *Phys. Rev.* **118**, 1100 (1960); **121** 1556 (1961); **122**, 997 (1961).

[2]   Y. Choquet-Bruhat, Cauchy problem, In *Gravitation, an Introduction to Current Research*, (L. Witten, Ed.), New York, 1962.

[3]   S. W. Hawking and G. F. R. Ellis, *The Large Scale Structure of Space-Time*. Cambridge U. P., Cambridge, 1973.

[4]   P. S. Jang, On the positivity of energy in general relativity, (Preprint).

[5]   R. Schoen and S. T. Yau, On the proof of the positive mass conjecture in general relativity. *Comm. Math. Phys.* **65** 45–76. (1979).

[6]   R. Schoen and S. T. Yau, *Proc. Natl. Acad. Sci. U.S.A.* **75** #(6), 2567 (June 1978).

[7]   R. Schoen and S. T. Yau, Existence of incompressible minimal surfaces and the topology of three dimensional manifolds with non-negative scalar curvature, *Ann. Math.* **110** 127–142 (1979).

[8]   R. Schoen and S. T. Yau, On the strucutre of manifolds with positive scalar curvature, *Manuscripta Math.* **28** 159–183 (1979).